普通高等教育“十四五”规划教材

有机合成及设计
Organic Synthesis & Design

吴亚　吴丽　主编

中国石化出版社

内 容 提 要

本书为中英文双语教材，首先介绍有机合成基础知识，包括有机分子骨架的构建和官能团的转换方法，在此基础上讨论逆合成分析方法和策略，最后补充适应时代发展的有机合成新技术。本书将有机合成反应方法和有机合成路线设计进行融合，按照学生的认知规律循序渐进，而且新的有机合成技术和应用拓展了学生的想像力。

本书内容涵盖有机合成基本原理和方法、逆合成分析策略、有机合成新技术及应用，可作为应用化学、化学、化学工程等专业的本科、研究生教材，也可供有机化学、药物化学等相关领域研究人员参考。

图书在版编目(CIP)数据

有机合成及设计 = Organic Synthesis & Design / 吴亚，吴丽主编．—北京：中国石化出版社，2022.10
ISBN 978-7-5114-6903-8

Ⅰ．①有… Ⅱ．①吴… ②吴… Ⅲ．①有机合成-设计-研究 Ⅳ．①O621.3

中国版本图书馆 CIP 数据核字(2022)第 193144 号

中国石化出版社出版发行

地址：北京市东城区安定门外大街 58 号
邮编：100011　电话：(010)57512500
发行部电话：(010)57512575
http://www.sinopec-press.com
E-mail:press@sinopec.com
北京科信印刷有限公司印刷
全国各地新华书店经销

*

787×1092 毫米 16 开本 19 印张 454 千字
2022 年 10 月第 1 版　2022 年 10 月第 1 次印刷
定价：69.00 元

前言

PREFACE

随着高等院校教学改革的深化，双语教学在各个层次的教育体制中开展和推广，但有机合成及设计双语课程可供选择的教材很少。各个学校根据具体教学情况有的选用全英文教材，有的选用内部讲义。公开出版的教材基本分两类，一类侧重于有机合成，主要课程内容偏重原理和反应方法；另一类侧重于有机合成路线设计，主要课程内容偏重合成策略和路线。

鉴于上述情况，西安石油大学与中国科学院大学联合编写了本书。本书为双语内容，以英文为主。首先介绍有机合成基础知识，包括有机分子骨架的构建和官能团的转换方法，在此基础上讨论逆合成分析方法和策略，最后补充适应时代发展的有机合成新技术。有机合成反应方法是路线设计的基础，有机合成路线设计是有机合成的重要思维方法。本书将二者进行融合，有机合成原理、方法、策略和路线编排循序渐进，符合学生的认知规律，新的有机合成技术和应用拓展了学生的想像力。同时考虑到学生的语言、知识信息，文字内容简明扼要。

本书内容涵盖有机合成基本原理和方法、逆合成分析策略、有机合成新技术及应用，适合应用化学、化学、化学工程本、硕专业需求，也可供有机化学、药物化学研究生和从事有机合成化学、药物化学等相关领域的研究人员作参考书使用。

本书第 1~9 章由西安石油大学吴亚编写；第 10 章及第 7 章部分内容由中国科学院大学吴丽编写；全书由吴亚统稿。本书出版得到了西安石油大学校级教材建设项目的支持，特此感谢。

由于编者水平所限，不足之处在所难免，敬请专家读者批评指正。

前言

目录
CONTENTS

第二部分　分子骨架修饰
Part Ⅱ　Molecular Skeleton Modification

第三部分　设计有机合成
Part Ⅲ　Designing Organic Synthesis

第四部分　有机合成技术
Part Ⅳ　Organic Synthesis Techniques

1 有机合成设计概论

Chapter 1 Introduction to Designing Organic Synthesis

1.1 有机合成设计：一般问题
Designing Organic Synthesis：General Issues

1.1.1 有机合成定义与历史
Definition and History of Synthetic Organic Chemistry

The term *synthesis* means in Greek "put together". Synthetic organic chemistry is the "art" of building-up complex molecular structures of organic compounds putting together smaller, easily accessible (commercially available) compounds. This art has a relatively recent story.

In 1828 Friedrich Wöhler produced the *organic* chemical urea (carbamide), a constituent of urine, from *inorganic* starting materials (the salts potassium cyanate and ammonium sulfate), in what is now called the Wöhler synthesis. These types of foundational studies marked the birth of synthetic organic chemistry.

After the very first example of organic synthesis, acetic acid was performed by Kolbe in 1845. From around 1900, a great number of synthetic efforts have been made, and more complex structures such as camphor (Komppa, 1903 and Perkin, 1904) or the complex structure of haemin (Fisher, 1929) have been produced. By the 1930s, several large classes of natural products were known. Important milestones in natural products research have included several notable Nobel Prize awards[1]. From 1940 to 1980, new reactions appear stereochemistry is understood, complex structure fall. Woodward and Doering made quinine in 1944 and Sheehan made penicillin V in 1957. Kishi made monensin in 1979. In the setting, Woodward arrived at the chemistry scene and began to reshape the synthetic paradigm. This period was known as the Woodward era. In late 1960's biological synthesis was developed and biologically active insulin (a hormone with 51 amino acids) and DNA fragments are made. In order to simplify and fasten the synthetic procedures, a solid phase approach has been developed. This method allows automation of some repetitive synthetic procedures, such as peptide or oligonucleotide synthesis, and shortens the others by simplifying the purification steps. In 1980's automated peptide and oligonucleotide synthesizers are common. Bruce Merrifield invented solid phase synthesis and he was awarded by Nobel Prize in 1984. In Mid-1990's, molecules get bigger, more complex and less stable. The solid phase technique allowed the development of combinatorial synthesis, an approach that generates a high number of organic compounds, based on the combinatorial disposition of different building blocks in the construction of the products. In 1995, K. C. Nicolaou made brevetoxin B. He developed "total synthesis". Total synthesis involves:

(ⅰ) choosing a target molecule;

(ⅱ) conducting a "retrosynthetic analysis" to formulate a plan (called as a strategy);

(ⅲ) selecting a sequence of reactions and reagents (called as a tactics);

(ⅳ) executing the synthesis in the lab. Finally, nowadays there is an effort to render synthetic chemistry more environmentally friendly. This effort is mainly based on the use of reaction media such as water or fluorous recyclable biphasic systems. Recently the integration of biological technology and chemical engineering was focused. All these studies make organic synthesis more and more efficient, economic and safe.

Organic synthesis is considered, to a large extent, to be responsible for some of the most exciting and important discoveries of the 20th century in chemistry, biology, and medicine, and continues to fuel the drug discovery and development process with myriad processes and compounds for new biomedical breakthroughs and applications[2]. The gradual sharpening of this tool is demonstrated by considering its history along the lines of pre-World War Ⅱ, the Woodward and Corey eras, and the 1990's, and by accounting major accomplishment along the way. Today, natural product total synthesis is associated with prudent and tasteful selection of challenging and preferably biologically important target molecules; the discovery and invention of new synthetic strategies and technologies; and explorations in chemical biology through molecular design and mechanistic studies. Future strides in the field are likely to be aided by advances in the isolation and characterization of novel molecular targets from nature, the availability of new reagents and synthetic methods, and information and automation technologies[3]. Such advances are destined to bring the power of organic synthesis closer to, or even beyond, the boundaries defined by nature, which, at present, and despite our many advantages, still look so far away.

1.1.2 有机合成的重要性 Importance of Organic Synthesis

A very large number of compounds with unimaginably complicated structures are produced by plants, animals or microorganisms. Examples are antibiotics, alkaloids, rubber, chlorophyll, steroids, proteins, carbohydrates, fats, vitamins, dyes and perfumes. These are called natural products. The chemist isolates, purifies, analyses and determines the structures of these compounds. Then people find various applications for these substances such as in medicine, plastics, paints, textiles, electronics etc.

Examples：

Natural products： e. g. Steroids（类固醇）， prostaglandins（前列腺素）， alkaloids（生物碱）.

15-Methyl $PGF_{2\alpha}$ (prostaglandin)
(前列腺素)

Epibatidine (South American frog alkaloid)
(南美蛙生物碱)

< 15 mg isolated from 750 frogs

Industrially important compounds： such as pharmaceuticals， agrochemicals， flavours， dyes， cosmetics， monomers and polymers.

Naproxen (painkiller)
萘普生(止痛药)

Carbaryl (insecticide)
甲萘威(杀虫剂)

Sarin (nerve gas)
沙林(神经毒气)

Isobutavan (异丁酸香兰酯)

(smells of mint chocolate) (薄荷巧克力的味道)

Methylenedioxymethamphetamine
(MDMA亚甲二氧甲基苯丙胺)

(Ecstasy) (迷魂药)

"5 CB" (liquid crystal) (液晶)

Kevlar (fancy polymer) (花式聚合物)

Theoretically interesting molecules：

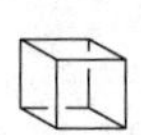

Cubane (立方烷)

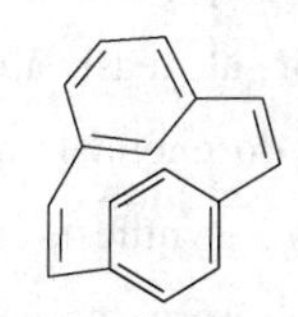

meta para Cyclophane (环芳)

Structure proof: while spectroscopy and crystallography are used to determine molecular structures, unambiguous total synthesis is still important.

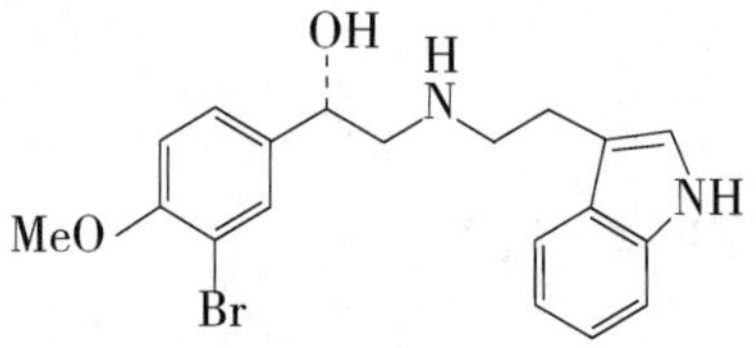

S-(+)-Chelonin B (蛇头宁) (marine sponge alkaloid) (海绵生物碱)

New methodology: new ways to make molecules, improvement of existing ways, ways of doing what was previously impossible.

Science and Technology: materials with special applications; molecular switches, non-linear optics, nanotechnology.

The synthesis of these compounds becomes important for the following reasons:

(ⅰ) Synthesis from simple starting materials of known structure using familiar reactions with predictable regioselectivity (in which part of the molecule will the reaction take place?) and stereochemistry (what will be the spatial arrangement of the atoms and groups?) and demonstrating that the product obtained is identical with the one in nature becomes the ultimate proof of the structure of the compound.

(ⅱ) The natural product is available only in very small quantities and will be very costly. Synthesis provides a means for the cheap mass production of the compound and makes it available for research and use for the benefit of man, for example in medicines.

(ⅲ) Structural variations made in the natural product may provide molecules which are more active, more useful or with lesser side effects.

(ⅳ) Syntheses of complicated molecules require good planning derived from a deep knowledge of the various reactions and their mechanism in great detail. It is extremely challenging and provides immense intellectual satisfaction to the chemist as a test and proof of his ability.

Till the 1950s, most syntheses were developed by selecting an appropriate starting material after a trial and error search of a few commercially available starting materials having structural resemblance to the molecule to be synthesized. Attempts were then made to convert them to the required product through a handful of well-known reactions using common laboratory reagents.

The total synthesis of structurally complex compounds is a challenging undertaking, in intellectual as well as practical respects. Whereas simple compounds can usually be made by synthetic routes comprising of a few reaction steps (say two to five), complicated molecules may require a lengthy sequence of reactions, usually more than twenty. Most such multi-step syntheses are executed, or at least attempted, according to a plan designed beforehand on paper or blackboard. How do chemists arrive at such synthetic plans?

Traditionally, synthesis design was based upon associative thinking processes, the most important of which were: association with existing syntheses of similar compounds; association with known starting material (s); association with a hypothetical advanced intermediate. The last of these represents an attempt to reduce the complexity of the design problem, but the selection of a suitable intermediary structure is still a highly intuitive process. The associative approach to synthesis design becomes less practical as the complexity of the problem, and hence the number of steps required, increases. There had been no systematic planning or visualization of the synthetic steps required to reach the target material economically. But a look at some of the complex molecules that were synthesized in the last century (given on the previous page) will indicate the difficulty in recognizing any available starting material. This called for a more systematic method for recognizing simpler molecules from which the required product could be synthesized in a small number of steps. Such a systematic method was developed by Prof. Elias J. Corey of the Harvard University and was called retrosynthetic analysis (RA).

The method was so logical and efficient that not only students of chemistry, but even machines (computers) could be taught (programmed) to find out reasonable synthetic routes to obtain complex molecules. A few of these programs such as LHASA [Logic and Heuristics (= empirical rules) Applied to Synthetic Analysis], AIPHOS (Artificial Intelligence for Planning and Handling Organic Synthesis), COMPASS (Computer Assisted Organic Synthesis), OSET (Organic Synthesis Exploration Tool), SECS (Simulation and Evaluation of Chemical Synthesis), IGOR (Interactive Generation of Organic Reactions) and CHIRON (Chiral Synthon) are now commercially available.

1.1.3 有机合成的定位 Orientation of Organic Synthesis

The efforts of synthetic organic chemists therefore are devoted not only to the total synthesis of complex organic compounds (target oriented synthesis), but also to the development of new synthetic methods (method oriented synthesis).

(ⅰ) 目标导向的合成

Target Oriented Synthesis

The goal of target oriented synthesis is the obtainment of a more or less complex organic molecule. It can be a natural bioactive compound, or a compound derived from rational design as potentially bioactive, or a compound of commercial relevance, or even a compound of theoretical interest. Drugs, flavors, nutraceuticals, new materials are examples of the most common and interesting targets[4].

A target oriented synthesis must be as much efficient as possible in terms of yield, cost and time. The target can be a new molecule, the properties of which must be tested, or a known molecule that already found industrial application. In this last case it is interesting to note that the new synthetic procedure make sense only if it is more efficient than the previously reported[5].

The orientation of organic synthesis is shown in Fig. 1-1.

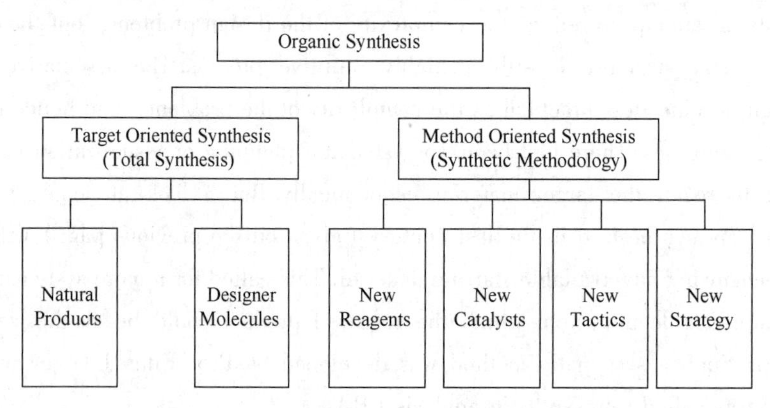

Fig. 1-1 Orientation of organic synthesis

(ⅱ) 方法导向的合成

Method Oriented Synthesis

The method oriented synthesis is devoted to the development of new reagents, new catalysts, new reaction and work-up procedures, in general to any innovation that can improve a synthetic procedure. Particular attention is devoted to the yields, the stereochemical outcome, the atomic economy of the reactions (in terms of atoms of the reagents that are not inserted in the products, and therefore lost), and more generally to the environmental impact of the process. In order to improve the synthetic methods (and to show their ability) synthetic chemists have chosen very often quite complex synthetic targets such as palytoxin and taxol(紫杉醇). Despite the synthesis of these targets will require several years and will never be industrially applicable, the efforts to solve the synthetic problems encountered during the synthesis are a fantastic "practice field" for the methodological innovation[6].

Synthetic organic chemistry also concerns polymerization processes, which are treated in Polymer Chemistry and Environmentally Degradable Polymers (化学和环境可降解聚合物), and structural modifications which require only one reaction, which can be deduced from the topic devoted to the organic chemical reactions (see Organic Chemical Reactions). This topic is mainly devoted to multi-step syntheses of complex molecular architectures. In this context, two main categories must be considered:

(a) syntheses that require the reiterative junction of bifunctional monomers, such as amino acids, carbohydrates and nucleotides.

(b) syntheses that require the construction of a complex skeleton made mainly by carbon atoms. In both cases a synthetic strategy is required.

1.1.4 合成策略 Synthetic Strategy

C—C Bond Formation: this is a strategy-level consideration. For example, the Diels-Alder and aldol reactions are strategy - level transformations; they result in a significant increase in complexity of the synthetic intermediate.

Functional Group Interconversions (FGIs): These may be regarded as largely tactical in nature, and in a sense are the "glue" which holds the strategy together. For example, a functional group arising from a strategy level transformation involving C—C bond formation may have to be converted into another functional group before the next strategy-level, C—C bond-forming transformation is possible. FGIs leave the carbon skeleton unchanged (almost always!), and usually involve exchange of heteroatoms.

Stereochemistry: A molecule with n stereocentres/double bonds may be in principle exist as up to 2^n stereoisomers, so we must always be aware of the likely stereochemical outcome of a transformation. Stereochemical factors often profoundly influence decisions about strategy[7].

When the target molecule presents a complex skeleton mainly made by carbon atoms chains, no constituent monomers or building blocks can be immediately envisaged. In this case the identification of the starting materials for the synthesis requires much more fantasy and some rules: in other word a retrosynthetic analysis.

The concept of retrosynthetic analysis has been developed by E. J. Corey who received for this reason the Nobel Prize in chemistry in 1990. In Corey's words, "Retrosynthetic analysis is a problem solving technique for transforming the structure of a synthetic target (TGT) molecule to a sequence of progressively simpler structures along a pathway which ultimately leads to simple or commercially available starting materials for a chemical synthesis". The retrosynthetic analysis is based on a sequence of disconnections. Each disconnection is a mental process in which a molecule is fragmented into two pieces that in the mind of the synthetic organic chemist can generate the molecule under examination by known reactions.

1.2 有机合成基本原理 The Principles of Synthetic Organic Chemistry

1.2.1 热力学和化学动力学 Thermodynamics and Chemical Kinetics

(ⅰ) 热力学 Thermodynamics

All chemical reactions are in principle reversible: reactants and products eventually reach equilibrium. From the point of view of devising an organic synthesis it is necessary to know whether the position of equilibrium will favor the desired product. The factors which determine the equilibrium constant of a reaction and its variation with changes in conditions follow from the principles of thermodynamics.

Chemical systems are subject to opposing influences: the tendency to minimize their enthalpy, H, is opposed by the tendency to maximize their entropy, S. Chemical reactions are usually carried out at constant pressure and under these conditions the compromise between the two trends is determined by the value of the function ($H-TS$): the function tends to decrease, and the compromise situation-equilibrium-corresponds to its minimum value. It is convenient to define a new function, G, the Gibbs free energy, as $G=H-TS$. A process will occur spontaneously if, as a result, G decreases and it will continue until G reaches a minimum[3].

A fundamental thermodynamic equation relates the equilibrium constant for a reaction to the free-energy change associated with the reaction:

$$A+B \Rightarrow C+D$$

The free energy contains both enthalpy and entropy terms:

$$\Delta G = \Delta H - T\Delta S$$

The principles of thermodynamics lead to an equation that relates the equilibrium constant, K, to the change in free energy accompanying a reaction. If the sums of the free energies of reactants and products in their standard states at temperature T are $G_2{}^0$ and $G_1{}^0$ and given by

$$RT \ln K_{eq} = -\Delta G^0 \quad (R=8.314\text{J} \cdot \text{K}^{-1} \cdot \text{mol}^{-1})$$

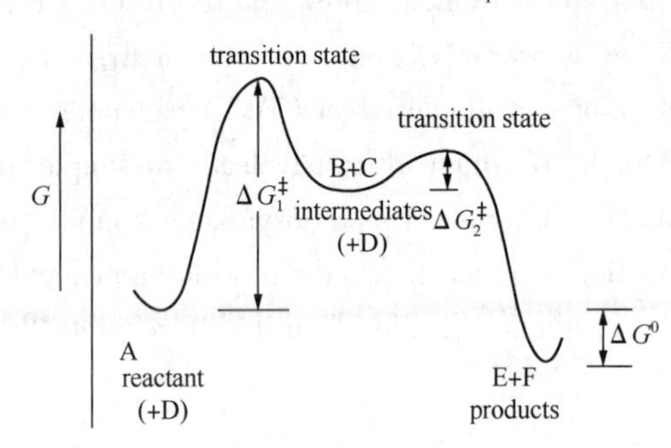

Fig. 1-2 Energy profile for a two-step reaction

Thermodynamic data give us a means of quantitatively expressing stability. Thermodynamic stability, as measured by free energy, places a limit on the extent of a chemical reaction. However, it does not directly give the information about the rates of reactions. The energy profile for a two-step reaction is shown in Fig. 1-2.

(ii) Chemical kinetics

Kinetics(动力学) is the field of chemistry that describes the rates of chemical reactions and the factors that affect those rates. Namely, the quantitative description of reactivity is called chemical kinetics. It is the energy of the transition state (TS) relative to the reactants that determines the reaction rate.

$$\Delta G^{\ddagger} = -RT \ln K^{\ddagger}$$

The free energy $\Delta G^{\ddagger}$ is called as the free energy of activation (Ea). The rate of a reaction step is then given by the free energy.

Transition State Theory（过渡态理论）

In transition state theory, a reaction is assumed to involve the attainment of an *activated complex* that goes on to product at an extremely rapid rate. Transition state(过渡态 TS) represents the highest-energy structure involved in the reaction. It is unstable and cannot be isolated, but we can imagine it[8].

Transition State: A transition state (TS) possesses a defined geometry and charge delocalization (电荷离域) but has no finite existence. At TS, energy usually higher and although many reactant bonds are broken or partially broken, the product bonds are not yet completely formed.

(iii) 动力学控制与热力学控制

Thermodynamic Versus Kinetic Control

For competitive reactions:

In most reactions which can proceed by two or more pathways each of which gives a different set of products, the products isolated are those derived from the pathway of lowest free energy of activation, regardless of whether this path results in the greatest decrease in the free energy of the system. These reactions are described as being kinetically controlled and most of the reaction product will be B (kinetic product). It is shown in Fig. 1-3.

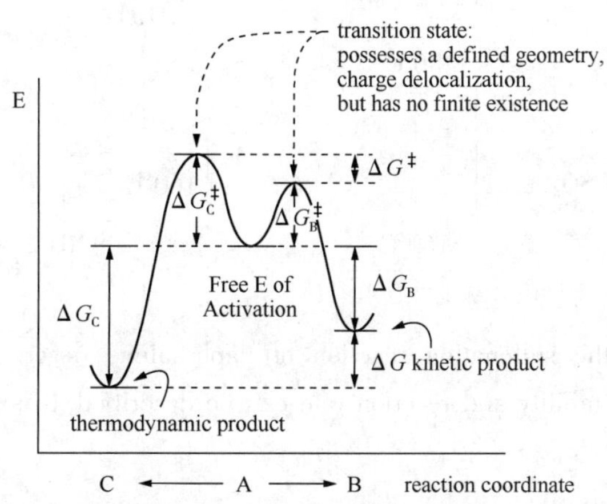

Fig. 1-3 Reaction energy profile via thermodynamic and kinetic control

If, however, the reaction conditions are suitable for equilibrium to be established between the reactants and the kinetically controlled products, a different set of products, formed more slowly but corresponding to a lower free energy for the system, can in some instances be isolated. Such reactions are described as being thermodynamically controlled and most of the product will be C (more stable, thermodynamic product).

A beautiful example of this was observed in the kinetic versus thermodynamic 1,4-addition of diene illustrated below. The most stable product (lower ΔG) was observed upon conducting the reaction under equilibrating conditions for the reversible reaction while the alternative kinetic product (lower $\Delta G^{\ddagger}$) was observed when the reaction was conducted under lower temperature and nonequilibrating conditions (kinetic conditions).

$$H_2C=CH-CH=CH_2 \xrightarrow{HBr} H_3C-\underset{\displaystyle Br}{CH}-CH=CH_2 + H_3C-\overset{\displaystyle H}{C}=CH-\underset{\displaystyle Br}{CH_2}$$

−80℃	81%	19%
r.t.	44%	56%

The example for the enolization of ketone in different basic conditions also displays the relationship.

LDA (TMSCl quench) → OTMS 84% kinetic enolate + OTMS + OTMS

Et$_3$N, TMSCl, DMF, Δ 60h → 13% ; 58% ; 29% (thermodynamic enolate)

Considering the reaction of naphthalene with concentrated sulfuric acid, two monosulfonated products are in principle obtainable:

+ H_2SO_4 — 80℃ → SO_3H 93% 1-sulfonic acid

— 160℃ → SO_3H 85% 2-sulfonic acid (1-sulfonic acid ⇌ 2-sulfonic acid, 160℃)

Energy profile for the sulfonation reaction of naphthalene describes general relationships between thermodynamic stability and reaction rate can be described. It is shown in Fig. 1-4.

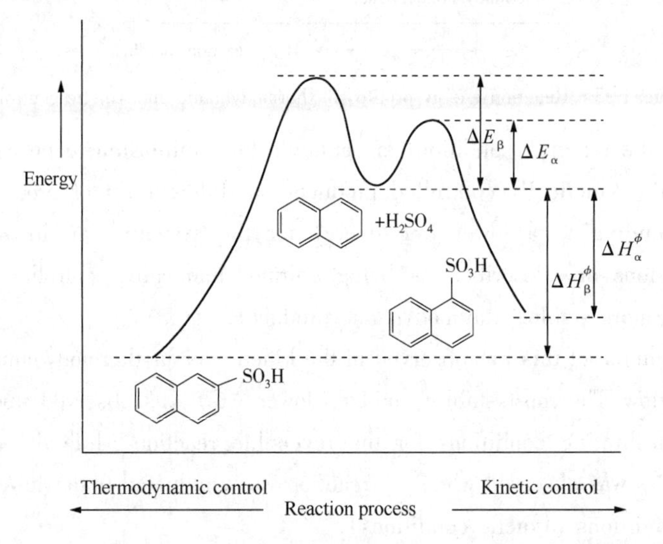

Fig. 1-4 Energy profile for the sulfonation reaction of naphthalene

The 1-derivative is formed the faster of the two but the 2-derivative is thermodynamically the more stable. At low temperatures (80℃), sulfonation at the 1-position occurs fairly rapidly whereas that the 2-position is very slow. The free energy of activation for the desulfonation (脱磺化) of the 1-sulfonic acid is such that in these conditions this product is essentially inert and is therefore isolated. At higher temperatures (160℃), desulfonation of the 1-sulfonic acid becomes important and equilibrium is fairly rapidly established between this product and the reactants. The rate of formation of the 2-sulfonic acid is now also greater so that gradually most of the naphthalene is converted into the 2-derivative and this becomes the major product. Product composition is usually governed by stability of products, which is called thermodynamic control.

Product composition at the end of the reaction may be governed by the equilibrium thermodynamics of the system. When this is true, the product composition is governed by thermodynamic control and the stability difference between the competing products, as given by the free-energy difference, determines the product composition. Alternatively, product composition may be governed by competing rates of formation of products, which is called kinetic control.

1.2.2 反应的选择性 Reactive Selectivity of Organic Reactions

(ⅰ)化学选择性 Chemoselectivity

The reaction steps are the actual synthesis, when we form or break C—C, C—O, C—X bonds etc. These reactions can often occur at more than one possible place. This would lead to other products that we are not interested in and are called side-products.

Functional groups are so called because they impart specific types of reactivity to organic molecules. In general, the characteristic reactions of functional groups are observed, irrespective of the precise molecular environment in which the functional group is situated. It should be obvious that the synthesis of a complicated molecule containing several functional groups depends on the chemoselectivity of the individual reaction steps. Reagents must be chosen which react only at the desired functional group or groups, and if necessary other functionality in the molecule must first be protected in order to prevent unwanted side reactions[9].

If the reaction of one functional group in preference to others, the reaction is chemoselective.

$$H_2C{=}CH{-}CH_2{-}C{\equiv}CH \xrightarrow[-20℃]{Br_2\text{-}CCl_4} H_2C(Br){-}CH(Br){-}CH_2{-}C{\equiv}CH \quad 90\%$$

If the reaction rate of the functional group in one site is ten times than others, the chemoselective reaction is ready to achieved.

Chemoselective reagents were employed in the following conditions[10].

$$\text{1,3-}C_6H_4(NO_2)_2 \xrightarrow{NH_4HS} \text{3-}NO_2C_6H_4NH_2$$

$$\text{3-}HOC_6H_4CH_2OH \xrightarrow[CH_2Cl_2,\ 25℃]{CrO_3,\ Py} \text{3-}HOC_6H_4CHO$$

(ⅱ) 区域选择性

Regioselectivity

There are certain functional groups which incorporate more than one reactive site. The most familiar of these are alkenes, alkynes and arenes. Unsymmetrically substituted alkenes and alkynes may in principle undergo addition reactions in either of two directions. One of these directions normal predominates(as determined, for example, by Markownikoff' s rule for ion addition). The reaction occurs at dominantly at one place and such reactions are thus said to be regioselective since the initial attack occurs at one " end" of the alkene or alkyne group in preference to the other.

Regioselectivity is also observed in electrophilic aromatic and heteroaromatic substitution:

Br + Cl_2 $\xrightarrow{AlCl_3}$ Br Cl (30%) + Br Cl (5%) + Br Cl (65%)

For example, in benzene derivatives the effect of existing substituents is to direct the incoming electrophile either to the *ortho*- and *para*-positions or to the *meta*-position.

In later chapters regioselectivity will be encountered again, for example in addition to an α, β-unsaturated carbonyl compound in the formation of specific enolates from unsymmetrical ketones.

However, if the reaction occurs at on place only (in 100% yield) the reaction is: **regiospecific**. The term regiospecific is, strictly speaking, reserved for reactions where only one of the possible products is formed, although in common parlance the term is used to refer to reactions where the regioselectivity is very high (but not necessary 100%).

(ⅲ) 立体选择性

Stereoselectivity

Where a particular reaction leads to a product that is capable of exhibiting stereoisomerism, it is not unusual for one stereoisomer to predominate; such a reaction is described as stereoselective and if only one isomer is formed the reaction is described as stereospecific[9].

The stereoisomerism in question may be geometrical (*E*-and *Z*-isomers).

O $\xrightarrow{LiAlH_4}$ H OH 90% + OH H 10%

Alternatively, the stereoselectivity of a reaction may result from the presence of stereogenic centres in the product or the starting compound or both. In every reaction which generates a stereogenic centre, the new stereogenic centre may have either the *R*- or *S*-configuration[11]. If the starting compound contains no stereogenic centre, the product then consists of a pair of enantiomers (usual in equal proportions: a racemate). However, if the starting compound already contains a stereogenic centre, the reaction may be expected to produce a pair of diastereomers (either or both of which may be chiral), not necessarily in equal proportions[12,13], this situation is commonly encountered, for example, in additions to carbonyl compounds.

$$\text{H}\text{—C(CH}_3\text{)(Ph)—CHO} + \text{CH}_3\text{MgI} \longrightarrow \text{erythro } 67\% + \text{threo } 33\%$$

erythro 67%   threo 33%

Enantioselective reactions represent a special case in which the conditions employed in the reaction lead to a product in which one configuration (R or S) at the new stereogenic centre or centres predominates over the other[14,15]. This process is often referred to as an asymmetric synthesis.

参考文献
References

1. Seebach D. Organic Synthesis—Where now? *J. Angewandte Chemie International Edition in English*. 1990, 29 (11): 1320-1367.
2. Mal D. Anionic Annulations in Organic Synthesis. Amsterdam: Elsevier, 2019. 1-29.
3. Carey F A. Advanced Organic chemistry (4th). 2000.
4. Pavlinov I, et, al. Next generation diversity-oriented synthesis: a paradigm shift from chemical diversity to biological diversity. *J, Organic & Biomolecular Chemistry*. 2019, 17 (7): 1608-1623.
5. Ringsdorf H, et, al. Molecular Architecture and Function of Polymeric Oriented Systems: Models for the Study of Organization, Surface Recognition, and Dynamics of Biomembranes. *J, Angewandte Chemie International Edition in English*. 1988, 27 (1): 113-158.
6. Kitayama T, Utaka Y. Chapter 14 - Synthesis of Valuable Optically Active Compounds From Versatile Natural Products Using Biocatalyst: Examples of Natural Materials - Related Diversity - Oriented Synthesis "NMRDOS". Amsterdam: Elsevier, 2017. 297-312.
7. D' Angelo J, Smith M B. Hybrid Retrosynthesis. Boston: Elsevier, 2015. 63-66.
8. Tantillo D J. Advances in Physical Organic Chemistry. New York: Academic Press, 2021. 1-16.
9. Mandal D K. Stereochemistry and Organic Reactions. New York: Academic Press, 2021. 247-265.
10. 徐家业等．高等有机合成．北京：化学工业出版社，1994，1-4.
11. Timms N, Daniels A D, et, al. Stereoselective Biocatalysts. Amsterdam: Elsevier, 2012. 21-45.
12. 张滂．有机合成进展．北京：科学出版社，1992. 45.
13. S. 特纳．有机合成设计[M]. 罗宣德译．北京：化学工业出版社，1984. 69.
14. 袁履冰等．有机反应动力学．大连：大连理工大学出版社，1989. 11.
15. 黄培强等．有机化学进展(2011—2012). 化学通报，2014，77 (07): 586-622.

第一部分 分子骨架的形成

Part Ⅰ Formation of Molecular Skeleton

2 通过亲核试剂形成碳碳键

Chapter 2 Carbon-Carbon Bonds Formation: Nucleophiles

Carbon-carbon bond formation is the basis for the construction of the molecular framework of organic molecules by synthesis. Carbon-carbon bond formation is a strategy-level consideration, and the decisions made here will influence both of the other main considerations set out below[1]. As was mentioned in the section, the strategy-level transformations would result in a significant increase in complexity of the synthetic intermediate. The methods of carbon-carbon bond formation are divided by nucleophilic reagents, electrophilic reagents and free radicals.

One of the fundamental processes for carbon-carbon bond formation is a reaction between a nucleophilic and an electrophilic carbon. The focus in this chapter is on organometallic compounds, enolates, imine anions, and enamines, which are carbon nucleophiles, and their reactions with alkylating agents. The nucleophilic reagents can react with various carbonyl compounds, including ketones, esters, and amides.

The crucial factors that must be considered include:

(ⅰ) the conditions for generation of the carbanions.

(ⅱ) the effect of the reaction conditions on the structure and reactivity of the nucleophile.

(ⅲ) the regio- and stereo-selectivity of the alkylation reaction. The reaction can be applied to various carbonyl compounds, including ketones, esters, and amides.

2.1 有机金属化合物 Organometallic Compounds

2.1.1 有机镁试剂 Organomagnesium Reagents

Victor Grignard reported what we now call the Grignard reaction in 1900, and he was awarded the Nobel Prize in 1912. The Grignard reaction is now one of the most important, and powerful in all of organic chemistry, and it set the stage for much of the organometallic chemistry that is so important in modern organic chemistry[2].

2.1.1.1 格氏试剂的制备与性质

Preparation and properties of Grignard reagents

The Grignard reagent (RMgX) is formed by the reaction of magnesium [Mg(0)] with an alkyl or aryl halide, usually in an ether solvent. A simple example is the reaction of bromoalkane with magnesium in ether to give RMgBr.

$$\mathrm{R{-}Br+Mg(0)\longrightarrow R{-}Br^{\cdot -}+Mg(I)}$$

$$\mathrm{R{-}Br^{\cdot -}\longrightarrow R^{\cdot}+Br^{-}}$$

$$\mathrm{R^{\cdot}+Mg(I)+Br^{-}\longrightarrow R{-}Mg{-}Br}$$

Grignard reagents can be formed from primary, secondary and tertiary halides. Alkyl chlorides, bromides or iodides can be used. The order of reactivity of the halides is RI>RBr>RCl.

In all cases, an ether solvent stabilizes the Grignard reagent by forming a Lewis acid–Lewis base charge–transfer complex such as $RMgBr\cdot(OEt_2)_2$. Coordination with ether also assists in the initial magnesium insertion reaction, and coordination minimizes decomposition of the Grignard reagent via disproportionation.

monomer(单体)　　　　dimer(二聚体)

Aryl and vinyl halides react with magnesium to form aryl magnesium halides such as phenylmagnesium bromide (PhMgBr), or vinyl Grignard reagents such as 1–propenylmagnesium bromide ($CH_3CH{=}CHMgBr$). In both cases, the C—Br bond of the aryl or vinyl bromide is stronger than the analogous C—Br bond of an alkyl halide. A stronger Lewis base is required, both to assist the insertion and to stabilize the organometallic. Therefore, a more basic solvent THF is used when aryl or vinyl Grignard reagents must be prepared. Grignard formation occurs in ether, but may be sluggish, and the yields can be poor[3].

Organomagnesium compounds are strongly basic and nucleophilic. The basicity of Grignard reagents is an important factor in their reactions with electrophilic compounds. Grignard reagents react with the slightly acidic hydrogen of water to give the hydrocarbon (R–H) and a magnesium hydroxide. Alcohols react in a similar manner. Another electrophilic reagent that reacts with Grignard reagents is molecular oxygen, which gives a hydroperoxide anion as an initial product. This anion reacts with additional Grignard reagent to give an alkoxide, and hydrolysis liberates the alcohol as the final product. The use of anhydrous solvents and exclusion of air is essential for optimum yields in the formation and reactions of Grignard reagents on small scale.

2.1.1.2 格氏试剂的反应

Reactions of Grignard reagents

The importance of the Grignard reagent is seen when the C—Br bond is contrasted with the C—Mg—Br bond. The normal bond polarization in the C—Br bond leads to a positive dipole on carbon, so it is an electrophilic carbon. The C—Mg bond in the Grignard reagent, however, generates a negative dipole on carbon, so it is a nucleophilic carbon. This ability to *invert* the polarity of a carbon atom means the Grignard reagent can function as a carbon nucleophile, usually in reactions with a carbonyl derivative.

(ⅰ) 与烷基化试剂的反应

Reaction with Alkylating Reagents

Good yields can be obtained in the reactions with allyl or benzyl halide. However, alkylations of Grignard reagents frequently proceed in very low yield (especially when the leaving group Y is a halogen) because of the intervention of side reactions (e. g. Elimination of hydrogen halide or redox processes giving radicals)[4]. One generally useful alkylation is the reaction with ROTs, Me_2SO_4.

$$RMgX + R'X' \longrightarrow R{-}R' + MgXX' (X=Cl, Br, I)$$

$$RMgX + C_6H_5CH_2Cl \longrightarrow C_6H_5CH_2R$$

$$C_6H_5MgBr + \text{2-ClC}_6H_4CH_2Cl \longrightarrow \text{2-ClC}_6H_4CH_2Ph \quad 88\%$$

$$CH_3MgI + ClCH_2C_6H_5 \longrightarrow CH_3CH_2C_6H_5 + MgICl \quad 35\%$$

$$CH_3CH{=}CH_2MgBr + BrCH_2CH{=}CH_2 \longrightarrow CH_3CH{=}CH{-}CH_2CH{=}CH_2 \quad 73\%$$

$$C_6H_5CH_2MgCl + CH_3(CH_2)_3OTs \xrightarrow{THF} C_6H_5CH_2(CH_2)_3CH_3 \quad 61\%$$

$$ArCH_2MgCl \xrightarrow{Me_2SO_4} ArCH_2Me + MgCl(OSO_2OMe)$$

$$ArCH_2MgCl \xrightarrow{n\text{-}C_4H_9OTs} ArCH_2{-}C_4H_9\text{-}n + MgCl(OTs)$$

(ⅱ) 与醛和酮的反应

Reaction with Aldehydes and Ketones

The most important reactions of Grignard reagents for synthesis involve addition to carbonyl groups. Grignard reagents react with aldehydes and ketones to give alcohols, in what is commonly called the Grignard reaction. With aldehydes and ketones, the reaction is addition: formaldehyde is converted into primary alcohols, other aldehydes into secondary alcohols and ketones into tertiary alcohols.

This reaction is shown for ketone R^1COR^2, which reacts with the nucleophilic RMgX at the electrophilic carbonyl carbon to give alkoxide. Hydrolysis is required to liberate the alcohol product.

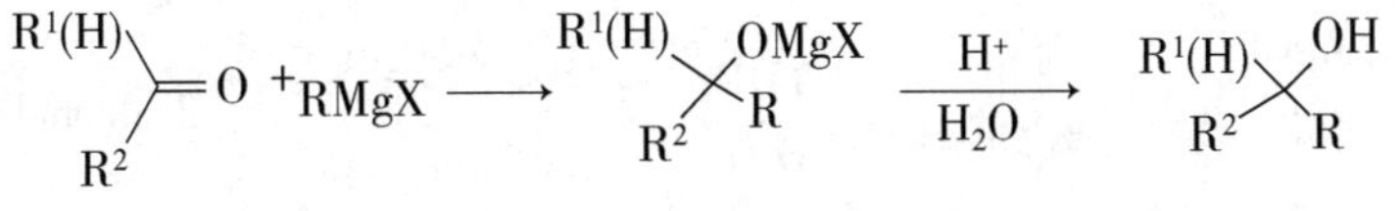

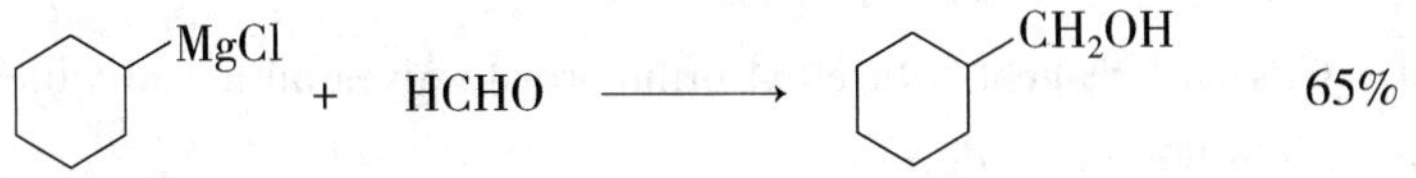

$$R^1(H)(R^2)C{=}O + RMgX \longrightarrow R^1(H)(R^2)C(OMgX)R \xrightarrow[H_2O]{H^+} R^1(H)(R^2)C(OH)R$$

$$C_6H_{11}MgCl + HCHO \longrightarrow C_6H_{11}CH_2OH \quad 65\%$$

$$(CH_3)_2CHMgBr + CH_3CHO \longrightarrow (CH_3)_2CH{-}CH(CH_3){-}OH \quad 52\%$$

$$CH_3(CH_2)_3MgBr + CH_3COCH_3 \longrightarrow CH_3(CH_2)_3{-}C(CH_3)_2{-}OH \quad 65\%$$

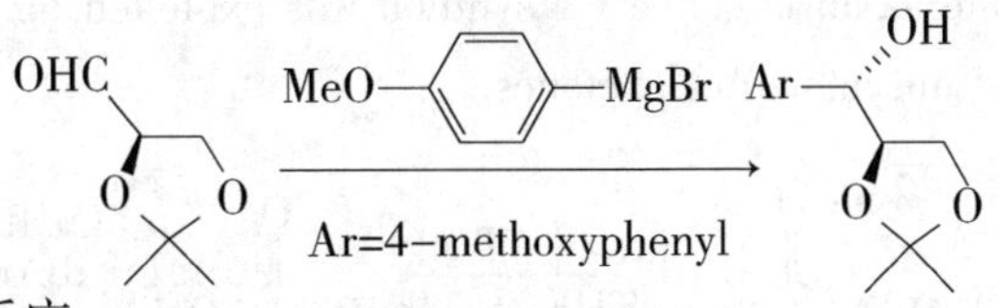

OHC–(2,2-dimethyl-1,3-dioxolan-4-yl) + MeO–C₆H₄–MgBr ⟶ Ar–CH(OH)–(2,2-dimethyl-1,3-dioxolan-4-yl)

Ar=4-methoxyphenyl

（ⅲ）与酯类的反应

Reaction with Esters

Although esters are much less reactive than acid chlorides (ethoxy is a poorer leaving group than Cl), they are only slightly less reactive than the ketone products resulting from reaction with a Grignard reagent. With esters, the first stage of the reaction is acylation, giving a ketone. It is seldom possible to stop these reactions at half-way stage, but carbonyl compounds have occasionally been isolated under special conditions e. g. when the temperature is low or when the product is sterically hindered[5].

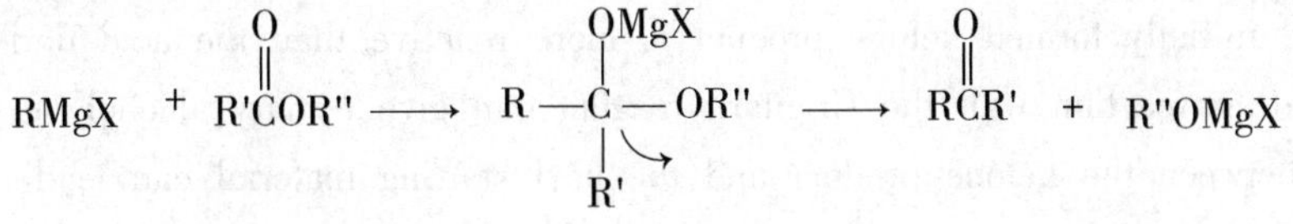

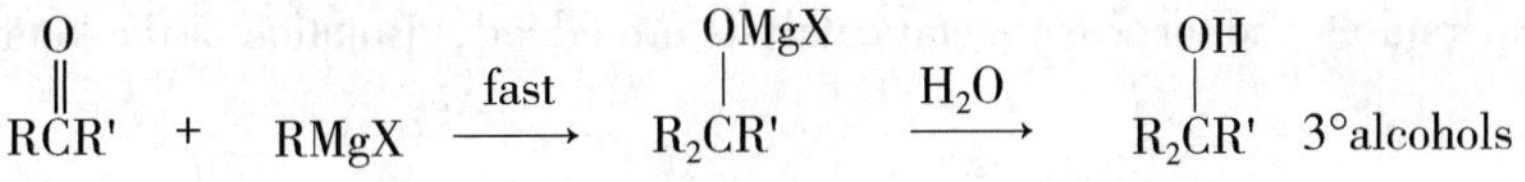

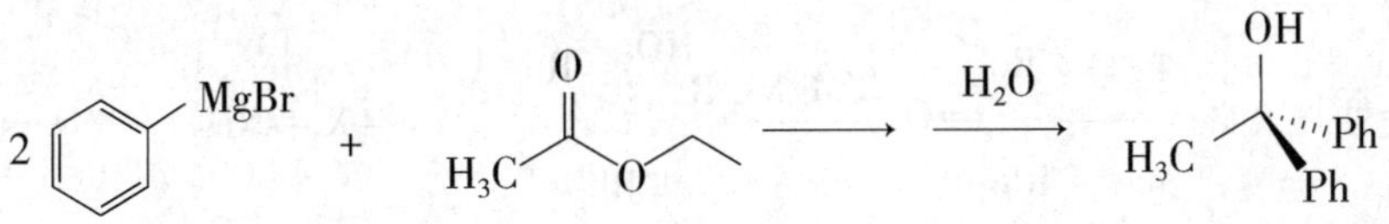

$$RMgX + R'COR'' \longrightarrow R{-}C(OMgX)(R'){-}OR'' \longrightarrow RCOR' + R''OMgX$$

$$RCOR' + RMgX \xrightarrow{fast} R_2C(OMgX)R' \xrightarrow{H_2O} R_2C(OH)R' \quad 3°\text{alcohols}$$

$$2\ C_6H_5MgBr + CH_3COOCH_2CH_3 \longrightarrow \xrightarrow{H_2O} H_3C{-}C(OH)Ph_2$$

Esters are less reactive than the intermediate ketones, therefore the reaction is only suitable for synthesis of tertiary alcohols using an excess of Grignard reagents.

$$C_6H_5MgCl + HCOC_2H_5 (O) \xrightarrow[H_3O^+]{THF} (C_6H_5)_2CH\text{-}OH \quad 2°\text{alcohols}$$

Formate esters, of course, give secondary alcohols.

The reaction of a Grignard reagent with ethyl orthoformate gives an acetal which is converted by mild acid hydrolysis into the aldehyde.

$$HC(OEt)_3 \xrightarrow[-EtOMgR]{RMgX} [EtO—CH^+—OEt \; X^-] \xrightarrow{RMgX} [Et\overset{+}{O}=C—OEt \; X^-]$$

$$RCH(OEt)_2 + MgX_2 \xrightarrow{H_3O^+} RCHO \quad 50\%$$

aldehydes

Sodium methyl carbonate as an effective C1 synthon was exploited for synthesis of carboxylic acids, benzophenones, and unsymmetrical ketones.

$$R—MgBr + \text{2eq. } MeOC(O)ONa \xrightarrow{THF,r.t.,24h} RC(O)OH \quad 53\%\sim99\%$$

X: Br,Cl
R:Ar,alkyl, alkynyl

(ⅳ) 与酰氯的反应

Reaction with Acyl Chlorides

Although the reaction with aldehydes and ketones is the most common, Grignard reagents react with most other carbonyl derivatives. Initial addition of a Grignard reagent to the acyl carbon of an acid derivative generates an alkoxide intermediate, called a tetrahedral intermediate, and displacement of the leaving group (Y) leads to a ketone[6]. Acyl halides react with nucleophilic RMgX via acyl substitution. With acyl halides, the first stage of the reaction is acylation, giving a ketone. If the initially formed ketone product is more reactive than the acid derivative starting material, further reaction with the Grignard reagent can give tertiary alcohol via alkoxide. The competition between the ketone product and the acyl starting material can lead to mixtures of products, and there maybe ignificant amounts of unreacted starting material. When the reaction is done at low temperatures and/or when metal catalysts are added, isolation of the ketone product is often possible.

$$RC(=O)Y + R'MgX \longrightarrow RC(=O)R' \xrightarrow{R'MgX} R_2R'C\text{-}OH \qquad Y=—X, —NH_2, —O—C(=O)CH_3$$

$$CH_3(CH_2)_5MgBr + CH_3CH_2CH_2\overset{O}{\overset{\|}{C}}Cl \xrightarrow{-30℃} CH_3(CH_2)_5\overset{O}{\overset{\|}{C}}(CH_2)_2CH_3 \quad 92\%$$

$$PhCH_2CH_2MgCl + H\overset{O}{\overset{\|}{C}}-N\text{(piperidine)} \longrightarrow \xrightarrow{H_2O} PhCH_2CH_2CH=O \quad 66\%\sim76\%$$

Cl + BrMg O O ⟶ O O O 78%

（ⅴ）与 CO_2的反应

Reactions with CO_2(by adding dry ice to the reaction mixture)

Carbon dioxide (CO_2, O ═C ═O) is actually a carbonyl derivative, and reacts with a Grignard reagent to give a carboxylate salt, that gives the corresponding carboxylic acid upon hydrolysis. Conversion of benzylmagnesium bromide to phenylacetic acid is an example of this transformation. This functional group interchange of a halide to an acid proceeds with a one-carbon extension of the chain.

MgBr + CO_2 $\xrightarrow[H_3O^+]{THF}$ COOH

（ⅵ）与 Schiff 碱或腈的反应

Reaction with Schiff Base or Nitriles

Cyano compounds also behave as electrophiles, nucleophilic addition with Grignard reagents to such compounds giving anions of imines. The imines themselves are frequently not stable and undergo hydrolysis to carbonyl compounds.

CH_3CH_2MgBr + CN N $\xrightarrow{Et_2O}$ H_3CH_2C C ═NMgBr N $\xrightarrow{H^+/H_2O}$ O C CH_2CH_3

$$RMgX + R'C\equiv N \longrightarrow R\overset{NMgX}{\overset{\|}{C}}R' \xrightarrow{H_2O} R\overset{O}{\overset{\|}{C}}R'$$

—C≡N + MgX ⟶ —C ═NMgX $\xrightarrow{H_3O^+}$ O —C—

Chemoselective synthesis of aryl ketones from amides and Grignard reagents via C (O)—N bond cleavage under catalyst-free conditions was exploited.

$$\mathrm{ArC(=O)CH(R)Boc} + \mathrm{BrMg{-}Ar'} \xrightarrow[-30℃,30\,min]{THF} \mathrm{ArC(=O)Ar'}$$

R: Me, Boc

Schiff base react with Grignard reagents give a final product alcohol.

$$\mathrm{C_6H_5{-}CH_2MgCl} + \mathrm{CH_3O{-}C_6H_4{-}CH{=}N{-}C_2H_5} \xrightarrow{Et_2O} \xrightarrow{H_3O^+} \mathrm{MeO{-}C_6H_4{-}CH(OH){-}CH_2{-}C_6H_5}$$

(vii) 与 α, β-不饱和酮或醛的反应

Reaction with α, β-Unsaturated Ketones or Aldehydes

α, β-Unsaturated carbonyl derivatives have two reactive sites, and nucleophilic addition of a Grignard reagent can occur at either the acyl carbon (1,2-addition) or the alkenyl carbon (1,4-addition).

Conjugated Addition:

$$\mathrm{R^1MgX} + \mathrm{{>}C{=}C{-}C{=}O} \begin{cases} \xrightarrow{1,\ 2\text{-addition}} \mathrm{{>}C{=}C{-}C(R^1)(H){-}OH} \\ \xrightarrow{1,\ 4\text{-addition}} \mathrm{{>}C(R^1){-}C(H){-}C{=}O} \end{cases}$$

When a Grignard reagent reacts with a conjugated carbonyl compound, the 1, 4-addition product is an enolate anion. The 1, 2-addition product is the usual alkoxide. In general, Grignard reagents undergo 1, 2-addition with conjugated aldehydes and relatively unhindered conjugated ketones, although conjugate alkylation often competes in the latter case. The course of this reaction varies with the steric bulk of R^1 in the Grignard reagent, and R in the carbonyl compound. As the size of R group at carbonyl carbon increases, the amount of 1, 4-addition increases, although α, β-unsaturated aldehydes usually give only 1, 2-addition. Conversely, as the R^1 group of the Grignard increases in size, less conjugate addition is observed for a given R^7.

$$\mathrm{C_2H_5MgBr+CH_3CH{=}CHCHO} \xrightarrow[(2)H_3O^+]{(1)干醚} \mathrm{CH_3CH{=}CHCH(C_2H_5)OH} \quad \text{1,2-addition}$$

$$\mathrm{C_2H_5MgBr+CH_3CH{=}CHC(=O)CH_3} \xrightarrow[(2)H_3O^+]{(1)干醚} \mathrm{CH_3CH(C_2H_5){-}CH_2C(=O)CH_3} + \mathrm{CH_3CH{=}CHC(OH)(C_2H_5)CH_3}$$

$$\mathrm{C_6H_5CH{=}CHC(=O)C(CH_3)_3 + C_2H_5MgBr} \xrightarrow[(2)H_3O^+]{(1)干醚} \mathrm{C_6H_5CH(C_2H_5){-}CH_2C(=O)C(CH_3)_3} \quad \text{1,4-addition}$$

$$C_6H_5MgBr + C_6H_5CH{=}CH\overset{O}{\overset{\|}{C}}OC_2H_5 \xrightarrow[(2)H_3O^+]{(1)干醚} (C_6H_5)_2CH{-}CH\overset{O}{\overset{\|}{C}}OC_2H_5$$ 1,4-addition

$$CH_3CH{=}CHCO_2CH_3 \xrightarrow[乙醚，Cu_2I_2]{CH_3MgI} (CH_3)_2CHCH_2CO_2CH_3$$ 1,4-addition

In general, addition of cuprous [Cu(I)] salts to a Grignard regeant facilitates conjugate addition, partly because a more highly reactive species (R_2MgCu or RCu) is formed, and partly because copper coordinates better to the carbonyl, which facilitates the six-center transition state.

$$\xrightarrow[CuCl,(CH_3)_2S]{CH_2{=}CHMgCl}$$ 1,4-addition

（viii）与环氧化物的反应

Reaction with Epoxides

Grignard reagents react with epoxides to form a new carbon-carbon bond with opening of the three-membered ring to give an alcohol after hydrolysis. There are two electrophilic carbon atoms, and the Grignard reagent attacks the less hindered carbon in an S_N2 like reaction. If the carbons of the epoxide moiety are primary or secondary, attack can occur at either carbon leading to a mixture of alcohol products. When 2-ethyloxirane reacts with methylmagnesium bromide, attack at the primary carbon gave 3-pentanol with only trace amounts of attack at the secondary carbon. When the epoxide has substituents that provide equal or close to equal steric hindrance at each carbon, as in 2-ethyl-3-propyloxirane the reaction produces a mixture of regioisomeric alcohol products[8]. When a Grignard reagent opens an epoxide, the new substituent is on the β-carbon relative to the OH.

$$RMgX + \text{(propylene oxide)} \longrightarrow H_3C{-}\underset{OMgX}{\underset{|}{CH}}{-}CH_2R + H_3C{-}\underset{R}{\underset{|}{CH}}{-}CH_2OMgX$$

main ↓ subsidiary

$$H_3C{-}\underset{OH}{\underset{|}{CH}}{-}CH_2{-}R$$

$$C_6H_5MgBr + \text{(ethylene oxide)} \xrightarrow{THF} C_6H_5CH_2CH_2OMgBr \xrightarrow{H_3O^+} C_6H_5CH_2CH_2OH$$

$$RMgX + \text{(oxetane)} \xrightarrow{干醚} RCH_2CH_2CH_2OMgBr \xrightarrow{H_3O^+} RCH_2CH_2CH_2OH$$

Synthetic procedures involving Grignard reagents are shown in the following.

A. Primary alcohols from formaldehyde.

$$\text{C}_6\text{H}_{11}\text{MgCl} + CH_2O \xrightarrow[H^+]{H_2O} \text{C}_6\text{H}_{11}\text{CH}_2\text{OH} \quad 64\%\sim69\%$$

B. Primary alcohols from ethylene oxide.

$$CH_3(CH_2)_3MgBr + H_2C\overset{O}{—}CH_2 \xrightarrow[H^+]{H_2O} CH_3(CH_2)_5OH \quad 60\%\sim62\%$$

C. Secondary alcohols from aldehydes.

$$PhCH{=}CHCH{=}O + HC{\equiv}CMgBr \xrightarrow[H^+]{H_2O} HC{\equiv}C\overset{OH}{C}HCH{=}CHPh \quad 58\%\sim69\%$$

$$\text{3-ClC}_6\text{H}_4\text{MgBr} + CH_3CH{=}O \xrightarrow{H_2O} \text{3-ClC}_6\text{H}_4\text{CHOHCH}_3 \quad 82\%\sim85\%$$

$$CH_3CH{=}CHCH{=}O + CH_3MgCl \xrightarrow{H_2O} CH_3CH{=}CH\overset{OH}{C}HCH_3 \quad 81\%\sim86\%$$

$$(CH_3)_2CHMgBr + CH_3CH{=}O \longrightarrow (CH_3)_2CH\overset{OH}{C}HCH_3 \quad 53\%\sim54\%$$

D. Secondary alcohols from formate esters.

$$2CH_3(CH_2)_3MgBr + HCO_2C_2H_5 \xrightarrow[H^+]{H_2O} (CH_3CH_2CH_2CH_2)_2CHOH \quad 83\%\sim85\%$$

E. Tertiary alcohols from ketones, esters, and lactones.

$$3C_2H_5MgBr + (C_2H_5O)_2CO \xrightarrow[NH_4Cl]{H_2O} (C_2H_5)_3COH \quad 82\%\sim88\%$$

$$2PhMgBr + PhCO_2C_2H_5 \xrightarrow{H_2O} Ph_3COH \quad 89\%\sim93\%$$

$$\text{(1,3-dioxolan-2-yl)}CH_2MgBr + \text{(diacetone ketofuranose)} \xrightarrow{LiBr} \text{product (HO, }CH_2\text{-dioxolanyl)}\quad 89\%$$

$$CH_3(CH_2)_4\text{-}\gamma\text{-lactone} + 2CH_3MgBr \xrightarrow[H^+]{H_2O} CH_3(CH_2)_4CH(OH)(CH_2)_2C(OH)(CH_3)_2 \quad 57\%$$

F. Aldehydes from triethyl orthoformate.

$$\text{9-phenanthryl-MgBr} + HC(OC_2H_5)_3 \xrightarrow[H^+]{H_2O} \text{9-phenanthryl-}CH{=}O \quad 40\%\sim42\%$$

$$CH_3(CH_2)_4MgBr + HC(OC_2H_5)_3 \xrightarrow[H^+]{H_2O} CH_3(CH_2)_4CH{=}O \quad 45\%\sim50\%$$

G. Ketones from nitriles, thioesters, amides, and anhydrides.

$$\text{9-phenanthryl-}C{\equiv}N + CH_3MgI \xrightarrow[HCl]{H_2O} \text{9-phenanthryl-}C(=O)CH_3 \quad 52\%\sim59\%$$

$$CH_3OCH_2C{\equiv}N + PhMgBr \xrightarrow[HCl]{H_2O} PhC(=O)CH_2OCH_3 \quad 71\%\sim78\%$$

$$\text{2-methyl-1,3-dithiolan-2-yl-}CH_2CH_2C(=O)S\text{-2-pyridyl} + BrMgCH_2CH{=}CHCH_2CH_3 \longrightarrow \text{2-methyl-1,3-dithiolan-2-yl-}CH_2CH_2C(=O)CH_2CH{=}CHCH_2CH_3 \quad 93\%$$

$$PhMgBr + ClCH_2C(=O)N(OCH_3)CH_3 \longrightarrow PhC(=O)CH_2Cl \quad 92\%$$

$$HC{\equiv}CCH_2CH_2C(=O)N(OCH_3)CH_3 + CH_2{=}CHMgBr \longrightarrow HC{\equiv}CCH_2CH_2C(=O)CH{=}CH_2$$

$$PhC{\equiv}CMgBr + (CH_3CO)_2O \longrightarrow PhC{\equiv}CC(=O)CH_3 \quad 80\%$$

H. Carboxylic acids by carbonation.

$$2,4,6\text{-}(CH_3)_3C_6H_2MgBr + CO_2 \xrightarrow[H^+]{H_2O} 2,4,6\text{-}(CH_3)_3C_6H_2CO_2H \quad 86\%\sim87\%$$

$$CH_3CH_2CH(MgBr)CH_3 + CO_2 \xrightarrow[H^+]{H_2O} CH_3CH_2CH(CO_2H)CH_3 \quad 76\%\sim86\%$$

$$\text{1-Cl-norbornane} \xrightarrow[(3)H^+,H_2O]{(1)\text{active Mg} \quad (2)CO_2} HO_2C\text{-norbornane} \quad 60\%\sim70\%$$

I. Amines from imines.

$$PhCH{=}NCH_3 + PhCH_2MgCl \xrightarrow{H_2O} PhCH(NHCH_3)CH_2Ph \quad 96\%$$

J. Alkenes after dehydration of intermediate alcohols.

$$PhCH{=}CHCH{=}O + CH_3MgBr \xrightarrow{H_2SO_4} PhCH{=}CHCH{=}CH_2 \quad 75\%$$

$$2PhMgBr + CH_3CO_2C_2H_5 \xrightarrow[H^+]{H_2O} Ph_2C{=}CH_2 \quad 67\%\sim70\%$$

2.1.2 C—Li 有机锂试剂 C—Li Organolithium Reagents

Grignard reagents are clearly important in synthesis. If MgX is replaced with Li, another class of organometallic reagents is available, the organolithium reagents. Organolithium reagents are also potent nucleophiles and they are more basic than Grignard reagents. Both features can be exploited in synthesis.

2.1.2.1 有机锂试剂的制备与性质 Preparation and Properties of Organolithium Reagents

Organolithium reagents, characterized by a C—Li bond, are as important in organic synthesis as the Grignard reagents[9]. Lithium is less electronegative than carbon, and the carbon of the C—Li bond is polarized, as in Grignard reagents. Organolithium reagents are expected to behave both as a nucleophile and as a base. It is important to understand the chemical properties of organolithium reagents, and to note differences with Grignard reagents before discussing their reactions.

These two reactions (lithium-halogen exchange, and lithium-hydrogen exchange) describe the main methods for preparing organolithium reagents and the latter also describes the most useful synthetic application, removal of an acidic hydrogen to generate a new organolithium species.

(ⅰ) 用金属锂制备

Preparation Using Metallic Lithium

Like Grignard reagents, organolithium compounds can often be made by treating an organic halide in ether with lithium metal.

$$R—X + 2Li \longrightarrow RLi + LiX$$

$$ClCH{=}C(OC_2H_5)_2 \xrightarrow[DTBB]{Li, 5equiv} LiCH{=}C(OC_2H_5)_2$$

4,4-di-t-butylbiphenyl(DTBB)

$$\text{(ethylene ketal)}\text{-}CH_2CH_2CH_2Cl \xrightarrow[\text{naphthalene}, -78℃]{\text{excess Li, THF}} \text{(ethylene ketal)}\text{-}CH_2CH_2CH_2Li$$

Because of the reactivity of organolithium with oxygen, the reaction is carried out in an atmosphere of dry nitrogen or argon.

For applications where the organolithium reagent is not commercially available, the reagent can be prepared by direct synthesis (by reaction of lithium metal with the appropriate alkyl halide).

(ⅱ) 锂-氢交换法制备

Preparation by Lithium-hydrogen Exchange

$$H—C\equiv C—R + R'MgBr \longrightarrow BrMgC\equiv C—R + R'H$$

$$H—C\equiv C—R + R'Li \longrightarrow LiC\equiv C—R + R'H$$

Lithium-hydrogen exchange describes the most useful synthetic application, removal of an acidic hydrogen to generate a new organolithium species.

(ⅲ) 锂-卤交换法制备

Preparation by Lithium-halogen Exchange

Lithium metal does not always react well with aryl and vinyl halides and the corresponding lithium compounds are then conveniently prepared by the metal-halogen exchange reaction using, e. g. preformed butyllithium. Coupling with alkyl halides is an important reaction of the relative stable aryllithium reagents and vinyllithium reagents.

$$\text{PhCH=CHBr} \xrightarrow[\text{pentane-THF-Et}_2\text{O}]{t\text{-BuLi}, -120℃} \text{PhCH=CHLi}$$

$$m\text{-MeOC}_6\text{H}_4\text{Br} \xrightarrow[-78℃]{n\text{-BuLi}} m\text{-MeOC}_6\text{H}_4\text{Li}$$

The degree of association is related to the solvent and structure of the organolithium, but tends to be higher for straight-chain than for branched (secondary and tertiary) organolithium reagents. The extent of association of the organolithium is important since it can affect the rate of metal-halogen or metal-hydrogen exchange, as well as the product distribution. The general reactivity of organolithium reagents in cleavage reactions of diethyl ether via Li-H exchange is following:

t-BuLi > sec-alkyl-Li > i-PrLi > cyclohexyl-Li > alkyl-Li > n-alkyl-Li > EtLi > n-alkyl-Li > n-alkyl-Li > 1-naphthyl-Li > 2-naphthyl-Li > Li-C₆H₄-C₆H₄-NMe₂ > Li-C₆H₄-C₆H₅ > CH_3Li

Organolithium reagents are more strongly nucleophilic than the corresponding Grignard reagents. Reactivity: RNa>RLi>RMgX.

2.1.2.2 有机锂化合物的反应 Reactions of Organolithium Compounds

The reactions of organolithium reagents (RLi) parallel those of Grignard reagents (RMgX) in many ways. Organolithium reagents are more reactive than the corresponding Grignard reagent[10]. Commercially available organolithium reagents can be added to carbonyl derivatives, and often provide the most efficient route to substituted alcohols. Organolithium reagents react with ketones and aldehydes via acyl addition, as expected of a carbon nucleophile. Unlike similar reactions with Grignard reagents, reduction is not a serious problem. *n*-butyllithium, for example, added readily to the hindered ketone di-*tert*-butyl ketone to give alcohol by direct addition to the carbonyl. Organolithium reagents usually prefer 1,2-addition in conjugated systems due primarily to their enhanced nucleophilicity and poor coordinating ability relative to Grignard reagents. Indeed, reaction of methyllithium with 3-penten-2-one gave < 1% of conjugate addition, with 2-methyl-3-penten-2-ol being the major product. Note that addition of cuprous iodide to the reaction led to>99% of conjugate addition.

(i) 烷基化反应

Reactions with Alkylating Agents

Organolithium compounds are more reactive than Grignard reagents towards alkyl halides and the reaction is called Wurtz coupling reaction.

$$PhCH_2Li + H_3C\text{-}C(H)(CH_2CH_3)\text{-}Br \longrightarrow PhCH_2\text{-}C(H)(CH_2CH_3)\text{-}CH_3$$

58% yield
100% inversion

$$CH_3(Br)C{=}C(H)CH_3 \xrightarrow[(2)CH_3(CH_2)_3I]{(1)Li} CH_3(CH_3(CH_2)_3)C{=}C(H)CH_3$$

$$\text{2-F-}C_6H_4Li + C_2H_5Br \xrightarrow[-70℃]{KOtBu,\ THF} \text{2-F-}C_6H_4C_2H_5$$

75%

Alkenyllithium reagents can be alkylated in good yields by alkyl iodides and bromides. However, the reactions of aryllithium reagents are accelerated by inclusion of potassium alkoxides.

（ⅱ）与酮的反应

Reactions with Ketones

$$(CH_3)_2CH\text{—}\overset{O}{\overset{\|}{C}}\text{—}CH(CH_3)_2 \xrightarrow{(CH_3)_2CHLi} (CH_3)_2CH\text{—}C(OH)(CH(CH_3)_2)\text{—}CH(CH_3)_2$$

Organolithium reagents are less subject to steric hindrance.

（ⅲ）与 α，β-不饱和酮或醛的反应

Reactions with α,β-Unsaturated Ketones or Aldehydes

Whereas Grignard reagents often react with α,β-unsaturated ketones predominantly by 1,4-addition, lithium reagents prefer 1,2-addition rather than conjugate addition in conjugated systems[11].

$$PhCH{=}CH\text{—}\overset{O}{\overset{\|}{C}}\text{—}Ph \xrightarrow{PhMgX} Ph_2CHCH_2\overset{O}{\overset{\|}{C}}\text{—}Ph \quad \text{1,4-addition}$$

$$PhCH{=}CH\text{—}\overset{O}{\overset{\|}{C}}\text{—}Ph \xrightarrow{PhLi} PhCH{=}CH\text{—}C(OH)(Ph)\text{—}Ph \quad \text{1,2-addition}$$

（ⅳ）与 CO_2 或 RCOOH 的反应

Reactions with CO_2 or RCOOH

Carbon dioxide, which reacts with Grignard reagents to give carboxylic acids, react with organolithium compounds to give ketones. The lithium compounds are more strongly nucleophilic than Grignard reagents and they are able to react with the intermediate resonance - stabilized carboxylate anion.

$$RLi + CO_2 \longrightarrow RCOOLi \xrightarrow{RLi} R_2C(OLi)-OLi \xrightarrow{H_3O^+} R_2CO$$

$$R-COOH \xrightarrow[-RH]{RLi} RCOOLi \xrightarrow{R'Li} RC(OLi)(R')-OLi \xrightarrow[-2LiOH]{H_3O^+} R_2CO$$

$$\text{C}_6\text{H}_{11}\text{-COOH} \xrightarrow[(2)TMS\text{-}Cl \quad (3)H^+/H_2O]{(1)4eq.CH_3Li} \text{C}_6\text{H}_{11}\text{-C(=O)-CH}_3 \quad 92\%$$

The reaction of a Grignard reagent and a carboxylic acid gives the carboxylate salt. Reaction of an organolithium reagent with a carboxylic acid gives the expected lithium carboxylate salt, but a second equivalent adds to the carboxylate to give a ketone.

（V）有机锂试剂与其他官能团的反应

Reactions of Organolithium Reagents with Other Functional Groups

Organolithium reagents react with carbonyl compounds other than aldehydes and ketones analogous to Grignard reagents. Esters usually react to give a tertiary alcohol, and acid chlorides give mixtures of unreacted starting material and tertiary alcohol. Adding an excess of the organolithium reagent gives the tertiary alcohol exclusively, which is formally analogous to the acyl substitution reactions noted for Grignard reagents.

Me, MeO, Me, O, CO_2Me $\xrightarrow[0\sim5℃,3h]{MeLi,ether}$ Me, MeO, Me, O, OH 95%

If an ester is treated with excess methyllithium, a tertiary dimethyl carbinol results as in the conversion of ester to alcohol. Careful selection of RLi and the reaction conditions often allow the synthesis of ketones rather than tertiary alcohols, however.

$$PhCH_2C(=O)NH_2 \xrightarrow[THF\text{-}hexane]{3n\text{-}C_4H_9Li} [PhCH(Li)CH=N-Li] \longrightarrow [PhCH(Li)C\equiv N] \xrightarrow{H_3O^+} PhCH_2C\equiv N \quad 72\%$$

Primary amides react with excess organolithium to give a nitrile, as in the conversion of phenylacetamide to benzonitrile in 72% yield. Reaction of phenylacetamide with three equivalents of butyllithium gave the intermediate trilithiated species, fragmentation and hydrolysis gave benzonitrile.

Nitriles can be hydrolyzed to or prepared from carboxylic acids, and nucleophiles can attack the electrophilic carbon, so they are considered to be acid derivatives. 3-Thienyllithium reacted with 4-methylpyridine carbonitrile, for example, to give a *N*-lithio imine, analogous to the reaction of Grignard reagents[12]. Subsequent acid hydrolysis converted the initially formed imine to the ketone.

(ⅵ) 与环氧化物的反应

Reactions with Epoxides

Epoxides are opened to the alcohol as expected, untroubled by the MgX_2 catalyzed rearrangements often observed with Grignard reactions. Methyllithium reacted with epoxide to give the alcohol, in 68% yield. The methyllithium added selectively to the allylic position to give the (*E*)-carbinol. In general, the organolithium attacks the epoxide at the less sterically hindered (less substituted) carbon, as was noted with Grignard reagents.

(ⅶ) 与烯烃的反应

Reactions with Alkenes

Organolithium compounds, unlike Grignard reagents, react with C═C bonds. Secondary and tertiary alkyllithium react efficiently with ethylene at low temperature to give monomeric products.

$$H_3C\text{-}CH_2\text{-}CH(CH_3)\text{-}Li + CH_2{=}CH_2 \xrightarrow{-25℃} H_3C\text{-}CH_2\text{-}CH(CH_3)\text{-}CH_2\text{-}CH_2\text{-}Li$$

2.1.3 有机锂试剂 Organocopper Reagents

Two types of copper(Ⅰ) compound are of value in synthesis: organocopper compounds, RCu, and lithium organocuprates, R_2CuLi, where R is alkyl (primary, secondary, or tertiary), alkenyl, or aryl.

2.1.3.1 有机铜化合物与有机铜锂的制备与性质 Preparation and Properties of Organocopper Compounds and Lithium Organocuprates

Organocopper compounds and lithium organocuprates can be prepared from the corresponding lithium compounds with copper(Ⅰ) iodide in an aprotic solvent such as an ether and in an inert atmosphere.

$$RLi + CuI \longrightarrow RCu + LiI$$

$$2RLi + CuI \longrightarrow R_2CuLi + LiI$$

$$2\ n\text{-}BuLi + CuI \xrightarrow[-20℃]{ether} n\text{-}Bu_2CuLi$$

2.1.3.2 有机铜锂的反应 Reaction of Lithium Organocuprates

(ⅰ) 烷基化的偶联反应

Coupling Reactions with Alkylating Agents

$$R_2CuLi + R'X \longrightarrow R\text{-}R' + LiX$$

$$CH_3Cu + CH_3(CH_2)_9I \longrightarrow CH_3(CH_2)_9CH_3 \quad 68\%$$

$(CH_3)_2CuLi$ + 2-iodocyclohexanol (OH, I) ⟶ 2-methylcyclohexanol (OH, CH_3)

In addition to the expected displacement of alkyl halides, organocuprates react with vinyl halides and aryl halides to give good yields of coupling product.

$(CH_3)_2CuLi$ + (H, Br / Ph, H alkene) ⟶ (H, CH_3 / Ph, H alkene) 81%

$$Ph_2CuLi \xrightarrow[O_2]{THF,\ -78℃} Ph\text{—}Ph$$

(ⅱ) 与酮或醛的反应

Reactions with Ketones or Aldehydes

The reaction of organocuprates with aldehydes is fast. Dimethylcuprate reacted with benzaldehyde to give phenylethyl alcohol at < −90℃. Ketones react more slowly, as illustrated by the reaction of dimethylcuprate and 5-nonanone which occurred at about −10℃ to give 5-methyl-5-nonanol[13]. No reaction occurred at lower temperatures because ketones react very slowly with cuprates. Addition of chlorotrimethylsilane to the reaction can alleviate this problem.

(ⅲ) 与 α, β-不饱和酮或醛的反应

Reactions with α, β-Unsaturated Ketones or Aldehydes

The organocuprates derived from organolithium reagents (R_2CuLi) are extremely useful in conjugate addition reactions. Organocopper reagents derived from the reaction of a Grignard reagent with a cuprous salt, however, are also very useful. Reaction takes place specifically in the 1,4-manner. The organocuprates react more readily and usually give better yields than organocopper compounds. However, as the equation above shows, one of the R groups in the R_2CuLi is "wasted" as RH, and if this group is difficult to synthesize or particularly expensive to obtain, it is better to employ a mixed organocuprate.

$$R_2CuLi + CH_3CH{=}CH\overset{\overset{\Large O}{\|}}{C}CH_3 \longrightarrow CH_3\underset{\underset{\Large R}{|}}{C}HCH_2\overset{\overset{\Large O}{\|}}{C}CH_3$$

$(CH_3)_2CuLi$ + 3-methylcyclohex-2-enone ⟶ 3,3-dimethylcyclohexanone 98%

(ⅳ) 与酰氯的反应

Reactions with Acyl Chlorides

$$(CH_3)_2CuLi + (CH_3)_3C\overset{\overset{\Large O}{\|}}{C}Cl \longrightarrow (CH_3)_3C\overset{\overset{\Large O}{\|}}{C}CH_3$$

The reaction usually requires low temperatures to isolate the ketone product.

(ⅴ) 与环氧化合物的反应

Reactions with Epoixdes

$$(CH_3)_2CuLi + \text{methyloxirane} \longrightarrow CH_3\overset{\overset{\Large OH}{|}}{C}HCH_2CH_3$$

1-methylcyclohexene oxide (Me) $\xrightarrow[(2)H^+]{(1)R_2CuLi}$ product (Me, OH, R)

S_N2 mechanism

Organocuprates also react with epoxides, with addition occurring at the less sterically hindered carbon, typical of a nucleophilic addition.

2.1.4 有机锌试剂 Organozinc Reagents

The formation of ester - stabilized organozinc reagents and their addition to carbonyl compounds. Organozinc compounds are prepared from α-halogenesters in the same manner as Grignard reagents. This reaction is possible due to the stability of esters against organozincs. The enolate-like reaction is the Reformatsky reaction, which involves the formation of ester-stabilized organozinc reagents, generated from an α - halo carbonyl and zinc metal. The addition of organozinc reagent to carbonyl compounds gives the condensation product, β-hydroxyester.

An ester-stabilized organozinc reagent:

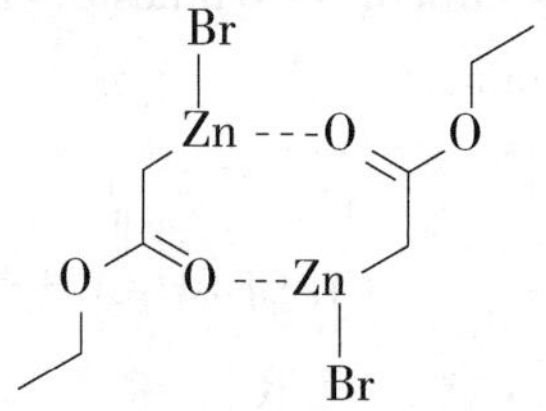

The organozinc reagent is generally less reactive than a Grignard or organolithium reagent, and condensation reactions proceed well with aldehydes and ketones but are sluggish with esters.

Preparation of the organozinc complex in the Reformatsky reaction can be a problem, and it often requires special preparation of the zinc (activated zinc). The use of ultrasound techniques produces a finely dispersed zinc that also facilitates the Reformatsky reaction.

$$C_6H_5CHO + BrCH_2COOC_2H_5 \xrightarrow[(2)H^+/H_2O]{(1)Zn} C_6H_5CH(OH)CH_2COOC_2H_5$$

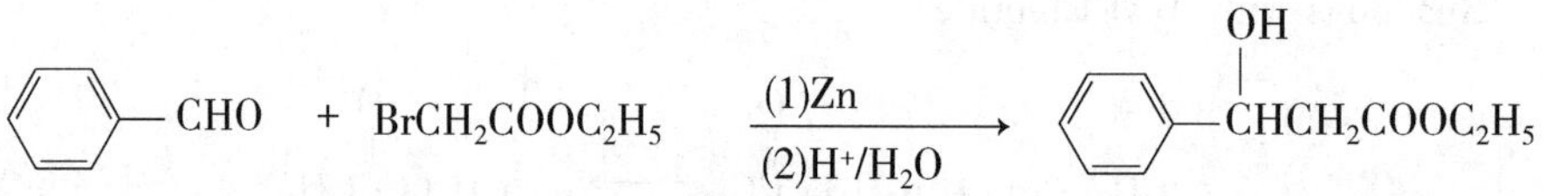

α-haloesters (bromoesters) react with aldehydes or ketones and metallic zinc to give β-hydroxyesters or α,β-unsaturated carbonyl compounds. Aldehydes generally show poorer selectivity than do ketones for formation of the anti-product. The bromozinc aldolate products from ketones were shown to equilibrate under reaction conditions, but those from aldehydes did not. As noted above, hydrolysis gives the hydroxy ester to complete this two-carbon chain extension process.

2.1.5 有机钯试剂
Organopalladium Reagents

Palladium is often used as a catalyst in the reduction of alkenes and alkynes with hydrogen. This process involves the formation of a palladium-carbon covalent bond. Palladium is also prominent in carbon-carbon coupling reactions, as demonstrated in tandem reactions. In contrast to its next-door neighbors the group 11 elements, the element palladium in organic chemistry does not involve preparation of organopalladium compounds itself but rather organopalladium reactive intermediates. On top of that in many reactions only catalytical amounts of the metal are used, known as Pd-catalyzed cross-coupling reactions.

The 2010 Nobel Prize in Chemistry was awarded by the Royal Swedish Academy of Sciences to Professor Richard F. Heck, Professor Ei-ichi Negishi and Professor Akira Suzuki for "palladium-catalyzed cross-couplings in organic synthesis". The discoveries made by the these three organic chemists have had a great impact on academic research and industrial applications. Their reactions have proved extremely powerful and are widely used for the synthesis of organic electronic materials, new drugs, pharmaceuticals and biologically active compounds.

2.1.5.1 Heck 反应 Heck Reaction

The palladium-catalyzed C—C coupling between aryl halides or vinyl halides and activated alkenes in the presence of a base is referred as the "Heck Reaction". Recent developments in the catalysts and reaction conditions have resulted in a much broader range of donors and acceptors being amenable to the Heck reaction.

One of the benefits of the Heck reaction is its outstanding trans selectivity.

RX + R'CH=CHR'' —(PPh_3, $Pd(OAc)_2$)→ R(R')C=CH(R'')

RPdX —(R'CH=CHR'')→ R'CH=CHR'' ⋯ RPdX —(Syn加成)→ R(R')CH–CH(PdX)R'' ≡ —(−HPdX, Syn消除)→ R(R')C=CHR''

Alkene + PhHal —(4; Solvent, Base, T℃)→ PhCH=CHR

Triethanolamine as an efficient and reusable base, ligand and reaction medium for phosphane-free palladium-catalyzed Heck reactions was developed in 2006.

Ar—X + CH₂=CHR —(1mol% $Pd(OAc)_2$, $N(CH_2CH_2OH)_3$, 100℃, 10h)→ ArCH=CHR

X: Br, I

A palladium-catalyzed decarbonylative alkenylation of various benzoic acids with terminal alkenes provides the corresponding internal alkenes in very good yields. The conversion of cinnamic acids and bioactive benzoic acids such as 3-methylflavone-8-carboxylic acid, probenecid, adapalin, and febuxostat demonstrates the synthetic value of this new reaction.

Ar—CO_2H + 2eq. $CH_2=CHC(O)NR_2$ —(5mol% $Pd(TFA)_2$, 5~10mol% dppb, 1.5eq. Piv_2O, 0.5eq. NaCl, dioxane or THF, 150℃, 10h)→ ArCH=CHC(O)NR_2

R: 1° alkyl

2.1.5.2 Suzuki 反应 Suzuki Reaction

Suzuki coupling, which is the palladium-catalysed cross coupling between organoboronic acid and halides. The scope of the reaction partners is not restricted to aryls, but includes alkyls, alkenyls and alkynyls. Potassium trifluoroborates and organoboranes or boronate esters may be used in place of boronic acids. Some pseudohalides (for example triflates) may also be used as coupling partners.

$$R{-}X + R^1{-}B(R^2)_2 \xrightarrow[NaOR^3]{L_2Pd(0)} R{-}R'$$

Br–C₆H₄–R + C₆H₅–B(OH)₂ → (2eq. K_2CO_3aq., 3mol% $Pd(PPh_3)_4$, benzene, Δ) → C₆H₅–C₆H₄–R

Mechanism:

$$R{-}X + L_2Pd(0) \xrightarrow{\text{oxidative addition}} (R)(L)Pd(L)(X)$$

$$R^1{-}BH(R^2)_2 + NaOR^3 \longrightarrow R^1{-}\underline{B}(OR^3)(R^2)_2$$

(both) $\xrightarrow{\text{transmetalation}}$

$$(R)(L)Pd(L)(R^1) + R^3O{-}B(R^2)_2 \xrightarrow[\text{elimination}]{\text{reductive}} R{-}R^1 + L_2Pd(0)$$

In part due to the stability, ease of preparation and low toxicity of the boronic acid compounds, there is currently widespread interest in applications of the Suzuki coupling, with new developments and refinements being reported constantly.

The Pd-catalyzed cross-coupling of phenyl esters and alkyl boranes with a Pd-NHC system give alkylketones. Use of a Pd-dcype catalyst enables alkylated arenes to be synthesized by a modified pathway with extrusion of CO via a Suzuki-Miyaura reaction proceeding by activation of the C(acyl)-O bond.

Note:dcype R_2P–CH₂CH₂–PR_2 *R*=Cyclohexyl

Ar–C(=O)–OPh + 1.5eq. 9-BBN–CH₂CH₂–R $\xrightarrow[\text{1.5eq.}Cs_2CO_3,\text{1.5eq.}H_2O\text{, toluene, 60℃, 16h}]{\text{5mol\% Pd(IPr)(cinnamyl)Cl}}$ Ar–C(=O)–CH₂CH₂–R

R:benzyl, alkyl

2.1.5.3 Negishi 反应 Negishi reaction

The Negishi coupling published in 1977, was the first reaction that allowed the preparation of unsymmetrical biaryls in good yields. The versatile nickel- or palladium-catalyzed coupling of organozinc compounds with various halides (aryl, vinyl, benzyl, or allyl) has broad scope, and is not restricted to the formation of biaryls.

$$R^1—X + R^2Zn—Y \xrightarrow[\text{solvent}]{NiL_n \text{ or } PdL_n} R^1—R^2$$

R^1=aryl, alkenyl, alkynyl, acyl
R^2=aryl, heteoaryl, alkenyl, allyl, homoallyl, homopropargyl
X=Cl, Br, I, OTf
Y=Cl, Br, I
L_n=PPh_3, dba, dppe

A silyl-Negishi reaction between secondary zinc organometallics and silicon electrophiles provides direct access to alkyl silanes.

$$R—ZnBr + I—SiR'_3 \xrightarrow[\text{1eq.NEt}_3\text{, dioxane,r.t.,1~4h}]{\text{1 mol\% (ligand)}_2\text{PdI}_2} R—SiR'_3$$

SiR'_3: $SiMe_2Ph$, $SiMe_3$, $SiEt_3$, $SiMe_2Bn$, $SiMePh_2$

ligand: $P(3,5\text{-}tBu_2\text{-}C_6H_3)_3$

2.1.5.4 Sonogashira 耦合反应 Sonogashira Coupling Reactions

This coupling of terminal alkynes with aryl or vinyl halides is performed with a palladium catalyst, a copper(Ⅰ) cocatalyst, and an amine base[14]. Typically, the reaction requires anhydrous and anaerobic conditions, but newer procedures have been developed where these restrictions are not important.

$$PhX + H—C\equiv C—R' \xrightarrow[\text{CuI,R}_3\text{N}]{(Ph_3P)_2PdCl_2} Ph—C\equiv C—R'$$

$$R'R''C=C(R''')X + H—C\equiv C—R \xrightarrow[\text{CuI,Et}_2\text{NH}]{Pd(PPh_3)_4} R'R''C=C(R''')—C\equiv C—R$$

Ynamides are moisture-sensitive and prone to hydration especially under acidic and heating conditions. An environmentally benign, robust Sonogashira coupling enables the synthesis of sulfonamide-based ynamides and arylynamines in water, using a readily available quaternary ammonium salt as the surfactant.

(1)2mol% $PdCl_2(PPh_3)_2$, 1mol% CuI, NEt_3, 55℃, 1-2h
(2)removal of volatiles
(3)0.1eq.CuI, DMF, 100℃, 1-5h

R-(2-X-aryl)-CH=N-t-Bu + 1,2eq. H-C≡C-R → R-isoquinoline (3-R')

R': alkyl, Ar, alkenyl

2.1.5.5 Stille 反应 Stille Reaction

$$R'—X + RSnBu_3 \xrightarrow{Pd-Cat} R'—R + XSnBu_3$$

The Stille coupling is a versatile C–C bond forming reaction between stannanes and halides or pseudohalides, with very few limitations on the R-groups. Well-elaborated methods allow the preparation of different products from all of the combinations of halides and stannanes depicted below. The main drawback is the toxicity of the tin compounds used, and their low polarity, which makes them poorly soluble in water. Stannanes are stable, but boronic acids and their derivatives undergo much the same chemistry in what is known as the Suzuki coupling. Improvements in the Suzuki coupling may soon lead to the same versatility without the drawbacks of using tin compounds.

$$\underset{X: Br, Cl}{Ar—X} + \underset{1.1eq.}{Bu_3Sn—R'} \xrightarrow[\text{DMF, } 45℃\ (-Br) \text{ or } 100℃\ (-Cl),\ 15h]{2mol\%PdCl_2,\ 4mol\%PtBu_3,\ 4mol\%CuI,\ 2\ eq.\ CsF} \underset{R': Ar, vinyl}{Ar—R'}$$

Stille coupling made easier – the synergic effect of copper(i) salts and the fluoride ion.

2.1.6 有机钠试剂 Organosodium Reagents

The Wurtz coupling is one of the oldest organic reactions, and produces the simple dimer derived from two equivalents of alkyl halide. The intramolecular version of the reaction has also found application in the preparation of strained ring compounds:

$$2R—X + 2Na \longrightarrow R—R + 2NaX$$

$$Br—\diamond—Cl + 2Na \longrightarrow \text{(bicyclobutane)} + 2NaX$$

Using two different alkyl halides will lead to an approximately statistical mixture of products. A more selective unsymmetric modification is possible if starting materials have different rates of reactivity.

Mechanism of the Wurtz reaction:

$$R—X + 2Na \longrightarrow R^-Na^+ + NaX$$

$$R^-Na^+ + R—X \longrightarrow R—R + NaX$$

Side products:

$$R'—CH_2—\bar{C}H_2 + R'CH(H)—CH(H)—Br \xrightarrow{-Br^-} RCH_2CH_3 + RCH{=}CH_2$$

Wurtz-Fittig Reaction:

This reaction allows the alkylation of aryl halides. The more reactive alkyl halide forms an organosodium first, and this reacts as a nucleophile with an aryl halide as the electrophile. Excess alkyl halide and sodium may be used if the symmetric coupled alkanes formed as a side product may be separated readily.

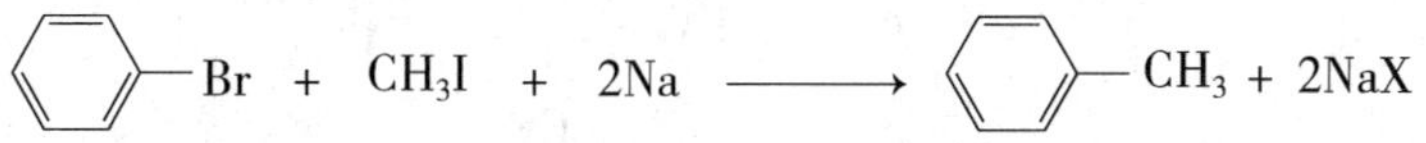

$$C_6H_5\text{-}Br + CH_3I + 2Na \longrightarrow C_6H_5\text{-}CH_3 + 2NaX$$

2.1.7 用于烯烃复分解的过渡金属卡宾配合物 Transition Metal Carbene Complexes for Olefin Metathesis

The word metathesis means "change-places". In metathesis reactions, double bonds are broken and made between carbon atoms in ways that cause atom groups to change places. This happens with the assistance of special catalyst molecules. Metathesis can be compared to a dance in which the couples change partners.

In 1971 Yves Chauvin was able to explain in detail how metathesis reactions function and what types of metal compound act as catalysts in the reactions. Now the "recipe" was known. The next step was, if possible, to develop the actual catalysts. Richard Schrock was the first to produce an efficient metal-compound catalyst for metathesis. This was in 1990. Two years later Robert Grubbs developed an even better catalyst, stable in air, that has found many applications.

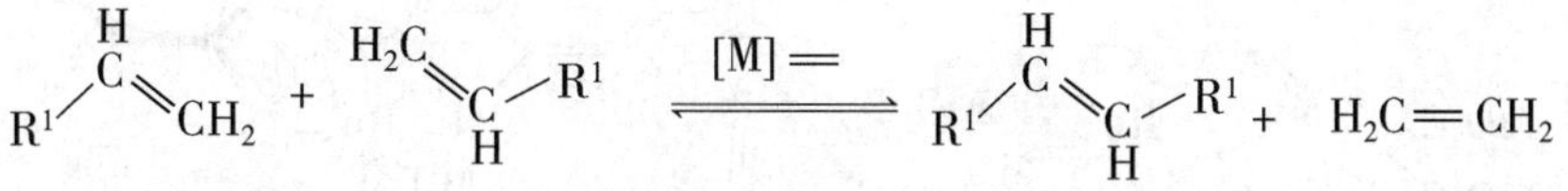

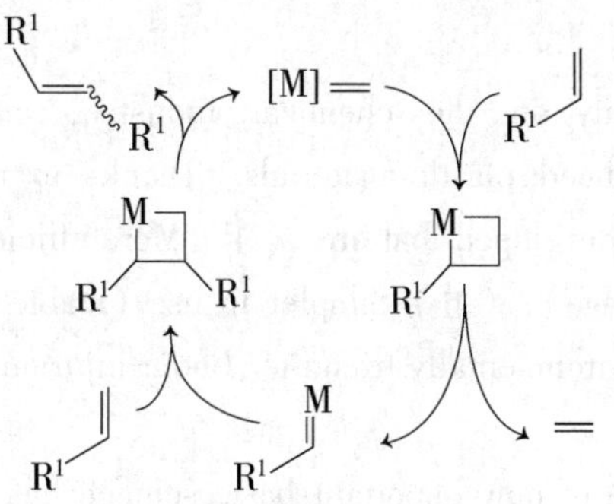

$$R^1CH{=}CH_2 + H_2C{=}CHR^1 \xrightleftharpoons{[M]=} R^1CH{=}CHR^1 + H_2C{=}CH_2$$

Chauvin catalytic cycle

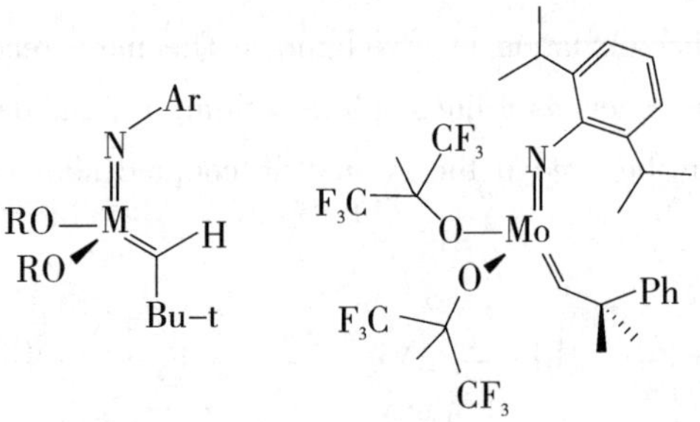

Schrock' s catalysts

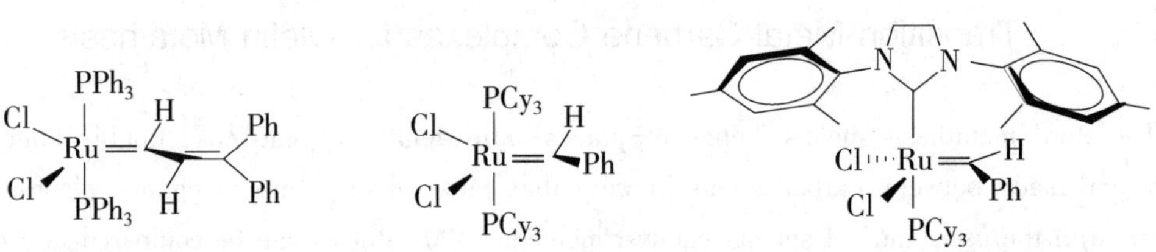

Grubbs,1992 Grubbs, 1995 Grubbs,1999

Grubbs' s catalysts

There is room for improvement in catalyst efficiency. Warsaw found that the following Ru complex is several times faster than the widely used G2 ruthenium catalyst, yielding a TON of 3200 in a ring-closing metathesis.

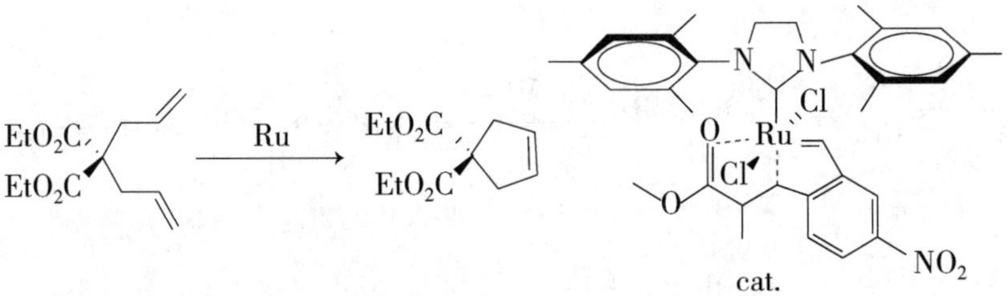

cat.

Metathesis is used daily in the chemical industry, mainly in the development of pharmaceuticals and of advanced plastic materials. Thanks to the Laureates' s contributions, synthesis methods have been developed that are: (ⅰ) More efficient (fewer reaction steps, fewer resources required, less wastage); (ⅱ) Simpler to use (stable in air, at normal temperatures and pressures) and (ⅲ) environmentally friendlier (non-injurious solvents, less hazardous waste products).

Metathesis is an example of how important basic science has been applied for the benefit of man, society and the environment. The 2005 Nobel Prize in Chemistry has been awarded to Yves Chauvin, Robert H. Grubbs and Richard R. Schrock for the development of the metathesis method in organic synthesis.

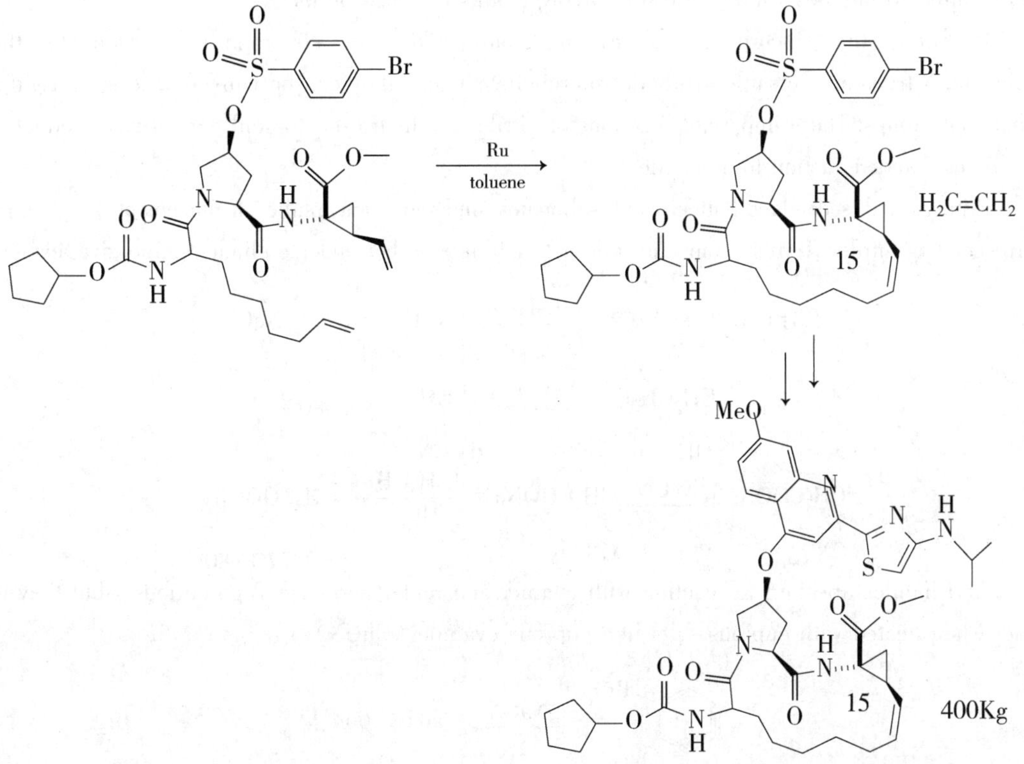

Ru
toluene
$H_2C{=}CH_2$
15
MeO
400Kg

Simeprevir（丙肝药物：司美匹韦）

2.2 稳定的碳负离子 Stabilized Carbanions

2.2.1 Cyanide (CN^-)

Cyanide ion is a powerful nucleophile that offers a simple but convenient method for generating carbon-carbon bonds. When cyanide displaces a leaving group in a S_N2 reaction, the carbon chain is extended by one carbon, and the nitrile can be converted to a variety of other functional groups.

$$RX+NaCN \longrightarrow RCN$$

$$X=Br,\ Cl,\ I,\ OSO_2R$$

Since the reaction with an alkyl halide is an S_N2 process, the best yields of nitrile are obtained with primary and secondary substrates, whereas tertiary halides such as 2-chloro-2-methylbutane sometimes react via elimination to give the alkene. Cyanide displacement of secondary halides often gives poor yields, however, the use of polar, aprotic solvents such as DMSO, DMF or THF gives the best yields of nitriles such as hexanenitrile (capronitrile, 97%

yield) after 20 minutes. Sulfonate ester leaving groups can also be used.

Reactions that incorporate a cyano group add one electrophilic carbon to that molecule. Nitriles are versatile synthetic intermediates since they can be converted to several other functional groups. These important reactions of nitriles are hydrolysis to acid derivatives, reduction to amines, and reduction to aldehydes.

Primary and secondary halides and sulfonates undergo nucleophilic displacement to give the corresponding nitriles. Tertiary compounds do not yield nitriles but undergo elimination to give alkenes.

$$C_2H_5CH_2Cl + NaCN \xrightarrow{C_2H_5OH} C_6H_5CH_2CN + NaCl \quad 90\%$$

$$\underset{OH}{CH_2}CH_2Cl \xrightarrow{NaCN} \underset{OH}{CH_2}\underset{CN}{CH_2} \quad 79\%\sim80\%$$

$$\underset{CN}{CH_2}COONa \xrightarrow{NaCN} \underset{CN}{CH_2}COONa \xrightarrow[HCl]{C_2H_5OH} \underset{CN}{CH_2}COOC_2H_5 \quad 77\%\sim80\%$$

Aryl halides are inert to reaction with cyanide ion under normal S_N2 conditions, but they do react when heated with cuprous salts like cuprous cyanide (CuCN).

$$C_6H_5Br \xrightarrow{NaCN} \text{no reaction}$$

$$C_6H_5NH_2 \xrightarrow[0℃]{NaNO_2,HCl} C_6H_5N_2Cl \xrightarrow[KCN]{CuCN} C_6H_5CN \quad 64\%\sim70\%$$

Reactions of cyanide are not limited to alkylation of halides and sulfonate esters. Epoxides are also opened at the less sterically hindered carbon by cyanide via an S_N2-like process.

Other acid derivatives can also be prepared with HCN and carbonyl derivatives. Addition of HCN to a ketone or aldehyde gives a cyanohydrin in a reversible reaction, and hydrolysis gives the α-hydroxy acid.

$$R-\overset{O}{\overset{\|}{C}}H \xrightarrow[OH^-]{HCN} R-\underset{H}{\overset{OH}{C}}-CN \xrightarrow{H^+} R-\underset{H}{\overset{OH}{C}}-COOH$$

$$C_6H_5CH=CH\overset{O}{\overset{\|}{C}}C_6H_5 + HCN \longrightarrow C_6H_5\underset{CN}{HC}-CH\overset{O}{\overset{\|}{C}}C_6H_5$$

$$\xrightarrow{\text{1,4-addition}} \left[C_6H_5\underset{CN}{HC}-\underset{H}{C}=\overset{OH}{C}C_6H_5\right] \xrightarrow{\text{Rearrangement}}$$

Cyanide ion acts as a carbon nucleophile in the conjugate addition reaction.

$$Ph\text{-}CH{=}CH\text{-}C(O)Ph + HCN \longrightarrow NC\text{-}CH(Ph)\text{-}CH_2\text{-}C(O)Ph$$

$$\xrightarrow[\text{EtOH–H}_2\text{O}]{\text{KCN,NH}_4\text{Cl}}\quad (12\%) + (42\%)$$

2.2.2 Alkyne Anions (R−C≡C : ⁻)

The conjugate base of an alkyne is an alkyne anion, and it is generated by reaction with a strong base and is a carbanion. It functions as a nucleophile (a source of nucleophilic carbon) in S_N2 reactions with halides and sulfonate esters. Alkyne anions react with ketones, with aldehydes via nucleophilic acyl addition and with acid derivatives via nucleophilic acyl substitution. Alkyne anions are important carbanion synthons for the creation of new carbon−carbon bonds[15].

Acetylene and other terminal alkynes have an acidic hydrogen atom (C≡C—H), and they are weak acids. A strong base is required to remove that proton to give an alkyne anion, and this carbanion is a carbon nucleophile that reacts with alkyl halides or sulfonate esters (R^1−X where R^1=Me, 1°, 2° alkyl, and X=halogen or OSO_2R) via an S_N2 sequence to give disubstituted alkynes.

$$RC{\equiv}CNa + BrR' \xrightarrow[\text{Na}]{\text{NH}_3\text{(l)}} RC{\equiv}CR'$$

$$HO\text{(CH}_2)_n\text{C}{\equiv}CH \xrightarrow{\text{2LiNH}_2\text{(THF/NH)}_3} HO\text{(CH}_2)_n\text{C}{\equiv}CLi$$

$$\xrightarrow[\text{(2) diluteHCl } \triangle]{\text{(1)Br(CH}_2)_m\text{COOH}} HO\text{(CH}_2)_n\text{C}{\equiv}C\text{(CH}_2)_m\text{COOH}$$

$$H_3CC{\equiv}CH \xrightarrow[\text{Et}_2\text{O–78℃}]{n\text{–BuLi}} \xrightarrow[\text{Et}_2\text{O0℃}]{\text{ClCO}_2\text{Et}} H_3CC{\equiv}CCO_2Et \quad 95\%\text{~}97\%$$

As with cyanide, S_N2 reactions of alkyne anions can be done with substrates other than halides or sulfonate esters. Epoxides are opened by acetylides at the less sterically hindered carbon to give an alkynyl alcohol.

Alkyne anions react with ketones or aldehydes to form α−acetylenic alcohols, which is the other major synthetic use of alkyne anions.

$$(CH_3)_2C{=}O + H{-}C{\equiv}C{-}H \xrightarrow{\text{NH}_2^-} (CH_3)_2C(O^-)C{\equiv}CH \xrightarrow{\text{H}^+} (CH_3)_2C(OH)C{\equiv}CH$$

$$HC{\equiv}CH \xrightarrow[\text{–78℃THF}]{\text{n–BuLi}} HC{\equiv}CLi \text{ or } \xrightarrow[\text{–78℃THF}]{\text{EtMgBr}} HC{\equiv}CMgBr;\quad \xrightarrow[\text{–78℃}]{\text{RC(O)R'}} HC{\equiv}C{-}C(R)(R')OH \quad >90\%$$

A variety of polar substituents can be attached to the C=C unit. Alkyne anions show some propensity for 1,2-addition in conjugated systems, especially if the α-carbon of the double bond is hindered.

2.2.3 烯醇盐和其他稳定的负离子 Enolates and Other Stabilized Carbanions

2.2.3.1 烯醇盐 Enolates

（ⅰ）通过脱质子生成烯醇盐

Generation of Enolates by Deprotonation

$$RCH_2C(=O){-}R' + NH_2^- \rightleftharpoons R\bar{C}HC(=O){-}R' + NH_3$$

$$R'O{-}C(=O)CH_2C(=O){-}OR' + R'O^- \rightleftharpoons R'O{-}C(=O)\bar{C}HC(=O){-}OR' + R'OH$$

$$CH_3{-}C(=O)CH_2C(=O){-}OR' + R'O^- \rightleftharpoons CH_3{-}C(=O)\bar{C}HC(=O){-}OR' + R'OH$$

$$NCCH_2C(=O){-}OR' + R'O^- \rightleftharpoons NC\bar{C}HC(=O){-}OR' + R'OH$$

$$RCH_2NO_2 + OH^- \rightleftharpoons R\bar{C}HNO_2 + H_2O$$

A primary consideration in the generation of an enolate or other stabilized carbanion by deprotonation is the choice of base. In general, reactions can be carried out under conditions in which the enolate is in equilibrium with its conjugate acid or under which the reactant is completely converted to its conjugate base. The key determinant is the amount and strength of the base. For complete conversion, the base must be derived from a substantially weaker acid than the reactant.

Their ability to stabilize carbanions:

$$NO_2 > COR > CN \sim CO_2R > SO_2R > SOR > Ph \sim SR > H > R$$

The carbonyl group of an aldehyde or a ketone is an electron-withdrawing group by both induction and resonance. So α-hydrogens to the group possess increased acidity simply due to their proximity and the incipient anion can be stabilized through π-resonance with the carbonyl group.

base

H

acidic hydrogen

p*K*a=19~20

resonance structures

Metal tautomerism:

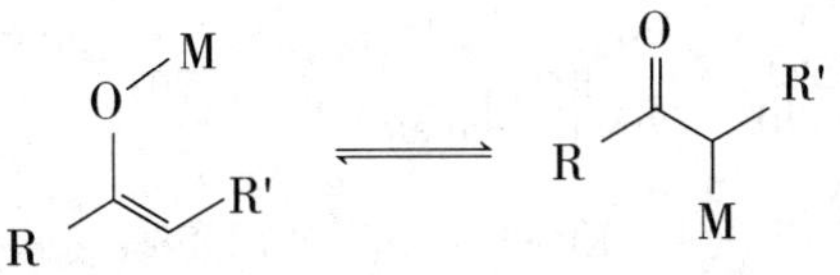

For alkali metal enolates (M=Li, Na, K, etc.) and alkaline earth enolates (Mg^{2+}, etc.), the O-metal tautomer is strongly favored. These are the generally useful enolate nucleophiles. For certain metal enolates from heavy metals, such as M = Hg^{2+}, the C-metal tautomer is sometimes favored.

Alkylating agents:

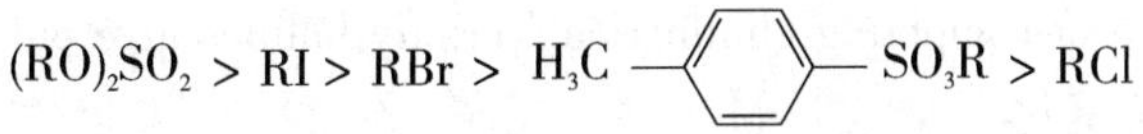

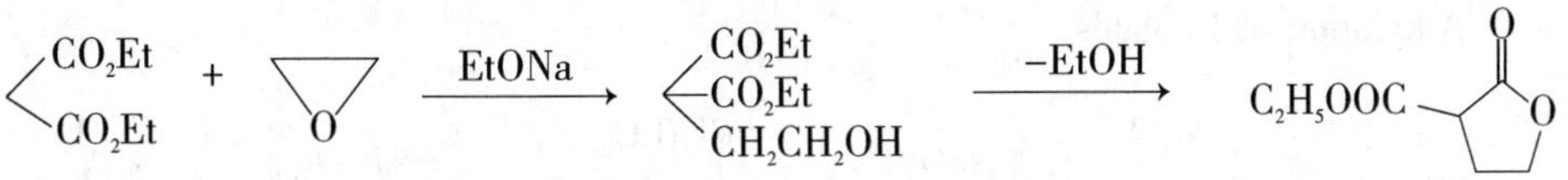

$$(RO)_2SO_2 > RI > RBr > H_3C-C_6H_4-SO_3R > RCl$$

Solvent: C_2H_5ONa/C_2H_5OH.

Polar aprotic solvents: Dimethyl sulfoxide (DMSO), *N*, *N*-dimethylformamide (DMF), NMP, HMPA.

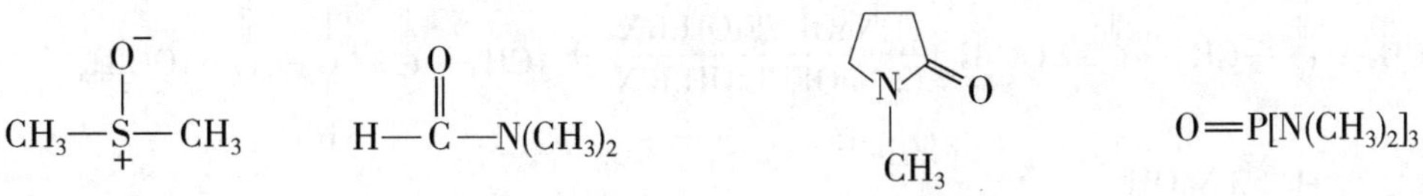

$O=P[N(CH_3)_2]_3$

dimethyl sulfoxide (DMSO) ε = 47

N,*N*-dimethylformamide (DMF) ε = 37

N-methylpyrrolidone (NMP) ε = 32

hexamethylphosphoric triamide(HMPA) ε = 30

The solvent and other coordinating or chelating additives also have strong effects on the structure and reactivity of carbanions formed by α-substituents.

These p*K* data provide a basis for assessing the stability and reactivity of carbanions. The acidity of the reactant determines which bases can be used for generation of the anion. Another crucial factor is the distinction between kinetic or thermodynamic control of enolate formation by deprotonation, which determines the enolate composition.

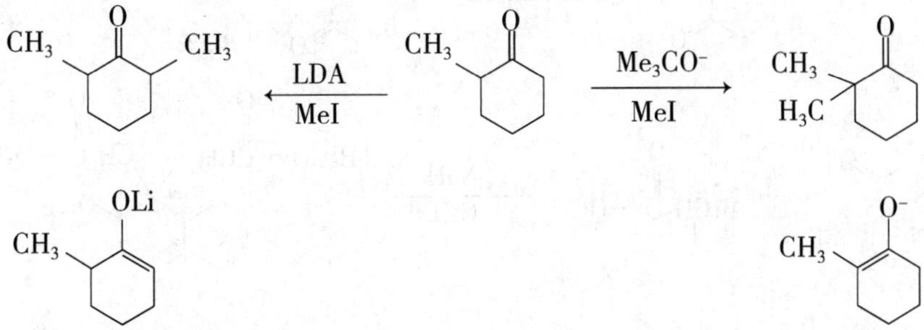

thermodynamic control

kinetic control

LDA (TMSCl quench): OTMS 84% kinetic enolate + OTMS 7% + OTMS 9%

Et_3N, TMXCl, DMF, △, 60h: 13%; 58%, 29% thermodynamic enolate

The presence of two electron-withdrawing substituents facilitates formation of the resulting enolate. Alkylation occurs by an S_N2 process, so the alkylating agent must be reactive toward nucleophilic displacement. Primary halides and sulfonates, especially allylic and benzylic ones, are the most reactive alkylating agents. Secondary systems react more slowly and often give only moderate yields because of competing elimination. Tertiary halides give only elimination products.

（ⅱ）烯醇酯的烷基化

Alkylation of Enolates

$CH_2(CO_2Et)_2$ + 3-chlorocyclopentene $\xrightarrow{NaOEt}$ 3-[$CH(CO_2Et)_2$]cyclopentene $\xrightarrow[\triangle]{H^+}$ 3-[$CHCO_2H$]cyclopentene

$$CH_3-\overset{O}{\overset{\|}{C}}-CH_2-\overset{O}{\overset{\|}{C}}-OC_2H_5 \xrightarrow[(2)NaOEt/EtOH,R'X]{(1)NaOEt/EtOH,RX} CH_3-\overset{O}{\overset{\|}{C}}-\underset{R'}{\overset{R}{C}}-\overset{O}{\overset{\|}{C}}-OC_2H_5$$

$$\xrightarrow[(2)H^+\ \triangle]{(1)\ 5\%NaOH} CH_3-\overset{O}{\overset{\|}{C}}-\underset{R'}{\overset{R}{CH}}$$

$$CH_3-\overset{O}{\overset{\|}{C}}-CH_2-\overset{O}{\overset{\|}{C}}-CH_3 \xrightarrow[\text{丙酮}]{K_2CO_3,\ CH_3I} CH_3-\overset{O}{\overset{\|}{C}}-\underset{CH_3}{CH}-\overset{O}{\overset{\|}{C}}-CH_3\quad 75\%\sim77\%$$

1,3-cyclohexanedione $\xrightarrow[CH_3I]{NaOEt}$ 2-methyl-1,3-cyclohexanedione (CH_3) 65%

2-tetralone + $BrCH_2\overset{O}{\overset{\|}{C}}-OCH_3$ $\xrightarrow[C_6H_6]{NaH}$ 1,1-bis($CH_2\overset{O}{\overset{\|}{C}}-OCH_3$)-2-tetralone ($H_3CO-\overset{O}{\overset{\|}{C}}CH_2$, $CH_2\overset{O}{\overset{\|}{C}}-OCH_3$)

Reactions of enolates with α-halo carbonyl derivatives:

$$CH_3-\overset{O}{\overset{\|}{C}}-CH_2-\overset{O}{\overset{\|}{C}}-OC_2H_5 + BrCH(CH_3)C(=O)CH_3 \xrightarrow{NaOEt} CH_3-\overset{O}{\overset{\|}{C}}-\underset{CH(CH_3)C(=O)CH_3}{\underset{|}{CH}}-\overset{O}{\overset{\|}{C}}-OC_2H_5$$

$$\xrightarrow[(2)H^+\quad\triangle]{(1)5\%NaOH} CH_3-\overset{O}{\overset{\|}{C}}-\underset{CH(CH_3)C(=O)CH_3}{\underset{|}{CH_2}}$$

$$\text{2-(CO}_2\text{CH}_3\text{)cyclopentanone} + BrCH_2(CH_2)_5CO_2Et \xrightarrow{NaH,DMF} \text{2-[CH}_2\text{(CH}_2\text{)}_5\text{CO}_2\text{Et]-2-(CO}_2\text{CH}_3\text{)cyclopentanone}$$

$$CH_3-\overset{O}{\overset{\|}{C}}-CH_2-\overset{O}{\overset{\|}{C}}-OC_2H_5 + ClCH_2CO_2Et \xrightarrow{NaOEt} CH_3-\overset{O}{\overset{\|}{C}}-\underset{CH_2CO_2Et}{\underset{|}{CH}}-\overset{O}{\overset{\|}{C}}-OC_2H_5$$

$$\xrightarrow[(2)H^+\quad\triangle]{(1)5\%NaOH} CH_3-\overset{O}{\overset{\|}{C}}-\underset{CH_2CO_2Et}{\underset{|}{CH_2}}$$

Rearrangement occurs α-halo allyl compounds.

$$CH_2(CO_2Et)_2 \xrightarrow[C_2H_5-\underset{Cl}{\underset{|}{C}}HCH=CH_2]{(1)\ NaOEt/EtOH} C_2H_5-CH-CH(CO_2Et)_2-CH=CH_2 + C_2H_5-CH_2CH=CH-CH(CO_2Et)_2\ (10\%)$$

In these procedures, an ester group is removed by hydrolysis and decarboxylation after the alkylation step. The malonate and acetoacetate carbanions are the synthetic equivalents of the simpler carbanions that lack the additional ester substituent. In the preparation of 2-heptanone, for example, ethyl acetoacetate functions as the synthetic equivalent of acetone.

$$RR'C(CO_2Et)_2 \xrightarrow{hydrolysis} RR'C(CO_2H)_2 \xrightarrow[-CO_2]{heat} RR'CH-CO_2H$$

Use of dihaloalkanes as alkylating reagents leads to ring formation. Malonic ester (丙二酸二乙酯) may also be used for the synthesis of the three- and four-membered alicyclic compounds from dibromides, e. g.

$$CH_2(CO_2Et)_2 \xrightarrow[(2)\,Br(CH_2)_3Br]{(1)NaOEt/EtOH} \text{cyclobutane-1,1-}(CO_2Et)_2$$

In the presence of a very strong base such as an alkyllithium, sodium or potassium hydride, sodium or potassium amide and LDA, 1, 3-dicarbonyl compounds can be converted to their dianions by two sequential deprotonations.

$$\text{2-acetylcyclohexanone} \xrightarrow[\text{液氨}]{KNH_2} \text{dianion} \xrightarrow[(2)\,H_2O,H^+]{(1)C_6H_5CH_2Cl} \text{2-}(COCH_2CH_2Ph)\text{cyclohexanone}$$

$$PhCH_2\overset{O}{\overset{\|}{C}}CH_2\overset{O}{\overset{\|}{C}}CH_3 \xrightarrow{2NaNH_2} PhCH{=}\overset{\bar{O}\,Li^+}{C}{-}CH{=}\overset{\bar{O}\,Li^+}{C}{-}CH_3 \xrightarrow{PhCH(Cl)CH_3} PhCH(PhCHCH_3)\overset{O}{\overset{\|}{C}}CH_2\overset{O}{\overset{\|}{C}}CH_3$$

Methylene groups can be dialkylated if sufficient base and alkylating agent are used. Dialkylation can be an undesirable side reaction if the monoalkyl derivative is the desired product. Sequential dialkylation using two different alkyl groups is possible.

(ⅲ) 酮、酯、羧酸、酰胺和腈的烷基化

Alkylation of Ketones, Esters, Carboxylic Acids, Amides, and Nitriles

$$(CH_3)_2CH\overset{O}{\overset{\|}{C}}{-}OC_2H_5 \xrightarrow[(C_2H_5)_2O]{(C_6H_5)_3CNa} (CH_3)_2\bar{C}\overset{O}{\overset{\|}{C}}{-}OC_2H_5 \longrightarrow$$

$$(CH_3)_2C{=}\overset{O^-}{C}{-}OC_2H_5 \xrightarrow{CH_3I} (CH_3)_3C\overset{O}{\overset{\|}{C}}{-}OC_2H_5$$

$$C_6H_5CH_2CN \xrightarrow[NH_3(l)]{NaNH_2} C_6H_5\bar{C}HCN \xrightarrow[\text{benzene,relux}]{C_6H_{11}Br} C_6H_5CH(C_6H_{11})CN$$

The cyano group also facilitates deprotonization.

$$(CH_3)_2CHCO_2H \xrightarrow{2LDA} (H_3C)_2C{=}C(O^-Li^+)_2 \xrightarrow[(2)H^+]{(1)\,CH_3(CH_2)_3Br} CH_3(CH_2)_3C(CH_3)_2CO_2H \quad 80\%$$

2.2.3.2 烯胺(烯醇和烯醇盐的氮类似物：烯胺和亚胺阴离子)

Enamines (The Nitrogen Analogs of Enols and Enolates: Enamines and Imine Anions)

(ⅰ) 烯胺的制备

Preparation of Enamines

$$R-\overset{O}{\overset{\|}{C}}-R + H_2NR' \longrightarrow R-\overset{N-R'}{\overset{\|}{C}}-R + H_2O \quad \text{imines}$$

$$R'C(=O)CH_2R + HNR''_2 \longrightarrow R'C(NR''_2)=CHR + H_2O \quad \text{enamine}$$

$$\underset{\text{enamine}}{R'_2N-\underset{R}{\underset{|}{C}}=CR_2} \longleftrightarrow \underset{\text{carboanion}}{R'_2\overset{+}{N}=\underset{R}{\underset{|}{C}}-\overset{-}{C}R_2} \xrightarrow{H^+} R'_2\overset{+}{N}=\underset{R}{\underset{|}{C}}-CHR_2 \quad \text{an iminium ion}$$

Enamines behave as nitrogen enolates in their reactions with alkyl halides, generating an iminium salt that can isomerize to the less substituted enamine. The behavior of a generic enamine reacting with a halide via the α-carbon gives α-alkyl iminium salt, via an S_N2 process.

(ⅱ) 烯胺的反应

Reactions of Enamines

$$R'_2\ddot{N}-\underset{R}{\underset{|}{C}}=CR_2 + H_2\underset{R''}{\underset{|}{C}}-X \longrightarrow R'_2\overset{+}{N}=\underset{R}{\underset{|}{C}}-\overset{R}{\underset{R}{C}}-CH_2R'' \xrightarrow{H_2O} R-\overset{O}{\overset{\|}{\underset{R}{C}}}-\overset{R}{\underset{R}{C}}-CH_2R''$$

The nucleophilicity of the β-carbon atoms permits enamines to be used synthetically for alkylation reactions. Alkylation gives the best yields with reactive primary halides, since it is essentially an S_N2 reaction. Methyl iodide, allyl and benzyl halides, α-halo esters, α-halo ethers, and α-halo ketones are the most successful alkylating agents. Enamines are usually formed by reaction of a secondary amine with a ketone, in the presence of an acid catalyst. Pyrrolidine, piperidine, morpholine or diethylamine are the most common amine precursors to enamines. It is noted that when aldehydes react with secondary amines, initial reaction gives an iminium salt.

Alkylation of enamines(烯胺烷基化反应):

$$\text{cyclohexanone} \xrightarrow[(3)H_2O]{(1)\ \text{pyrrolidine}\quad (2)CH_2=CHCH_2Br} \text{2-}(CH_2CH=CH_2)\text{cyclohexanone} \quad 66\%$$

$$\text{4-}OCH_3\text{-cyclohexanone} \xrightarrow[(3)H_2O]{(1)\text{pyrrolidine}\quad (2)CH_3I} \text{2-}CH_3\text{-4-}OCH_3\text{-cyclohexanone} \quad 60\%$$

$$C_2H_5\text{-tetralone} \xrightarrow[(3)H_2O]{(1)\text{pyrrolidine}\quad (2)CH_3CHICO_2C_2H_5} C_2H_5\text{-tetralone-2-}CH(CH_3)CO_2C_2H_5$$

$CH_3O_2CCH_2$ (cyclopentanone) $\xrightarrow[(3)H_2O]{(1)\text{pyrrolidine}\quad(2)CH_3COCHBrCH_3}$ $CH_3O_2CCH_2$ (cyclopentanone) $CH(H_3C)C(O)CH_3$ 31%

(enamine) H^+ $Br(CH_2)_4O_2CCH_3$ $\longrightarrow$ (cyclohexanone) O_2CCH_3 20%~40%

CH_3 CH_3 O O (ketal) O $\xrightarrow[(3)H_2O]{(1)\text{pyrrolidine}\quad(2)\ H_2C{=}C(Cl)CH_2Cl,\ NaI,\ iPr_2NEt}$ CH_3 CH_3 O O $CH_2{=}C(Cl)CH_2$ O $CH_2C(Cl){=}CH_2$ 91%

An enamine is essentially a nitrogen enolate, and it can react with alkyl halides such as iodomethane to give an equilibrium mixture of the iminium salt and the alkylated enamine. Hydrolysis with aqueous acid led to loss of pyrrolidine and formation of 2-methylcyclohexanone.

Enolization of enamine is very effective in controlling the kinetic-thermodynamic product. The intermediate iminium salt can isomerize under the reaction conditions to the less substituted enamine in an equilibrium process, and hydrolysis yields the corresponding α-alkyl carbonyl compound.

The enamine from an unsymmetrical ketone which can react at two sites is predominantly the less substituted one, e. g.

(pyrrolidine, N–H) + CH_3 (cyclohexanone) $\xrightarrow{-H_2O}$ CH_3 (enamine) 90% + CH_3 (enamine) 10%

If the steric interaction is greater, there is a preference for that kinetic nitrogen enolate. This is because an enamine is stabilized by interaction of the alkene π-system with the unshared electron-pair in a p orbital on nitrogen. This requires coplanarity of the bonds in unsaturated carbon atom and those to nitrogen, which is possible in (a) but, owing to steric repulsion, is not possible in (b). This can be applied in synthesis.

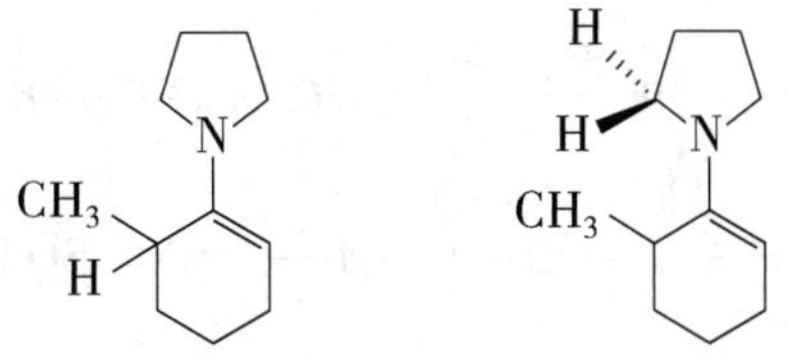

(a) kinetic control

(b) thermodynamic control

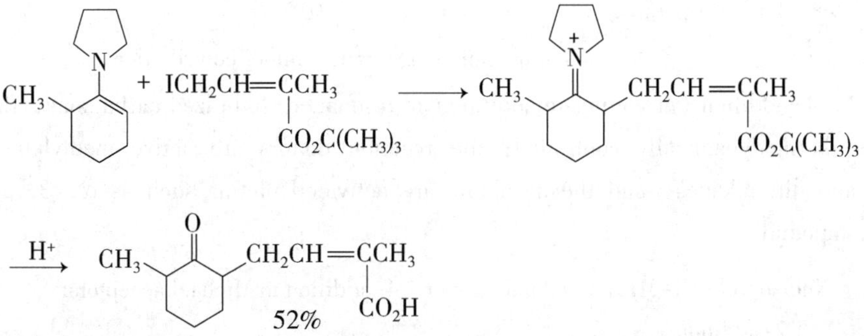

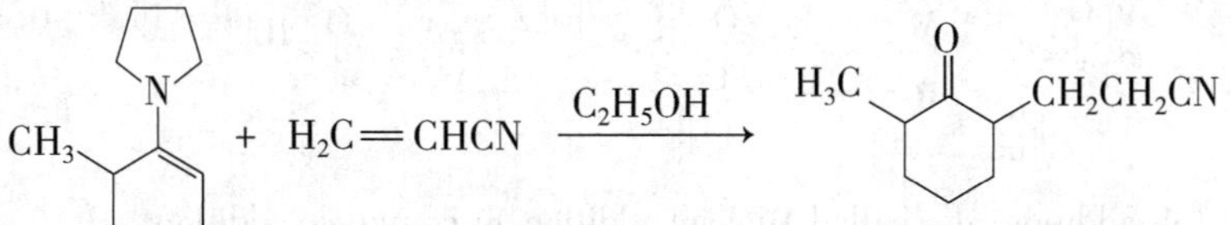

$$CH_3\text{-enamine} + ICH_2CH{=}C(CH_3)CO_2C(CH_3)_3 \longrightarrow \xrightarrow{H^+} \text{2-methyl-6-}(CH_2CH{=}C(CH_3)CO_2H)\text{cyclohexanone (52\%)}$$

$$CH_3\text{-enamine} + H_2C{=}CHCN \xrightarrow{C_2H_5OH} \text{2-methyl-6-}(CH_2CH_2CN)\text{cyclohexanone}$$

kinetic product

2.2.3.3 迈克尔加成(通过碳亲核试剂的共轭加成)

Michael Additions (Conjugate Addition by Carbon Nucleophiles)

(ⅰ) General Information

The 1,4-conjugate addition of a carbon nucleophile to an α,β-unsaturated carbonyl system is usually reversible, and referred to as Michael addition. The term Michael addition is also applied to the conjugate addition of amines to unsaturated carbonyl compounds. Enolates and carbanions are common partners in Michael additions. The electrophilic reaction partner is activated olefins. Solvent is important, and alcoholic solvents usually promote the equilibrium[16].

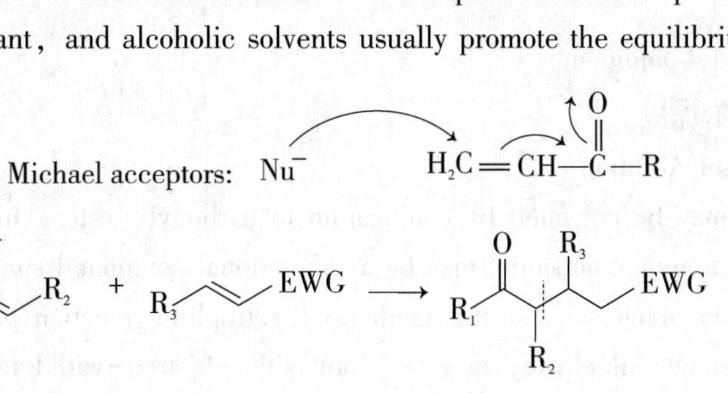

Michael acceptors: Nu^- $H_2C{=}CH{-}C({=}O){-}R$

$$R_1C(O^-){=}CHR_2 + R_3CH{=}CH{-}EWG \longrightarrow R_1C({=}O){-}CH(R_2){-}CH(R_3){-}CH_2{-}EWG$$

Review: What have we seen as $C^{\oplus}$ $C^{\ominus}$ so far?

Now, another broad class of $\overset{\oplus}{C}$: 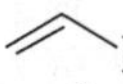 EWG called"Michael acceptors".

EWG can be: —C(=O)—R —C(=O)—OR —C(=O)—H —C≡N —SO_2Ph —NO_2

The electrophilicity of the C=O (or EWG) is translated down the p-system:

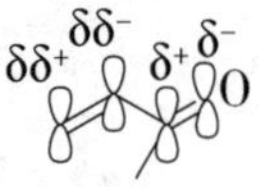

δδ⁻ δδ⁺ δ⁺ δ⁻ O or as resonance structures:

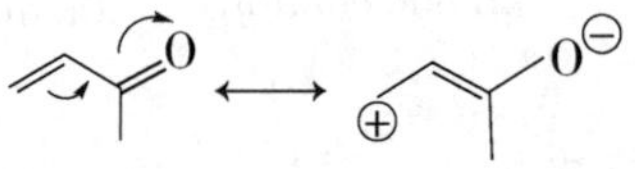

electron deficient C=C minor contribution

As a 1, 4-addition (or conjugate addition) of resonance-stabilized carbanions, the Michael addition is thermodynamically controlled; the reaction donors are active methylenes such as malonates and nitroalkanes, and the acceptors are activated olefins such as α, β-unsaturated carbonyl compounds.

Nucleophiles(R–M) can undergo1,2– or 1,4–addition to Michael acceptors:

1,2–addition:

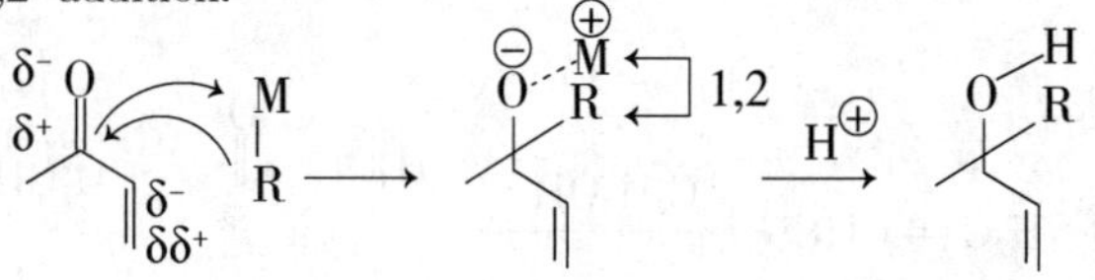

δ⁻ O δ⁺ δ⁻ δδ⁺ M R → 1,2 $H^{\oplus}$ →

1,4–addition, also called Michael addition, or conjuqate addition:

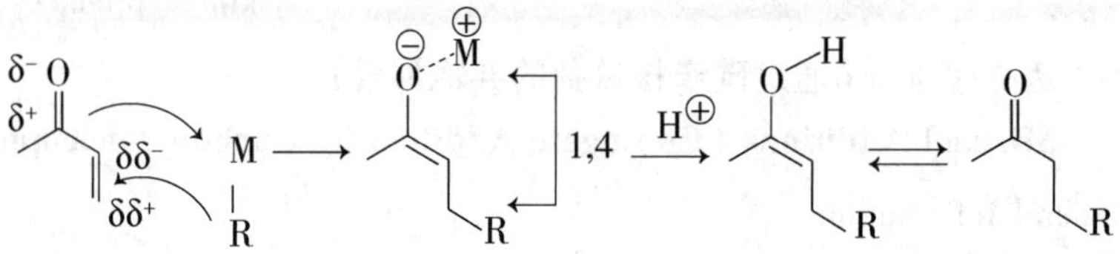

δ⁻ O δ⁺ δδ⁻ δδ⁺ M R → 1,4 $H^{\oplus}$ → ⇌

more stable

RC(=O)—C(R)(R)—C(R')HCH=CR″(O⁻) ⇌ RC(O⁻)=CR_2+ R'CH=CHC(=O)R″ ⇌ R'CH=CHC(O⁻)(R″)—C(R)(R)—C(=O)R

1,4–addition 1,2–addition

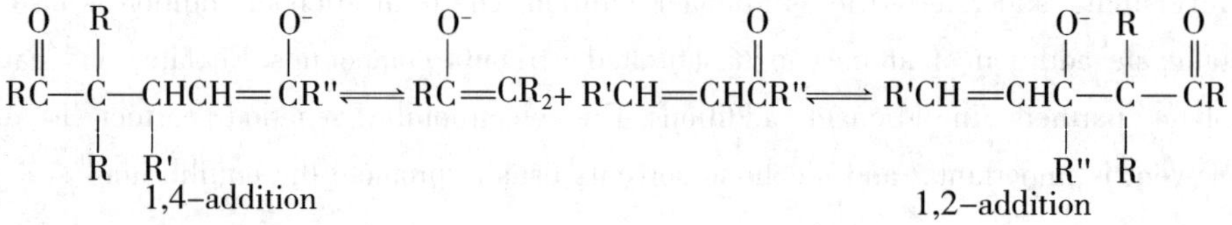

(ⅱ) 迈克尔加成

Michael Addition

The alkene may be activated by conjugation to carbonyl, ester, nitro, and nitrile groups, and the enolate-forming component may be a bifunctional compound such as malonic ester, or a monofunctional compound such as nitromethane. Electrophilic reaction partner is typically an α, β-unsaturated ketone, aldehyde, or ester, but other electron-withdrawing substituents such as nitro, cyano, or sulfonyl also activate carbon-carbon double and triple bonds to nucleophilic attack, e. g.

$$\text{2-methylcyclopentanone} + H_2C{=}CHCO_2CH_3 \xrightarrow[t\text{-BuOH}]{KOC(CH_3)_3\ 10mol\%} \text{2-}CH_3\text{-2-}(CH_2CH_2CO_2CH_3)\text{cyclopentanone}\quad 53\%$$

$$PhCH_2CH(CONH_2)CN + H_2C{=}CHCN \xrightarrow{NH_3(l)} PhCH_2C(CN)(CONH_2)CH_2CH_2CN\quad 100\%$$

$$CH_2(CO_2C_2H_5)_2 + H_2C{=}C(Ph)CO_2C_2H_5 \xrightarrow{NaOEt\ 12mol\%} (C_2H_5O_2C)_2CHCH_2CH(Ph)CO_2C_2H_5\quad 55\%\sim60\%$$

$$PhCH(CN)CO_2C_2H_5 + CH_2{=}CHCN \xrightarrow[(CH_3)_3COH]{KOH} PhC(CN)(CO_2C_2H_5)CH_2CH_2CN\quad 69\%\sim83\%$$

$$\text{methyl norbornadiene-2-carboxylate} + CH_3COCH_2CO_2C_2H_5 \xrightarrow{R_4N^+\text{–}OH^-} \text{norbornene-}CH(COCH_3)CO_2C_2H_5\text{, }CO_2CH_3\quad 86\%$$

$$(CH_3)_2CHNO_2 + CH_2{=}CHCO_2CH_3 \xrightarrow{PhCH_2\overset{+}{N}(CH_3)_3OH^-} O_2NC(CH_3)_2CH_2CH_2CO_2CH_3$$

$$(CH_3)_2CHNO_2 + CH_2{=}CHCOCH_2CH_3 \xrightarrow{\text{Amberlyst A27}} (CH_3)_2C(NO_2)CH_2CH_2COCH_2CH_3\quad 70\%$$

$$CH_3NO_2 + CH_2{=}C(CO_2CH_3)\text{–N(phthalimide)} \xrightarrow{KF} NO_2CH_2CH_2CH_2CH(CO_2CH_3)\text{–N(phthalimide)}\quad 80\%$$

$$BrZnCH_2CO_2C_2H_5 + Cl\text{–}C_6H_4\text{–}CH{=}CHNO_2 \longrightarrow Cl\text{–}C_6H_4\text{–}CH(CH_2CO_2C_2H_5)CH_2NO_2\quad 81\%$$

$$\text{1-(6-}CH_3\text{-4-}CH(CH_3)_2\text{-cyclohexenyl)pyrrolidine} + CH_2{=}CHCOCH_3 \xrightarrow[(2)NaOAc,HOAc,\ H_2O\ reflux]{(1)dioxane,16h} \text{2-}CH_3\text{-6-}(CH_2CH_2COCH_3)\text{-5-}CH(CH_3)_2\text{-cyclohexanone}\quad 66\%$$

Enamines react with α, β-unsaturated compounds almost exclusively by conjugate addition to give the corresponding substituted ester, ketone or nitrile. Reaction of enamines with acrylonitrile, for example, give a new enamine. Subsequent hydrolysis liberates the substitute ketone.

2. 2. 3. 4 Wittig 烯烃化
Wittig Olefination

Wittig discovered that triarylphosphonium ylides reacted with aldehydes and ketones to form an alkene product. This latter reaction has come to be called the Wittig olefination reaction, or simply the Wittig reaction. The reaction is particularly useful for incorporating exocyclic methylene groups. For this and other work, Wittig was awarded the Nobel Prize in 1979.

(i) 磷叶立德的制备

Preparation of Phosphorus Ylides

Ylides are defined as compounds in which a carbanionic carbon is immediately adjacent to an atom bearing a positive center. Many atoms can support a positive charge but phosphorus, sulfur or nitrogen leading to particularly useful and interesting compounds. This section will deal with ylides and their reactions, focusing on phosphorus ylides. Phosphonium salts are generated by an S_N2 reaction of a trialkylphosphine with an alkyl halide, and they react with strong bases to give the corresponding ylide, a resonance stabilized species.

$$Ph_3P + CH_3Br \xrightarrow{C_6H_6} Ph_3P^+CH_3Br^-$$

$$Ph_3P^+CH_3Br^- + PhLi \longrightarrow Ph_3\overset{+}{P}—\overset{-}{C}H_2 + C_6H_6 + LiBr$$

$$\updownarrow$$

$$Ph_3P=CH_2 \text{ Ylide}$$

$$Ph_3\overset{+}{P}—CH_2—CN\bar{X} \xrightarrow{NaOH} Ph_3\overset{+}{P}—\overset{-}{C}H—CN \longleftrightarrow Ph_3P=CH—CN$$

(ii) Wittig 烯烃化

Wittig Olefination

Triarylphosphonium or phosphorus ylides react with aldehydes and ketones to form an alkene product.

Solvent: Et_2O, THF, DME, DMSO, ROH.

Base: PhLi, *n*-BuLi, LDA, RONa, KHMDS.

$$H_3C—S(=O)—\overset{-}{C}H_2 \ \overset{+}{Na}$$

Substrates: ketones or aldehydes.

$$R'—C(=O)—R''$$

Wittig reaction:

$$Ph_3\overset{+}{P}-\overset{-}{C}H_2-R + R'C(=O)R'' \longrightarrow R'R''C=CRH$$

$$CH_3(CH_2)_4CH_2\overset{+}{P}Ph_3\overset{-}{Br} \xrightarrow[(2)\ THF,\ AcO(CH_2)_7CH_2CHO]{(1)\ NaHMDS} AcO(CH_2)_7CH=CH(CH_2)_4CH_3$$

79% yield, 98% cis

$$CH_3CH_2\overset{+}{P}Ph_3I^- \xrightarrow[(2)\ PhCHO]{(1)\ n\text{-}BuLi} PhCH=CHCH_3$$

76% yield, 58% cis

$$\text{cyclohexanone} + Ph_3\overset{+}{P}-\overset{-}{C}H_2 \longrightarrow \text{methylenecyclohexane}\ (C_6H_{10}=CH_2)$$

It is postulated that initial reaction of ylide with a ketone or aldehyde generates a labile intermediate called a betaine, which collapsed to an oxaphosphetane. Either of these intermediates can decompose to give an alkene, and a trisubstituted phosphine oxide. The P-O bond is particularly strong, with a bond dissociation energy of 130~140kcal · mol^{-1}(544. 2~586. 0kJ · mol^{-1}). Formation of this strong bond is largely responsible for decomposition of the intermediate betaine or oxaphosphetane to give the alkene.

Mechanism:

$$Ph_3\overset{+}{P}-\overset{-}{C}H-R \xrightarrow{solvent} Ph_3\overset{+}{P}-CH_2-R\ X^- \xrightarrow{base} Ph_3\overset{+}{P}-H\overset{-}{C}-R \xrightarrow{R'C(=O)R''} \text{betaine}$$

phosphonium salt; phosphorous ylide

p*Ka*≈18~20
(R=alky or H)

$$Ph_3\overset{+}{P}-H\overset{-}{C}-R \longleftrightarrow Ph_3P=HC-R$$

ylene

Betaine: $R'R''C(-\overset{-}{O})-CH(R)-\overset{+}{P}Ph_3$

betaine
a dipolar intermediate

⇕

Oxaphosphetane: four-membered ring $R'R''C-O-PPh_3-CH(R)$

oxaphosphetane

$$\text{oxaphosphetane} \longrightarrow Ph_3\overset{+}{P}-\overset{-}{O}\ (\longleftrightarrow Ph_3P=O) + R'R''C=CRH$$

triphenylphosphineoxide

Features:

(1) None-stabilized ylides (R=alkyl group or EDG) are sensitive to H_2O and O_2.

(2) Require strong base to generate ylide from the phosphonium salt.

(3) Cis-alkene is the predominant product.

e.g.

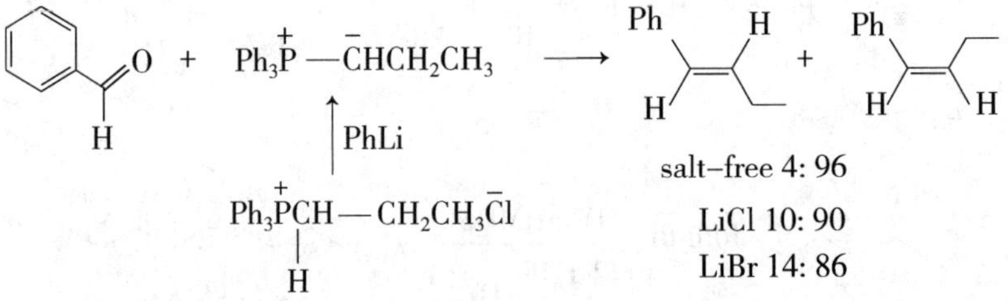

$$PhCHO + Ph_3\overset{+}{P}-\overset{-}{C}HCH_2CH_3 \longrightarrow$$ (Z)- and (E)-1-phenylbut-1-ene

$$Ph_3\overset{+}{P}CH_2-CH_2CH_3\overset{-}{Cl} \xrightarrow{PhLi} Ph_3\overset{+}{P}-\overset{-}{C}HCH_2CH_3$$

salt-free 4: 96

LiCl 10: 90

LiBr 14: 86

If R in the original alkyl halide are hydrogen or simply alkyl groups, α-hydrogen of the phosphonium salt is very weakly acidic and a very strong base (usually butyl-lithium or phenyl-lithium) is required to produce the ylide. The ylide, once formed, is a highly reactive compound and is not generally isolable; it is not only strongly basic (deprotonating acids as weak as water) is strongly nucleophilic in the manner of a Grignard reagent and react rapidly, under mild conditions, to give the adduct effectively irreversibly[17]. This then decomposes spontaneously to give the alkene. If stereoisomer of the product is possible, a mixture of E- and Z-isomers is generally obtained and cis-alkene is the predominant product.

Applications:

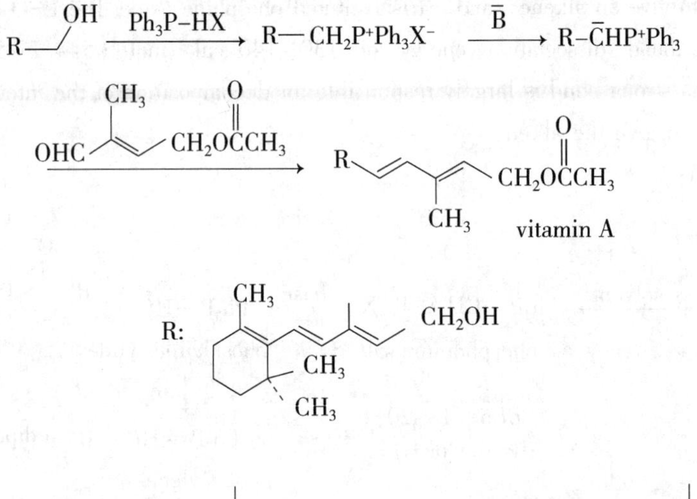

$$R-CH_2OH \xrightarrow{Ph_3P-HX} R-CH_2P^+Ph_3X^- \xrightarrow{\bar{B}} R-\bar{C}HP^+Ph_3$$

$$\xrightarrow{OHC-CH=C(CH_3)-CH_2OCOCH_3} R-CH=CH-C(CH_3)=CH-CH_2OCOCH_3$$

vitamin A

R: (2,6,6-trimethylcyclohexenyl)–CH=CH–C(CH₃)=CH– ; CH$_2$OH

6.5equiv CH_3PPh_3Br
5equiv n-BuLi
Et_2O, 25℃, 1h
reflux, 3h, 56%

OH → OH

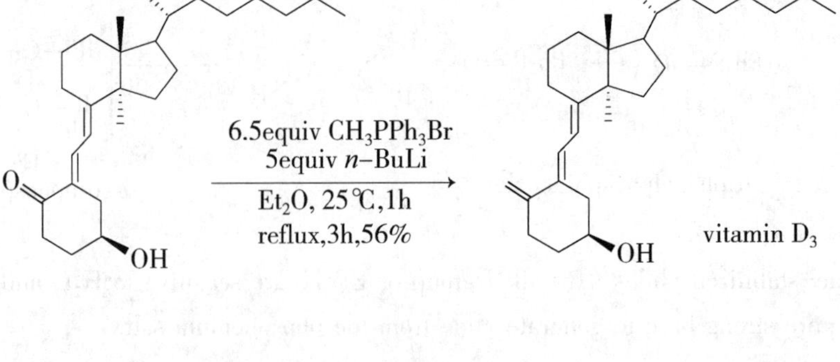

vitamin D_3

（ⅲ）稳定的叶立德

Stabilized Ylides

If R in the original alkyl halide is an electron-withdrawing group (e. g., an ester), deprotonation of the phosphonium salt is achieved under much less strongly basic conditions and the resulting ylide (or phosphorane) is often sufficiently stable to be isolated. It is also sufficiently stable that it may react reversibly with carbonyl groups and may not react at all readily with feebly electrophilic carbonyl groups. Where *E*- and *Z*-isomers of the final product can exist, it is the *E*-isomer which usually predominates.

$$Ph_3\overset{+}{P}-CH_2-EWG \xrightarrow{\text{weak base}} Ph_3\overset{+}{P}-\overset{-}{C}H-EWG$$

EWG: -COOR', -COPh, -CN, -Aryl.　　　Weak base: OR^-, OH^-, CO_3^{2-}.

Features of stabilized ylides:

(1) Has two EWGs, so the p*K*a is very low.

(2) Stabilized ylides are solid; stable to storage, not particularly sensitive to moisture.

(3) Easier to form, but less reactive (react with aldehydes and activated ketones).

(4) Usually gives trans-alkenes as the major product.

Horner-Wadsworth-Emmons Reaction:

$$(R'O)_2\overset{O}{\overset{\|}{P}}CH_2-EWG \xrightarrow{\text{base}} (R'O)_2\overset{O}{\overset{\|}{P}}\overset{-}{C}H-EWG \xrightarrow{R_2C=O} R_2C=CH-EWG$$

The reaction of aldehydes or ketones with stabilized phosphorus ylides (phosphonate carbanions) leads to olefins with excellent *E*-selectivity. The reaction mechanism is similar to the mechanism of the Wittig reaction. The stereochemistry is set by steric approach control, where the antiperiplanar approach of the carbanion to the carbon of the carbonyl group is favored when the smaller aldehydic hydrogen eclipses the bulky phosphoranyl moiety. This places the ester group syn to the aldehyde R group, but the incipient alkene assumes an *E*-orientation of these groups after rotation to form the oxaphosphetane. As the lithium counterion does not interfere with oxaphosphetane formation, use of BuLi is possible, but NaH and NaOMe are also suitable bases for forming the ylide. The resulting phosphate byproduct is readily separated from the desired products by simply washing with water.

$$(C_2H_5O)_3P + BrCH_2\overset{O}{\overset{\|}{C}}OC_2H_5 \longrightarrow (C_2H_5O)_2\overset{OC_2H_5}{\overset{|}{P^+}}-CH_2\overset{O}{\overset{\|}{C}}OC_2H_5\ Br^- \xrightarrow{-C_2H_5Br}$$

$$(C_2H_5O)_2\overset{O}{\overset{\|}{P}}CH_2\overset{O}{\overset{\|}{C}}OC_2H_5 \xrightarrow[\text{乙二醇二乙醚}]{NaH} \left[(C_2H_5O)_2\overset{O}{\overset{\|}{P}}\overset{-}{C}H\overset{O}{\overset{\|}{C}}OC_2H_5\right] Na^+ \xrightarrow{\text{cyclohexanone}}$$

磷酸酯

$$\text{cyclohexylidene}{=}CH\overset{O}{\overset{\|}{C}}OC_2H_5 + (C_2H_5O)_2\overset{O}{\overset{\|}{P}}ONa$$

e.g.

$$\text{4-}O_2N\text{-}C_6H_4\text{-}CHO + Ph_3P{=}C(Cl)C(O)Ph \longrightarrow \text{4-}O_2N\text{-}C_6H_4\text{-}CH{=}C(Cl)C(O)Ph \quad \text{major}$$

$$\text{6-hydroxybenzofuran-5-carbaldehyde} + Ph_3P{=}CHC(O)OEt \xrightarrow[\Delta]{\text{benzene}} \text{ethyl 3-(6-hydroxybenzofuran-5-yl)acrylate} \quad 86\%$$

Works of Wittig olefination:

$$Ph_3\overset{+}{P}CH_3I^- \xrightarrow[DMSO]{NaCH_2S(O)CH_3} Ph_3P{=}CH_2$$

$$\text{cyclohexanone} + Ph_3P{=}CH_2 \xrightarrow{DMSO} \text{cyclohexylidene}{=}CH_2 \quad 86\%$$

$$Ph_3\overset{+}{P}CH_2CH_2CH_2CH_2CH_3Br^- \xrightarrow[DMSO]{n\text{-}BuLi} Ph_3P{=}CHCH_2CH_2CH_2CH_3$$

$$CH_3\overset{O}{\overset{\|}{C}}CH_2 + Ph_3P{=}CH(CH_2)_3CH_3 \xrightarrow{DMSO} (CH_3)_2C{=}CH(CH_2)_3CH_3 \quad 56\%$$

$$CH_3CH_2\overset{+}{P}Ph_3Br^- \xrightarrow[NH_3]{NaNH_2} CH_3CH{=}PPh_3$$

$$C_6H_5CHO + CH_3CH{=}PPh_3 \xrightarrow{\text{benzene}} C_6H_5CH{=}CHCH_3 \quad 98\%\text{yield, } 87\%Z$$

$$CH_3CH_2\overset{+}{P}Ph_3I^- \xrightarrow{n\text{-}BuLi} CH_3CH{=}PPh_3$$

$$C_6H_5CH{=}O + CH_3CH{=}PPh_3 \xrightarrow{LiI} C_6H_5CH{=}CHCH_3 \quad 76\%\text{ yield, } 58\%\ Z$$

2.3 烯醇盐及相应的碳负离子和羰基化合物的反应 Reactions of Enolates and Related Nucleophiles with Carbonyl Compounds

The condensation reaction of enolate anions and carbonyl derivatives is one of the most useful in organic chemistry. The condensation is nothing more than an acyl addition reaction of the nucleophilic enolate to an electrophilic carbonyl carbon. There are many synthetic variations, and to further complicate matters the enolate may be generated under kinetic or thermodynamic conditions. The more common variations of this reaction are well known, and usually have a name associated with them (i. e. named reactions).

2.3.1 羟醛缩合 Aldol Condensation

The aldol condensation is one of the most commonly used methods for forming new C—C bonds. It consists of the reaction between two molecules of aldehydes or ketones, which may be the same or different: one molecule is converted into a nucleophiles by forming its enolate in basic conditions or its enol in acidic conditions and the second acts as electrophile. The aldol condensation is first reported by Chiozza in 1856 and later expanded by Wurtz and by Perkin. If acetophenone reacted with benzaldehyde in the presence of sodium ethoxide (in ethanol), the initial product was alkoxide. Hydrolysis provided the β-hydroxy ketone product (an aldol or aldolate). In general, the hydrolysis step provides the aldolate, but sometimes elimination of water (dehydration) accompanies the hydrolysis to give a conjugated ketone (α, β-unsaturated ketone)[18].

$$2H_3C-\overset{O}{\overset{\|}{C}}-CH_3 \xrightarrow[\text{reflux}]{\text{Soxhlet}} (H_3C)_2\underset{OH}{\underset{|}{C}}-H_2C-\overset{O}{\overset{\|}{C}}-CH_3 \quad 71\%$$

β-hydroxy carbonyl compounds

$$2H_3C-\overset{O}{\overset{\|}{C}}-H \xrightarrow{OH^-} H_3C-\underset{H}{\overset{O^-}{\overset{|}{C}}}-CH_2CHO \longrightarrow H_3C-\underset{H}{\overset{OH}{\overset{|}{C}}}-CH_2CHO$$

$$\xrightleftharpoons{OH^-} H_3C-\underset{H}{C}=CHCHO \xrightarrow[(2)CH_3CHO]{(1)OH^-} H_3C-\underset{H}{\overset{OH}{\overset{|}{C}}}-CH_2-\overset{H}{C}=CHCHO$$

$$\longrightarrow H_3C\left(CH=CH\right)_n CHO$$

$$HCHO + CH_3CHO \xrightarrow{OH^-} (HOCH_2)_3C-CHO \xrightarrow[conc.OH^-]{CH_3CHO} HOH_2C-C(CH_2OH)_2-CH_2OH + HCOO^-$$

If each of two aldehydes contains an α-hydrogen atom, aldol reaction can give each of four products. Such condensations are only of synthetic value if a required product can easily be separated from the resulting mixture. If, however, one of the two components has no α-hydrogen, only two products can be formed. Further, if this component has the more reactive carbonyl group of the two, one product predominates.

$$HCHO + CH_3CHO \xrightarrow{OH^-} HOCH_2CH_2-CHO$$

The mixed condensation of aromatic aldehydes with aliphatic aldehydes and ketones is often done under these conditions. If an aliphatic aldehyde is used, the aldol product is usually easy to isolate. When an aromatic aldehyde (such as benzaldehyde) is the reaction partner, the aqueous acid workup almost always leads to elimination of water from the aldol product. When an aromatic aldehyde is condensed with an enolate of an aliphatic aldehyde or ketone to give the α, β-unsaturated compound, the reaction is called the Claisen-Schmidt reaction.

$$ArCH{=}O + RCH_2\overset{O}{\overset{\|}{C}}R' \longrightarrow ArCH{=}\underset{R}{\underset{|}{C}}\overset{O}{\overset{\|}{C}}R'$$

α, β-unsaturated aldehydes and ketones(dimer)

$$C_6H_5-CHO + H_3C-\overset{O}{\overset{\|}{C}}-C(CH_3)_3 \xrightarrow[H_2O-EtOH]{NaOH} C_6H_5-CH{=}CH-\overset{O}{\overset{\|}{C}}-C(CH_3)_3 \quad 88\%\sim93\%$$

$$\text{furfural (2-furyl-CHO)} + H_3C-\overset{O}{\overset{\|}{C}}-CH_3 \xrightarrow[H_2O]{NaOH} \xrightarrow{H_3O^+} \text{2-furyl}-\underset{H}{C}{=}\overset{H}{C}-\overset{O}{\overset{\|}{C}}-CH_3 \quad 60\%\sim66\%$$

Other derivatives containing the electron-withdrawing groups can be used in the reaction such as nitro enolate.

$$C_6H_5—CHO + CH_3NO_2 \xrightarrow[H_2O\text{-}EtOH]{NaOH} (C_6H_5)(H)C=C(H)(NO_2) \text{ (E)} \quad 80\%\sim83\%$$

For unsymmetrical ketone, two carbon atom can be activated by the carbonyl group. For example, butanone might undergo reaction via either of the anions $^-CH_2COC_2H_5$ and $CH_3COCH^-CH_3$ leading to two isomers respectively.

$$CH_3COCH_2CH_3 + HOOC—CHO \xrightarrow[82\%]{H_3PO_4} CH_3COC(CH_3)=CHCOOH \quad \text{under thermodynamic control}$$

$$CH_3COCH_2CH_3 + HOOC—CHO \xrightarrow{Base} HOOCCH=CHCOCH_2CH_3 \quad \text{under kinetic control}$$

Mukaiyama aldol reaction:

This refers to Lewis acid-catalyzed aldol addition reactions of silyl enol ethers, silyl ketene acetals, and similar enolate equivalents.

$$PhCH=O + \text{1-(OTMS)cyclohexene} \xrightarrow[10mol\%]{Yb(O_3SCF_3)_3} PhCH(OH)\text{-2-oxocyclohexyl}$$

91% yield, 73:27syn:anti

Whereas at low temperature further aldol reactions occur until each of the three α-hydrogen of acetaldehyde has been replaced. This product, like formaldehyde, has no active hydrogen and next undergoes the crossed Cannizzaro reaction characteristic of such aldehydes.

$$HCHO + CH_3CHO \xrightarrow{OH^-} (HOCH_2)_3C—CHO \xrightarrow[浓OH^-]{HCHO} HOH_2C—C(CH_2OH)_2—CH_2OH + HCOO^-$$

Claisen-Schmidt reaction:

When an aromatic aldehyde is condensed with an enolate of an aliphatic aldehyde or ketone to give the α, β-unsaturated compound, the reaction is called the Claisen-Schmidt reaction.

$$PhCOCH_3 \xrightarrow[EtOH,reflux]{NaOEt,PhCHO} PhCOCH_2CH(O^-)Ph \xrightarrow{H_3O^+} PhCOCH_2CH(OH)Ph \xrightarrow[-H_2O]{heat} PhCOCH=CHPh$$

Robinson annulation:

The reaction using a Michael addition protocol to produce a bicyclic ketone are known as the Robinson annulation. The reaction conditions favor formation of a thermodynamic enolate, although this is irrelevant with ketone. Conversion of ketone to its enolate was followed by Michael addition to MVK to give a new enolate product. Under the equilibrium conditions of the reaction, an intramolecular aldol condensation is possible that generates a six-membered ring, and this is energetically more favorable than forming a four-membered ring.

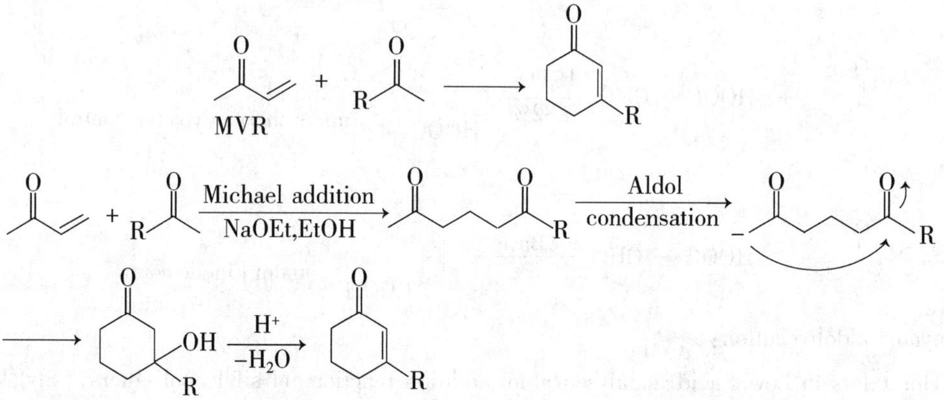

MVR + R–CO–CH₃ ⟶ 3-R-cyclohex-2-enone

Michael addition: NaOEt, EtOH; Aldol condensation; H^+, $-H_2O$

When a cyclic ketone such as cyclohexanone reacts with MVK, Robinson annulation generates a bicyclic ketone.

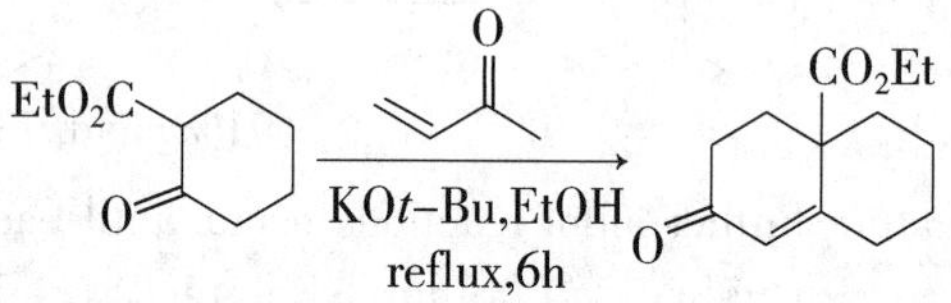

EtO_2C-cyclohexanone + MVK, KO*t*-Bu, EtOH, reflux, 6h ⟶ CO_2Et bicyclic enone

2.3.2 Claisen 和 Dieckmann 缩合反应 Claisen and Dieckmann Condensation Reactions

2.3.2.1 Claisen 酯缩合 Claisen Ester Condensation

Claisen ester condensation is the condensation of an ester enolate with an ester, illustrated by the self-condensation of ethyl acetate in the presence of sodium ethoxide to give β-keto-ester (acetoacetic ester). Initial reaction with the base under thermodynamic control in this case, generated the enolate anion. This anion attacked the carbonyl of a second molecule of ethyl acetate to give the tetrahedral intermediate. Displacement of ethoxide generated ketone, e. g.

OEt NaOEt EtOH O O O OEt acetoacetic ester

OEt EtO⁻ OEt OEt OEt OEt O -EtO⁻ O O OEt

O O OEt O⁻ O OEt O O⁻ OEt EtOH O O OEt acetoacetic ester

Reaction with base under thermodynamic conditions will give the 1, 3-dicarbonyl enolate, where deprotonation of a product is certain. The resulting anion is relatively stable, however, due to the increased resonance delocalization possible when two carbonyl groups are present. This means that the product enolate is less reactive than the enolate derived from the ester starting material, and generally does not give significant amounts of condensation product. Alcohol solvent is the conjugate acid of the alkoxide base (methanol with sodium methoxide, for example).

$C_6H_5COOC_2H_5 + CH_3CH_2COOC_2H_5 \xrightarrow{NaH} C_6H_5COCH(CH_3)COOC_2H_5$

Claisen condensation of two different esters can result in a mixture of products under thermodynamic control conditions. The reaction of two different esters is called the crossed-Claisen (or a mixed Claisen) ester condensation. However, mixed reactions are successful when only one of the two esters has α-hydrogen and the other has the more reactive of the two carbonyl groups. Diethyl oxalate and ethyl formate are particularly suitable.

EtO(O)C-C(O)OEt + PhCH_2CO_2Et $\xrightarrow[EtOH]{EtO^-}$ EtO(O)C-C(O)-CH(Ph)CO_2Et $\xrightarrow[-CO]{175℃}$ PhCH$(CO_2Et)_2$ 80%~85%

It is best to perform the enolate of one component with LDA and to treat it with the acid chloride corresponding to the other component.

OEt O $\xrightarrow{LDA}$ [OEt OH] $\xrightarrow{i\text{-}Pr\text{-}CO\text{-}OEt}$ i-Pr O O OEt

There are two important variations of this condensation. In the first, an ester enolate is condensed with a ketone or aldehyde. The acylation of ketones by esters has been called the Claisen reaction.

$$CH_3\overset{O}{\overset{\|}{C}}CH_3 + CH_3(CH_2)_4CO_2C_2H_5 \xrightarrow{NaH} CH_3\overset{O}{\overset{\|}{C}}CH_2\overset{O}{\overset{\|}{C}}(CH_2)_4CH_3 \quad 54\%\sim65\%$$

$$CH_3\overset{O}{\overset{\|}{C}}OC_2H_5 + CH_3\overset{O}{\overset{\|}{C}}CH_3 \xrightarrow{EtO^-} H_3C-\overset{O}{\overset{\|}{C}}CH_2\overset{O}{\overset{\|}{C}}-CH_3$$

$$PhCO_2C_2H_5 + H_3C-\overset{O}{\overset{\|}{C}}-Ph \longrightarrow Ph\overset{O}{\overset{\|}{C}}CH_2\overset{O}{\overset{\|}{C}}Ph$$

$$\text{2-methylcyclohexanone} + H\overset{O}{\overset{\|}{C}}OC_2H_5 \xrightarrow{CH_3O^-} \text{2-methyl-6-formylcyclohexanone (CHO)}$$

2.3.2.2 Dieckmann 缩合 Dieckmann Condensation

There is an intramolecular version of the Claisen condensation but it has been given a different name (the Dieckmann condensation)[19]. This reaction involves intramolecular cyclization of an α, ω-diester. The reaction is usually done under equilibrating conditions, although kinetic control conditions can also be used.

$$\text{diethyl adipate } (CO_2Et, CO_2Et) \xrightarrow[EtOH]{NaOEt} [\text{enolate}, O^-, OEt, OEt] \longrightarrow [CO_2C_2H_5, CO_2C_2H_5, O^-] \longrightarrow \text{2-}(CO_2C_2H_5)\text{cyclopentanone}$$

$$\xrightarrow[(3)\ heat(-CO_2)]{(1)\ aq.NaOH;\ (2)\ H_3O^+} \text{cyclopentanone}$$

$$\text{2-}CO_2C_2H_5\text{-2-}CH_3\text{-cyclohexanone} \not\longleftarrow C_2H_5O_2CCH_2(CH_2)_3\overset{CH_3}{\overset{|}{C}}HCO_2C_2H_5 \xrightarrow[xylene]{NaOEt} \text{enolate } (O^-, CH_3, CO_2C_2H_5)$$

(NaOEt, xylene: conversion of the 2-methyl-2-carbethoxycyclohexanone into the enolate)

$$\text{diethyl succinate } (CO_2Et, CO_2Et) \xrightarrow{EtONa} EtO_2C\text{-},\ \overset{O}{\overset{\|}{C}}OC_2H_5,\ CO_2Et,\ C{=}O \xrightarrow{EtONa} \text{cyclohexane-1,4-dione with } OC_2H_5\text{C=O and } EtO_2C$$

It is an important method for the formation of five－ and six－membered rings and has occasionally been used for formation of larger rings.

Synthetic procedures involving Claisen reaction and Dieckmann condensation：

$$CH_3(CH_2)_3CO_2C_2H_5 \xrightarrow{NaOEt} CH_3(CH_2)_3CO\underset{\displaystyle CH_2CH_2CH_3}{\underset{|}{C}}HCO_2C_2H_5 \quad 77\%$$

$$CH_3CH_2\underset{\displaystyle CH_3}{\underset{|}{C}}HCO_2C_2H_5 \xrightarrow{Ph_3C^-Na^+} CH_3CH_2\underset{\displaystyle CH_3}{\underset{|}{C}}H\overset{\displaystyle O}{\overset{\|}{C}}-\underset{\displaystyle CH_3}{\underset{|}{\overset{\displaystyle CH_2CH_3}{\overset{|}{C}}}}CO_2C_2H_5 \quad 63\%$$

$$C_2H_5O_2C(CH_2)_4CO_2C_2H_5 \xrightarrow{Na,\ toluene} \text{2-(ethoxycarbonyl)cyclopentanone } (CO_2C_2H_5) \quad 74\%\sim81\%$$

$$CH_3-N\begin{matrix} CH_2CH_2CO_2C_2H_5 \\ CH_2CH_2CO_2C_2H_5 \end{matrix} \xrightarrow[benzene]{NaOEt} \xrightarrow{HCl} \text{piperidinone ring: } CH_3N^+(H),\ CO_2C_2H_5,\ =O \quad 71\%$$

2. 3. 3 羧酸衍生物与醛酮的缩合反应 The Condensation of Carboxylic Enolates with Carbonyl Compounds

2. 3. 3. 1 Perkin 反应 Perkin Reaction

Condensation of an aldehyde (having no enolizable protons) with the enolate of an acid anhydride leads to an acetoxy ester and is called the Perkin reaction. The reaction consists of the condensation of an acid anhydride with an aromatic aldehyde catalyzed by a carboxylate ion. The mechanism is surprisingly complex, although each step is of a well－established type. The anhydride first provides the enolate under the influence of the basic carboxylate ion.

$$C_6H_5-CHO + \text{aliphatic anhydride} \xrightarrow{Cat.} \alpha,\beta\text{-unsaturated carboxylate}$$

Cat.: $RCOOK$, RCO_2Na, K_2CO_3, Et_3N.

$$C_6H_5-CHO + CH_3-\overset{\displaystyle O}{\overset{\|}{C}}-O-\overset{\displaystyle O}{\overset{\|}{C}}-CH_3 \xrightarrow{CH_3COONa} \xrightarrow{H_2O} C_6H_5-\underset{\displaystyle H}{C}=CH-\overset{\displaystyle O}{\overset{\|}{C}}-OH$$

cinnamon acid

$$CH_3COO^- + CH_3-\overset{O}{\overset{\|}{C}}-O-\overset{O}{\overset{\|}{C}}-CH_3 \rightleftharpoons {}^-CH_2COOCOCH_3 + CH_3COOH$$

$$ArCHO + {}^-CH_2COOCOCH_3 \rightleftharpoons \cdots \xrightleftharpoons{Ac_2O} \cdots \xrightleftharpoons{CH_3COO^-} \cdots \xrightleftharpoons{CH_3COO^-} ArCH{=}CHCOOCOCH_3 \xrightleftharpoons{CH_3COOH} ArCH{=}CHCOOH + (CH_3CO)_2O$$

When a OH or NH_2 group is adjacent the aldehyde group in the aromatic ring, a fused aromatic compound is formed by Perkin condensation.

$$o\text{-}HOC_6H_4CHO \xrightarrow[CH_3COONa]{(CH_3CO)_2O} o\text{-}HOC_6H_4CH{=}CHCOOH \xrightarrow{-H_2O} \text{coumarin}$$

Perkin reaction usually requires a long time and high temperature to achieve good yield.

2.3.3.2 Knoevenagel-Doebner 缩合反应 Knoevenagel-Doebner Condensation Reactions

Knoevenagel condensation involves malonate enolates in a condensation reaction with aldehydes, usually a non-enolizable aldehyde. An aldehyde or ketone can be condensed with an active methylene compound (H_2CX_2 or $HRCX_2$) such as a malonic ester using a primary or secondary amine as the base. To avoid a competing aldol condensation, the aldehyde should have no α-protons[20]. Diethyl malonate and benzaldehyde were condensed to form hydroxyl intermediate, for example. Spontaneous elimination of water gave the alkylidene product, saponification followed by decarboxylation gave the final Knoevenagel product, conjugated acid.

Contrary to aldol condensation without conjugation in the products, elimination commonly occurs upon hydrolysis of the Knoevenagel products. This is a facile route to substituted methylene malonic acids [$RCH{=}C(CO_2H)_2$] and acrylic acid derivatives. When the base is pyridine with a trace of piperidine (rather than diethylamine) and malonic acid is the enolate precursor, the reaction is called the Doebner condensation or the Doebner modification.

$$PhCHO + CH_2(COOEt)_2 \xrightarrow[\triangle]{AcOH, Et_2NH} PhCH{=}CH-COOH$$

Like malonic acid(丙二酸), succinic anhydride (丁二酸酐) is also involved in the reaction.

$$ArCHO + (CH_2CO)_2O \longrightarrow [ArCH(OH)-CH(CH_2CO)CO-O] \xrightarrow{-H_2O} [ArCH=C(CO)CH_2CO-O]$$

$$\xrightarrow{H_2O} [ArCH=C(COOH)CH_2COOH] \xrightarrow{150℃} ArCH=CH-CH_2-COOH + \text{ArCH-CH}_2\text{-CH}_2\text{-C(=O)-O (lactone)}$$

Other derivatives containing two electron-withdrawing groups can be used in the reaction, including nitro enolate, malononitrile, et. al.

$$ArCH=NC_4H_9 + \bar{C}H_2NO_2 \xrightarrow{H^+} ArCH(NHC_4H_9)CH_2NO_2 \longrightarrow ArCH(\overset{H^+}{N}HC_4H_9)-CHNO_2(H) \longrightarrow ArCH=CHNO_2$$

$$ArCH=C(CO_2H)_2 + C_5H_5NH^+ \longrightarrow ArCH(N^+C_5H_5)-CH(CO_2H)C(=O)OH \longrightarrow ArCH=CHCO_2H$$

Amine-catalyzed condensation reactions of the Knoevenagel type:

$$\text{cyclohexanone} + NCCH_2CO_2H \xrightarrow{NH_4OAc} \text{cyclohexylidene}=C(CN)(CO_2H) \quad 65\%\sim76\%$$

$$PhCH=O + CH_3CH_2CH(CO_2H)_2 \xrightarrow{pyridine} PhCH=C(CO_2H)(C_2H_5) \quad 60\%$$

$$CH_2=CHCH=O + CH_2(CO_2H)_2 \xrightarrow[60℃]{pyridine} CH_2=CHCH=CHCO_2H \quad 42\%\sim46\%$$

$$m\text{-}O_2NC_6H_4CHO + CH_2(CO_2H)_2 \xrightarrow{pyridine} m\text{-}O_2NC_6H_4CH=CHCO_2H \quad 75\%\sim80\%$$

$$C_6H_{11}CHO + CH_2(COOH)_2 \xrightarrow{pyridine} C_6H_{11}CH=CHCOOH$$

$$o\text{-}HOC_6H_4CHO + CH_2(COOC_2H_5)_2 \xrightarrow[HOAc]{pyridine} \text{3-COOEt-coumarin (2-oxo-2H-chromene)}$$

2.3.3.3 Darzen 甘油酯缩合 Darzens' Glycidic Ester Condensation

When a α-halo ester is treated with base and the resulting enolate anion condensed with a carbonyl derivative, the product is a halo-alkoxide. This nucleophilic species can displace the halogen intramolecularly to produce an epoxide, which forms the basis of a classical reaction known as the Darzens' glycidic ester condensation. Reaction of ethyl α-chloroacetate and sodium ethoxide in the presence of benzaldehyde generated the usual alkoxide. Intramolecular displacement of chloride, however, gave the α,β-epoxy ester (also called a glycidic ester). Saponification gave the α,β-epoxy acid, which was unstable and lost carbon dioxide to generate a substituted aldehyde or ketone such as phenyl acetaldehyde in this case. This sequence is a useful chain extension reaction in which an aldehyde (PhCHO) is converted to a longer chain aldehyde homolog ($PhCH_2CHO$). A ketone is converted to a longer chain aldehyde.

$$CH_2(Cl)CO_2C_2H_5 \xrightarrow{NaOBu\text{-}t} \bar{C}H(Cl)CO_2C_2H_5 \xrightarrow{R_2C{=}O} [R_2C(O^-)\text{—}CH(Cl)CO_2C_2H_5] \longrightarrow R_2C\overset{O}{\frown}CHCO_2C_2H_5$$

$$\xrightarrow{H_3O^+} R_2C\overset{O}{\frown}CHCO_2H \xrightarrow[\triangle]{-CO_2} R_2C{=}CH(OH) \xrightarrow{\text{tautomerizes}} R_2CH\text{—}CHO$$

Examples:

$$\text{cyclohexanone} + ClCH_2CO_2C_2H_5 \xrightarrow{KOC(Me)_3} \text{spiro-epoxy ester (}CO_2C_2H_5\text{, H)} \quad 83\%\sim95\%$$

$$PhCH{=}O + PhCH(Cl)CO_2C_2H_5 \xrightarrow{KOC(Me)_3} \text{epoxide (H, Ph; }CO_2C_2H_5\text{, Ph)} \quad 75\%$$

$$PhCOCH_3 + ClCH_2CO_2C_2H_5 \xrightarrow{KOC(Me)_3} \text{epoxide (}H_3C\text{, Ph; }CO_2C_2H_5\text{, H)} + \text{epoxide (Ph, }H_3C\text{; }CO_2C_2H_5\text{, H)} \quad 62\%$$

1∶1mixture of isomers

2.3.3.4 Stobbe 缩合 Stobbe Condensation

The Stobbe condensation is the condensation of diethyl succinate and its derivatives with non-enolizable carbonyl compounds in the presence of strong bases such as $t-C_4H_9OK$, RONa and NaH.

$$CH_2COOC_2H_5-CH_2COOC_2H_5 \xrightarrow{EtONa} \bar{C}HCOOC_2H_5-CH_2COOC_2H_5 \xrightarrow{RCOR'} RR'C(O^-)-CH(COOC_2H_5)-CH_2COOC_2H_5$$

$$\longrightarrow \text{cyclic intermediate} \longrightarrow \gamma\text{-lactone} \xrightarrow{H^+} RR'C{=}C(COOC_2H_5)-CH_2-COOH$$

γ-lactone

Condensation of succinic ester derivatives (such as diethyl succinate) with non-enolizable ketones or aldehydes and a base gives the condensation intermediate alkoxide. The alkoxide reacts with the distal ester via acyl substitution to give a γ-lactone intermediate. In the original version of this reaction, saponification of γ-lactone gave the α-alkylidene monoester[21]. The reaction is limited to those α, ω-diesters for which the Dieckmann condensation is not a competitive reaction so succinic acid derivatives are commonly used. This transformation is known as the Stobbe condensation. The aldehyde or ketone substrate is not limited to non-enolizable derivatives.

β, γ-unsaturated carboxylic acids can be obtained via Stobbe condensation.

$$C_6H_5CHO \xrightarrow[NaOEt]{(CH_2COOEt)_2} C_6H_5CH{=}C(CO_2Et)CH_2CO_2H \xrightarrow{H_3O^+} C_6H_5CH{=}C(CO_2H)CH_2CO_2H$$

$$\xrightarrow[-CO_2]{\triangle} C_6H_5CH{=}CH-CH_2COOH \quad \beta,\gamma\text{-unsaturated carboxylic acids}$$

$$\xrightarrow[cat.]{[H]} C_6H_5CH_2-CH_2-CH_2COOH \xrightarrow[-H_2O]{PPA} \alpha\text{-tetralone} \quad 75\%\sim80\%$$

2.3.3.5 Reformatsky 反应

Reformatsky reaction

Another enolate-like reaction is the Reformatsky reaction, which employs a nucleophilic organozinc intermediate, generated from an α-halo carbonyl and zinc metal (also see 2.1.4). Condensation of α-halogenesters with ketones and aldehydes in the presence of zinc metal in the same manner as Grignard reagents form β-hydroxyl carbonyl esters or α, β-unsaturated carbonyl esters.

$$ROOCCH_2Br \xrightarrow{Zn} ROOCCH_2-ZnBr \xrightarrow{>C=O} \text{(cyclic zinc chelate: } >C(-O-Zn(Br)-O=C(OR)-CH_2-)\text{)}$$

$$\xrightarrow{H_3O^+} >C(OH)-CH_2COOR \xrightarrow{-H_2O} >C=CHCOOR$$

β-hydroxyl carbonyl esters α, β-unsaturated carbonyl esters

2.3.3.6 曼尼希反应 Mannich reaction

Compounds which are enolic, or potentially enolic, and also certain alkynes, react with a mixture of an aldehyde (usually formaldehyde) and a primary or secondary amine in the presence of an acid to give, after basification, an aminomethyl derivative (Mannich base). For example, by refluxing a mixture of acetone, diethylamine hydrochloride, paraformaldhyde, methanol, and a little concentrated hydrochloride acid, and treating the product with base, 1-diethylamino-3-buttanone can be obtained in up to 70% yield.

$$RCH_2-C(=O)-R' + HCHO + HN(CH_3)_2 \longrightarrow RCH(CH_2N(CH_3)_2)-C(=O)-R'$$

Mannich base

$$C_6H_5COCH_3 + HCHO + HN(CH_3)_2 \xrightarrow{H^+} C_6H_5COCH_2CH_2NR_2 \xrightarrow[-R_2NH]{\triangle} C_6H_5CO-HC=CH_2$$

Mechanistically, the Mannich reaction is the condensation of an enolizable carbonyl compound with an iminium ion.

$$CH_2=O + HN(CH_3)_2 \rightleftharpoons HOCH_2N(CH_3)_2 \overset{H^+}{\rightleftharpoons} CH_2=\overset{+}{N}(CH_3)_2$$

It is usually done using formaldehyde and introduces an α-dialkylaminomethyl substituent. *N*, *N*-(Dimethyl) methylene ammonium iodide, preformed immonium salts, is commercially available and is known as Eschenmoser's salt. It advoids the need for an acid catalyst and thereby increases the scope of the reaction.

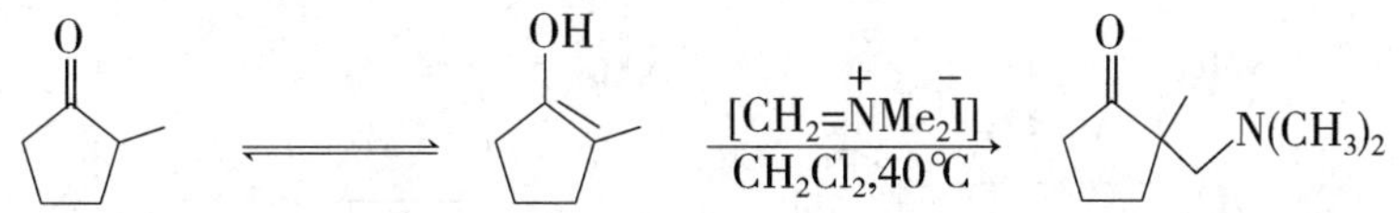

$$\text{[CH}_2\text{=}\overset{+}{\text{N}}\text{Me}_2\overset{-}{\text{I}}\text{]},\ \text{CH}_2\text{Cl}_2\text{, 40℃}$$

Phenylacetylene and some of its nuclear-substituted derivatives, although not enolic, also react, e. g.

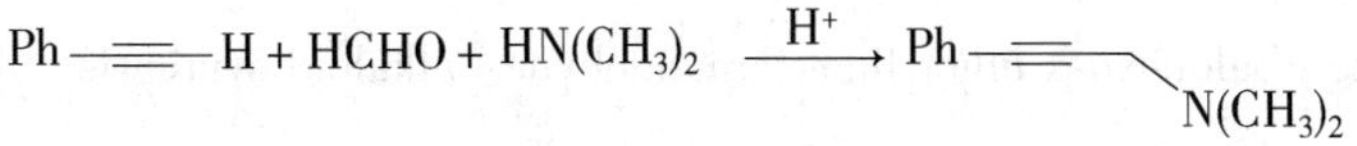

$$Ph-\equiv-H + HCHO + HN(CH_3)_2 \xrightarrow{H^+} Ph-\equiv-CH_2N(CH_3)_2$$

Mannich Bases as Intermediates in Synthesis:

（ⅰ）α, β-不饱和羰基化合物的形成

Formation of α, β-unsaturated carbonyl compounds

Mannich base

Robinson annulation

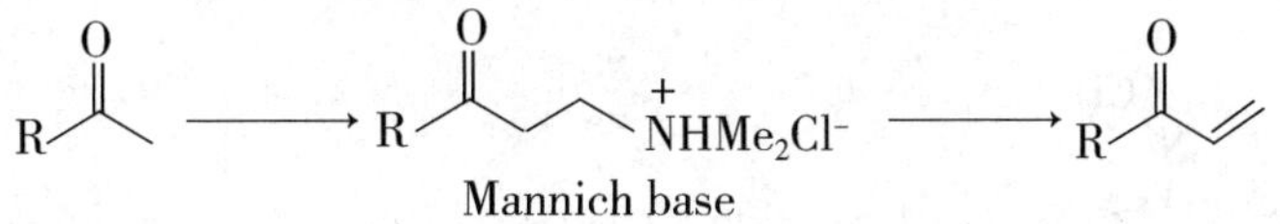

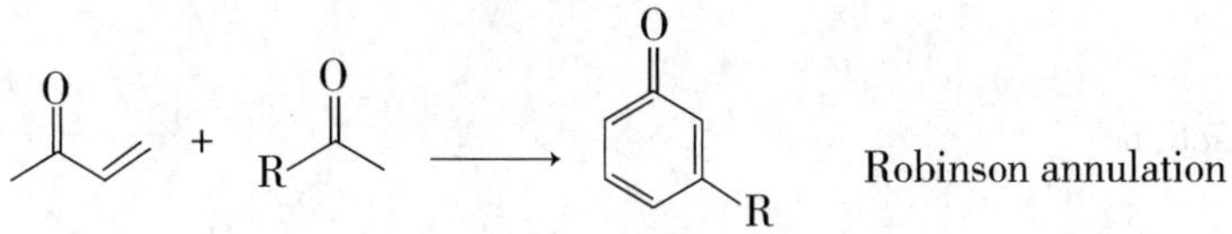

The Mannich reaction normally gives the hydrochloride of the Mannich base. These salts are usually stable at room temperature but those derived from aliphatic compounds undergo elimination on being heated, e. g.

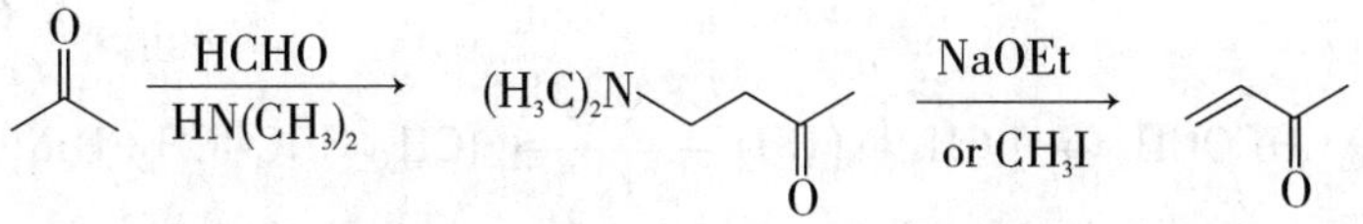

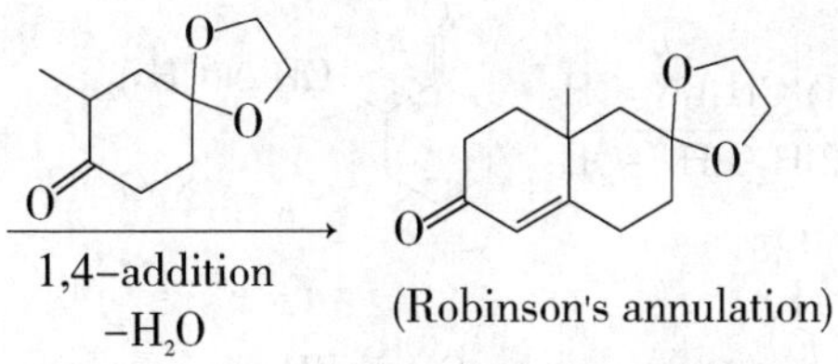

$$\xrightarrow[HN(CH_3)_2]{HCHO} (H_3C)_2N \ldots \xrightarrow[\text{or } CH_3I]{NaOEt}$$

1,4-addition
$-H_2O$

(Robinson's annulation)

The best known example of this application of α, β-unsaturated carbonyl compounds is Robinson ring extension, used in building up the ring system of steroids.

（ⅱ）氨基的置换

Replacement of the amino group

An alternative to the synthesis of heteroauxin described employs the Mannich base from indole[22]. Methylation with dimethyl sulfate followed by treatment with cyanide ion and then hydrolysis gives heteroauxin:

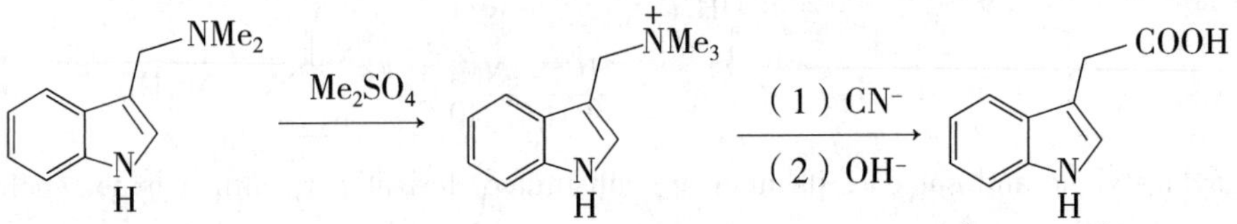

NMe_2 (indole, N-H) $\xrightarrow{Me_2SO_4}$ $\overset{+}{N}Me_3$ (indole, N-H) $\xrightarrow[(2)\ OH^-]{(1)\ CN^-}$ COOH (indole, N-H)

heteroauxin

(ⅲ) 甲醛以外的醛的使用：生物碱的合成

The use of aldehydes other than formaldehyde: alkaloid synthesis

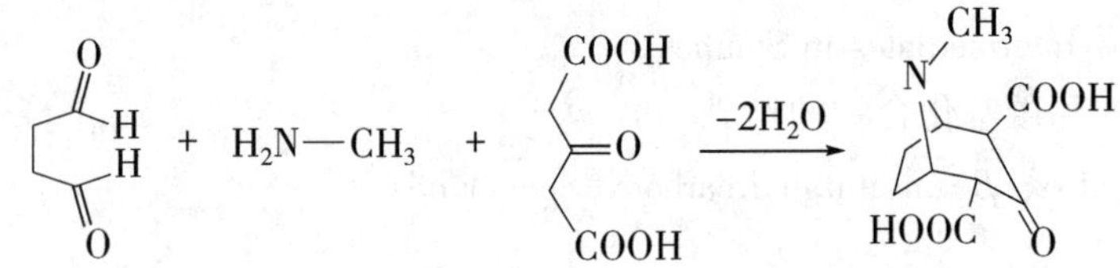

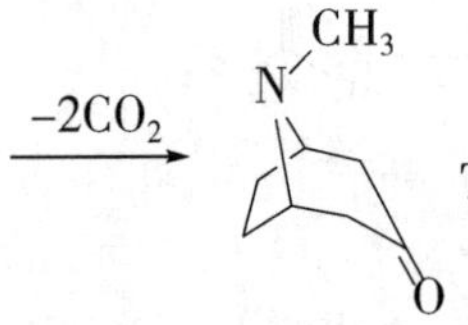

O, H, H, O + $H_2N—CH_3$ + COOH, =O, COOH $\xrightarrow{-2H_2O}$ CH_3, N, COOH, HOOC, O

$\xrightarrow{-2CO_2}$ CH_3, N, O Tropinone ----Alkaloid(used to prepare cocaine)

Mannich Reactions:

$$PhCOCH_3 + CH_2O + (CH_3)_2\overset{+}{N}H_2\overset{-}{Cl} \longrightarrow PhCOCH_2CH_2\overset{H}{\underset{+}{N}}(CH_3)_2\overline{Cl} \quad 70\%$$

$$CH_3COCH_3 + CH_2O + (CH_3CH_2)_2\overset{+}{N}H_2\overline{Cl} \longrightarrow CH_3COCH_2CH_2\overset{H}{N}(C_2H_5)_2\overline{Cl}$$

66%~75%

$$(CH_3)_2CHCOCH_3 + [(CH_3)_2N]_2CH_2 \xrightarrow{CF_3CO_2H} (CH_3)_2CHCOCH_2CH_2N(CH_3)_2$$

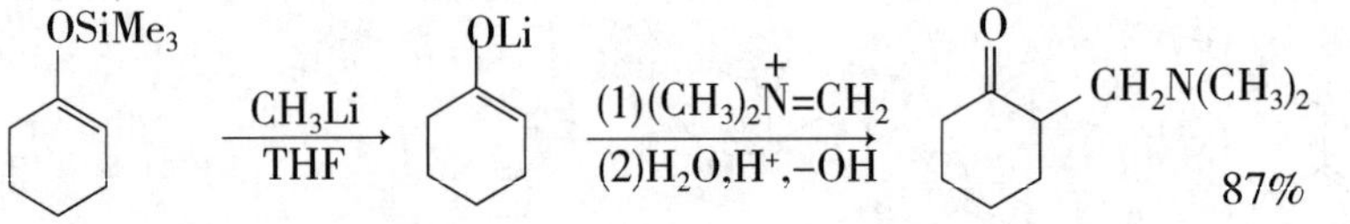

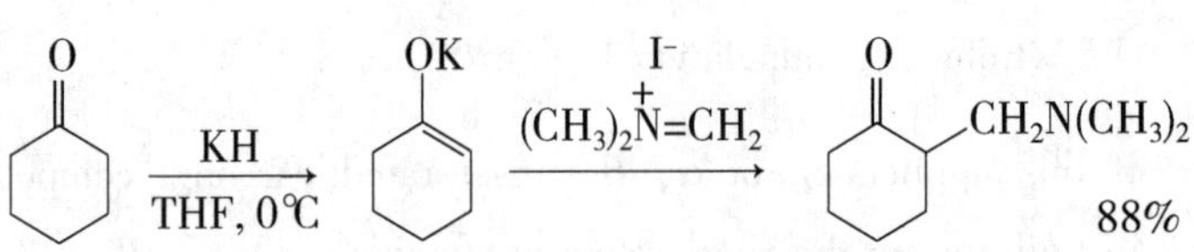

$OSiMe_3$ $\xrightarrow[THF]{CH_3Li}$ OLi $\xrightarrow[(2)H_2O,H^+,-OH]{(1)(CH_3)_2\overset{+}{N}=CH_2}$ O, $CH_2N(CH_3)_2$ 87%

O $\xrightarrow[THF,\ 0℃]{KH}$ OK $\xrightarrow{I^-\ (CH_3)_2\overset{+}{N}=CH_2}$ O, $CH_2N(CH_3)_2$ 88%

参 考 文 献

References

1. 张凤秀．有机化学．北京：科学出版社，2020.

2. 福尔哈特等．有机化学．北京：化学工业出版社，2020.

3. 陈艳．中级有机化学．北京：化学工业出版社，2019.

4. 李晓敏等 . 实用有机化学 . 北京：化学工业出版社，2019.
5. 江洪等 . 有机化学 . 北京：科学出版社，2019.
6. 曹晨忠 . 有机化学中的取代基效应 . 北京：科学出版社，2019.
7. 安琼等 . 有机化学学习指导 . 南京：东南大学出版社，2019.
8. 林晓辉等 . 有机化学概论 . 北京：化学工业出版社，2019.
9. 史密斯等 . March' s advanced organic chemistry reactions：mechanisms，and structure. 北京：化学工业出版社，2018.
10. 何树华等 . Secondary organic chemistry. 北京：化学工业出版社，2018.
11. 赵军龙等 . 基础有机化学反应 . 北京：高等教育出版社，2018.
12. 袁先友等 . 微波有机化学合成及应用 . 长沙：湖南大学出版社，2007.
13. Li Jie Jack. Carbocation Chemistry：Applications in Organic Synthesis. Boca Raton：CRC Press，2016.
14. Padwa Albert. Organic Photochemistry. Boca Raton：CRC Press，2017.
15. Frank Marken，Mahito Atobe. Modern Electrosynthetic Methods in Organic Chemistry. Boca Raton：CRC Press，2018.
16. Carey F A，Sundberg R J. Advenced Organic Chemistry. New York and London：Plenum Press，1990. 192.
17. Dean J Tantillo. Applied Theoretical Organic Chemistry. Singapore：World Scientific Publishing Company，2018.
18. McIntosh John M. Organic Chemistry：Fundamentals and Concepts. Berlin：De Gruyter，2018.
19. Quirk R P，Pickel D L. Polymer Science：A Comprehensive Reference. Amsterdam：Elsevier，2012. 351-412.
20. Smith M B. Organic Synthesis（Fourth Edition）. Boston：Academic Press，2017. 309-418.
21. Rice J E. Organic Chemistry Concepts and Applications for Medicinal Chemistry. Boston：Academic Press，2014. 51-65.
22. Ziffle V E，Fletcher S P. Comprehensive Organic Synthesis（Second Edition）. Amsterdam：Elsevier，2014. 999-1010.

3 通过亲电试剂形成碳碳键

Chapter 3 Carbon-Carbon Bonds Formation: Carbocations

Trivalent carbocations are one of the most fundamental classes of reactive intermediates. In this section, the emphasis is on carbocation reactions that modify the carbon skeleton, including carbon-carbon bond formation, rearrangements, and fragmentation reactions. The basic aspects of the structural and reactivity features of these intermediates will also be discussed[1]. The nucleophilic partners such as alkenes, alkynes, aromatic rings or compounds containing heteroatoms are included in this chapter. In addition, free carbocations and exploitation of their molecular rearrangements to generate carbon-carbon bonds and different structures are described in this chapter.

3.1 芳烃亲电取代 Electrophilic Aromatic Substitution

3.1.1 Friedel-Crafts 烷基化 Friedel-Crafts Alkylation

Friedel - Crafts alkylation reactions are an important method for introducing carbon substituents on aromatic rings. The reaction of an aromatic ring with a carbocation generates a Wheland-type intermediate that leads to a substituted aromatic system. Known as Friedel-Crafts reactions, they are among the most important reactions in organic chemistry. Alkylations usually involve alkyl halides and Lewis acids or reactions of alcohols or alkenes with strong acids[2].

3.1.1.1 Generation, structure and stability

There are many different methods for generating carbocations including the classical ionization by heterolytic bond cleavage, addition to unsaturated bonds, and exchange with other cations, and the rate-determining step can vary with each method.

The methods for generating carbocation:

(ⅰ) 通过异裂的经典电离

Classical ionization by heterolytic bond cleavage

Alkyl halides and Lewis acids: $R\text{–}X + AlCl_3 \rightleftharpoons R\text{–}X^+\text{–}\bar{A}lCl_3 \rightleftharpoons R^+ + X\bar{A}lCl_3$

Reactions of alcohols or alkenes with strong acids:

$$R\text{–}OH + H^+ \rightleftharpoons R\text{–}\overset{+}{O}(H)H \rightleftharpoons R^+ + H_2O$$

$$RCH{=}CH_2 + H^+ \rightleftharpoons R\overset{+}{C}HCH_3$$

Alcohols can serve as carbocation precursors in strong acids such as sulfuric or phosphoric acid. Alkenes can serve as alkylating agents when a protic acid, especially H_2SO_4, H_3PO_4, and HF, or a Lewis acid, such as BF_3 and $AlCl_3$, is used as a catalyst[3].

$$CH_3COF \xrightarrow{BF_3} CH_3CO^+ + {}^-BF_4$$

$$HO\text{—}C(CH_3)_2\text{—}CH_3 \xrightarrow[-60℃]{HSO_3F\text{–}SbF_5\text{–}SO_2} (H_3C)_2\overset{+}{C}\text{—}CH_3 + H_3O^+ + SO_3F^- + SbF_5 + SO_2$$

Treatment of a secondary or tertiary alcohol with an acid catalyst gives an onium salt, which loses water (a good leaving group) to give the cation, and 3° alcohols react faster than 2° alcohols. Heterolytic bond cleavage of a halide or sulfonate ester (C—X, X=Br, Cl, I, OMs, OTs, etc.) in aqueous solvents (solvolysis often requires heating) generates tertiary cations easily, secondary cations with difficulty and primary cations with extreme difficulty.

(ⅱ) Addition to unsaturated bonds with a suitable acid

$$Z{=}C< + H^+ \longrightarrow ZH\text{—}\overset{+}{C}<$$

Reactions of acids HX with alkenes give the more stable cation, where the order of stability is 3°>2°>1°.

$$>C{=}C< + HCl \longrightarrow >\overset{+}{C}\text{—}C(H)< + Cl^-$$

$$>C{=}O \xrightarrow{H^+} >C{=}\overset{+}{O}H \rightleftharpoons >\overset{+}{C}\text{—}OH$$

The reaction of aldehydes and ketones with an acid catalyst (Brønsted-Lowry or Lewis) also gives the oxygen-stabilized cation.

(ⅲ) Exchange with other cations

$$C_6H_5NH_2 \xrightarrow[HCl]{NaNO_2} C_6H_5N_2^+ \longrightarrow C_6H_5^+ + N_2$$

Reaction of amines with nitrous acid (HONO) initially gives a diazoalkane, which can generate carbenes when generated under different conditions, but in this case, it readily decompose to a carbocation.

$$C_7H_8 + Ph_3\overset{+}{C}\overset{-}{SbF_6} \longrightarrow C_7H_7^+ + SbF_6$$

Tropylium ion (环庚三烯正离子) shows the greatest stability because of aromaticity.

A carbocation is a sp^2 hybridized carbon bearing three substituents, with an empty p orbital perpendicular to the plane of the other atoms. The sp^2-hybridization leads to a trigonal planar geometry. This intermediate has been termed a carbenium ion by George Olah, who wan Nobel prize in 1994 for carbocations.

When R, R^1, and R^2 are alkyl groups, the carbon atoms of the alkyl substituents release electrons to the positive center. The hyperconjugation and inductive effect (超共轭与诱导效应) lead to greater stability with increasing numbers of alkyl groups. Stabilized carbocations can be generated from allylic and benzylic alcohols by reaction with $Sc(O_3SCF_3)_3$ and results in formation of alkylation products from benzene and activated derivatives. An important conclusion is that the order 3° > 2° > 1° applies to all cations, and 3° benzylic > 2° benzylic > 1° benzylic. It is not surprising that a 3° benzylic cation is more stable than a 3° alkyl cation since the former is resonance stabilized. Vinyl carbocations can be generated from a variety of sources, but they are generally less stable than alkyl carbocations.

3°benzylic>2°benzylic>3°alkyl>1°benzylic ≈ 1°allylic ≈ 2°alkyl>1°alkyl>methyl

3.1.1.2 Friedel-Crafts 烷基化反应

Friedel-Crafts alkylation reaction

The carbocation required for this reaction can be generated from an alkene, an alcohol, an alkyl halide, an epoxide, an aldehyde, a ketone or an ester.

Alkylating species: RX, $RCH=CH_2$, epoxide, ROH, RCHO, $RC(=O)OR'$, $RC(=O)R'$.

Alkyl fluorides are the most reactive of the alkyl halides. The order of reactivity for alkyl halides with aluminum chloride ($AlCl_3$):

$$R{-}F > R{-}Cl > R{-}Br > R{-}I$$

$$\text{1,3,5-}(CH_3)_3C_6H_3 + C_2H_5F \xrightarrow[-80℃]{BF_3} [\text{arenium ion } (CH_3)_3C_6H_3(CH_2CH_3)^+]\,BF_4^- \text{ (Crystal)} \longrightarrow \text{1-ethyl-2,4,6-trimethylbenzene} + HBF_4$$

Formation of a primary cation is somewhat slower and is, of course, subject to rearrangement(重排). Owing to the involvement of carbocations, Friedel-Crafts alkylations can be accompanied by rearrangement of the alkylating group. For example, isopropyl groups are often introduced when *n*-propyl reactants are used.

$$C_6H_6 + CH_3CH_2CH_2Cl \xrightarrow{AlCl_3} C_6H_5CH(CH_3)_2 \quad 89\%$$

$$\text{ethylene oxide} + C_6H_6 \xrightarrow{AlCl_3} C_6H_5CH_2CH_2OH$$

$$CCl_3CHO \xrightarrow{H^+} CCl_3CH{=}\overset{+}{O}H \longrightarrow CCl_3\overset{+}{C}H(OH) \xrightarrow{C_6H_5Cl}$$

$$[\sigma\text{-complex: } Cl_3CCH(OH){-}C_6H_5Cl^+] \longrightarrow Cl_3C{-}CH(OH){-}C_6H_4{-}Cl \xrightarrow{H^+} \xrightarrow{C_6H_5Cl} Cl_3C{-}CH(C_6H_4Cl)_2$$

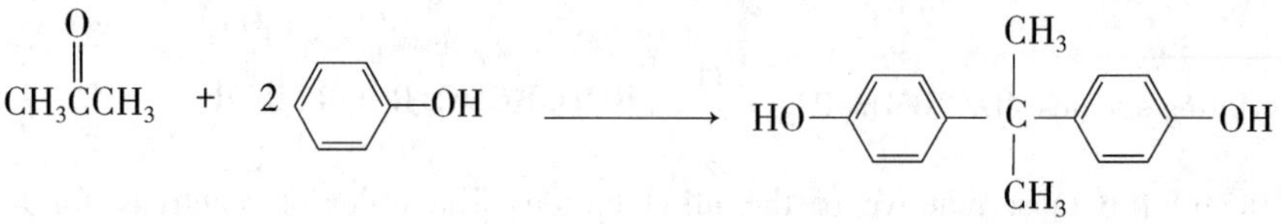

$$CH_3COCH_3 + 2\ C_6H_5OH \longrightarrow HO-C_6H_4-C(CH_3)_2-C_6H_4-OH$$

Bisphenol A

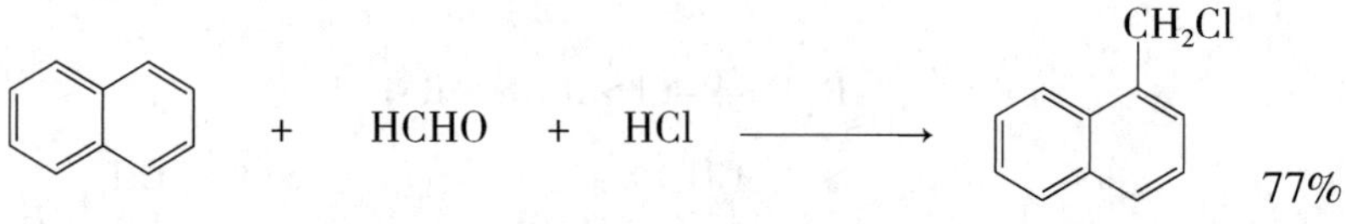

$$C_{10}H_8 + HCHO + HCl \longrightarrow C_{10}H_7CH_2Cl \quad 77\%$$

Chloromethylation (氯甲基化反应)

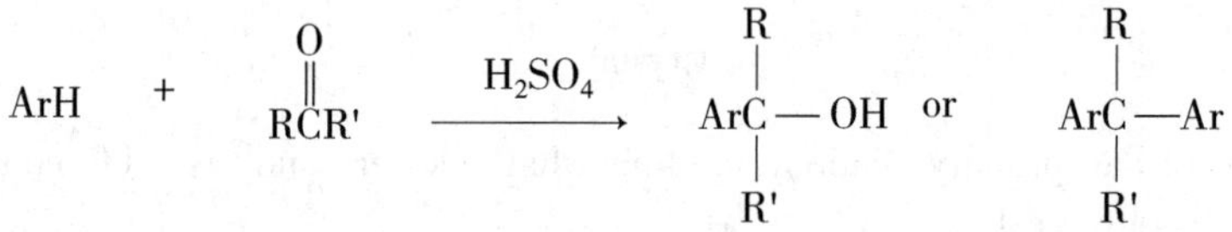

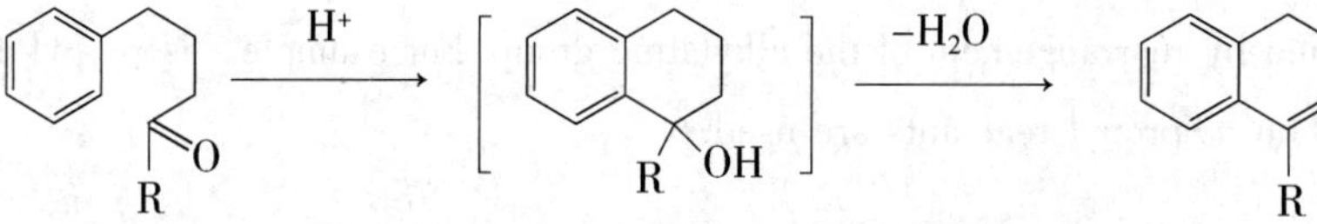

$$ArH + RCOR' \xrightarrow{H_2SO_4} ArC(R)(R')-OH \text{ or } ArC(R)(R')-Ar$$

$$\xrightarrow{H^+} \left[\ \text{R, OH}\ \right] \xrightarrow{-H_2O}$$

Amides and their derivatives can be used as alkylation reagents in the thermodynamic control conditions. For example:

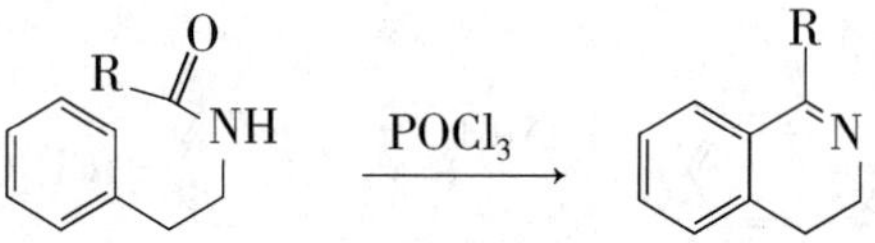

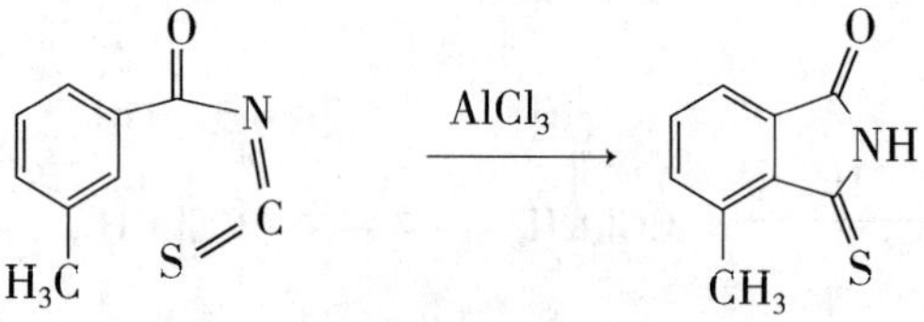

$$\xrightarrow{POCl_3}$$

$$\xrightarrow{AlCl_3}$$

Acyl isothiocyanate(异硫氰酸乙酰酯)

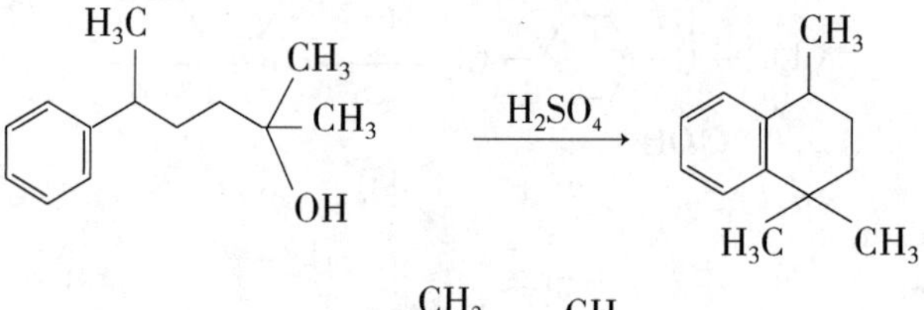

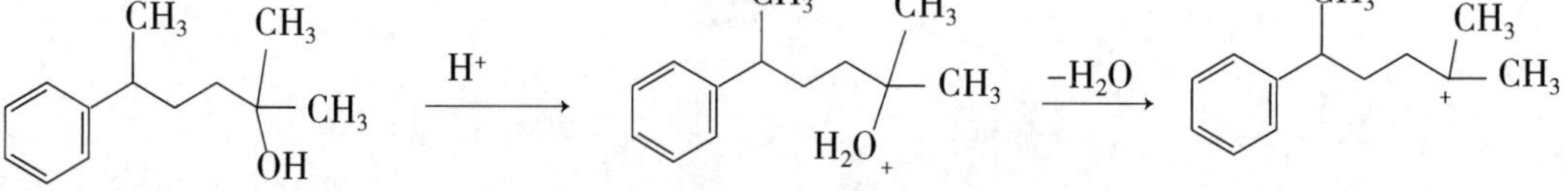

$$\xrightarrow{H_2SO_4}$$

$$\xrightarrow{H^+} \xrightarrow{-H_2O}$$

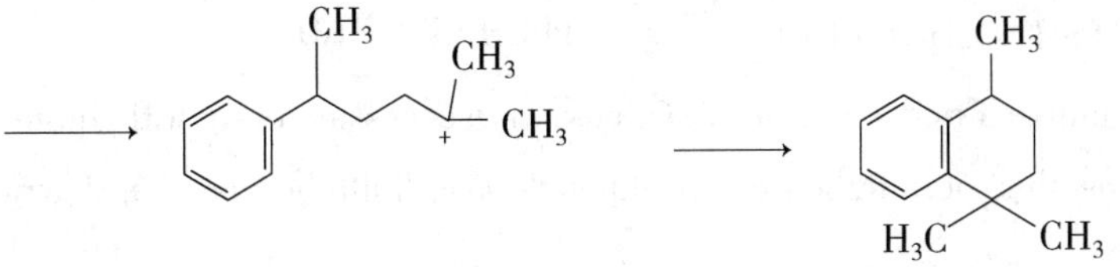

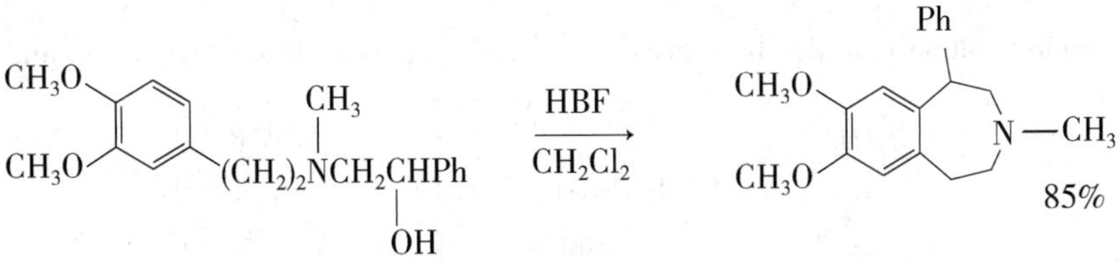

Internal Friedel-Crafts cyclization using alkene or alcohol substrates forms the basis of many natural product syntheses.

Friedel-Crafts alkylation can occur intramolecularly to form a fused ring. Intramolecular Friedel-Crafts reactions provide an important method for constructing polycyclic hydrocarbon frameworks. It is somewhat easier to form six-membered than five-membered rings in such reactions. Whereas 4-phenyl-1-butanol gives a 50% yield of a cyclized product in phosphoric acid, 3-phenyl-1-propanol is mainly dehydrated to alkenes.

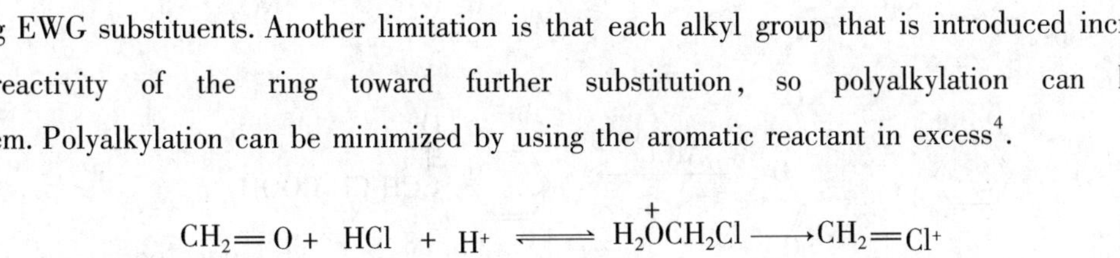

The Friedel-Crafts alkylation reaction does not proceed successfully with aromatic reactants having EWG substituents. Another limitation is that each alkyl group that is introduced increases the reactivity of the ring toward further substitution, so polyalkylation can be a problem. Polyalkylation can be minimized by using the aromatic reactant in excess[4].

$$CH_2{=}O + HCl + H^+ \rightleftharpoons H_2\overset{+}{O}CH_2Cl \longrightarrow CH_2{=}Cl^+$$

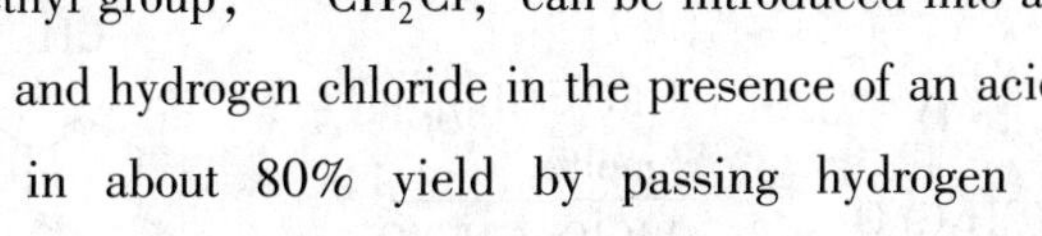

The chloromethyl group, $—CH_2Cl$, can be introduced into aromatic compounds by treatment with formaldehyde and hydrogen chloride in the presence of an acid. For example, benzyl chloride may be obtained in about 80% yield by passing hydrogen chloride into a suspension of paraformaldehyde and zinc chloride in benzene; the acid first liberates formaldehyde form paraformaldehyde and then takes part in the reaction.

$$PhH + CH_2O + HCl \xrightarrow{ZnCl_2} PhCH_2Cl + H_2O$$

Chloromethylation, unlike Friedel-Crafts reactions, is successful even with quite strongly deactivated nuclei such as that of nitrobenzene, although *m*-dinitrobenzene and pyridine are inert.

Chloromethylation can also be carried out using various chloromethyl ethers and $SnCl_4$.

$$\text{p-xylene} \xrightarrow[SnCl_4]{ClCH_2O(CH_2)_4OCH_2Cl} \text{2,5-dimethylbenzyl chloride}$$

3.1.2 Friedel-Crafts 酰基化 Friedel-Crafts Acylation

Friedel-Crafts acylation generally involves reaction of an acyl halide and Lewis acid such as $AlCl_3$, SbF_5, or BF_3, etc. Acid anhydrides can also be used in some cases. Mild Lewis acids can be used to affect Friedel-Crafts cyclization, which is particularly important with synthetic targets bearing other functional groups[5].

$$\text{2-nitroanisole} \xrightarrow[AlCl_3]{CH_3C(=O)-Cl} CH_3C(=O)-C_6H_3(OCH_3)(NO_2)$$

$$C_6H_6 + \text{succinic anhydride} \xrightarrow{AlCl_3} C_6H_5-C(=O)CH_2CH_2COOH \quad 92\%\sim95\%$$

A combination of hafnium (Ⅳ) triflate and $LiClO_4$ in nitromethane catalyzes acylation of moderately reactive aromatics by acetic anhydride in excellent yield.

$$\text{m-xylene} + (CH_3C(=O))_2O \xrightarrow[LiClO_4, CH_3NO_2]{Hf(O_3SCF_3)_4,\ 5mol\%} \text{2,4-dimethylacetophenone} \quad 91\%$$

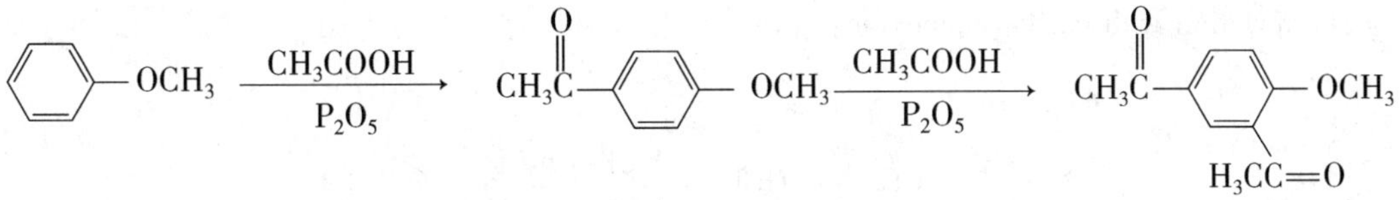

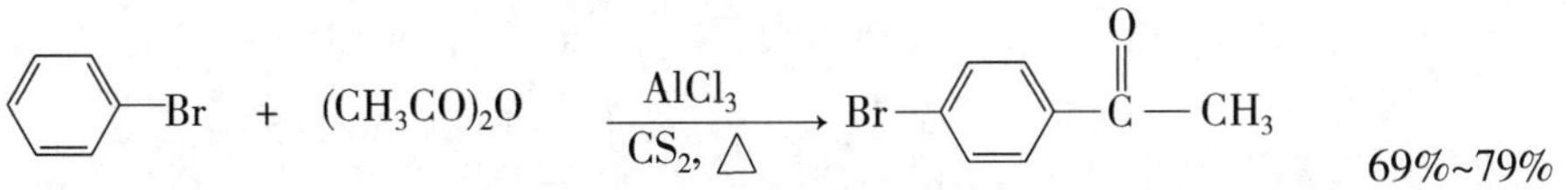

$$C_6H_5OCH_3 \xrightarrow[P_2O_5]{CH_3COOH} CH_3C(=O)C_6H_4OCH_3 \xrightarrow[P_2O_5]{CH_3COOH} CH_3C(=O)C_6H_3(C(=O)CH_3)OCH_3$$

$$C_6H_5Br + (CH_3CO)_2O \xrightarrow[CS_2,\ \triangle]{AlCl_3} Br{-}C_6H_4{-}C(=O){-}CH_3 \quad 69\%\sim79\%$$

The acylation is not limited to benzene, of course, but deactivated aromatic rings either do not undergo Friedel-Crafts reactions at all, or do so with difficulty.

Intramolecular Friedel-Crafts acylations are extremely valuable for the synthesis of polycyclic compounds, and the normal conditions involving an acyl halide and Lewis acid can be utilized. One useful alternative is to dissolve the carboxylic acid in polyphosphoric acid (PPA) and heat to effect cyclization.

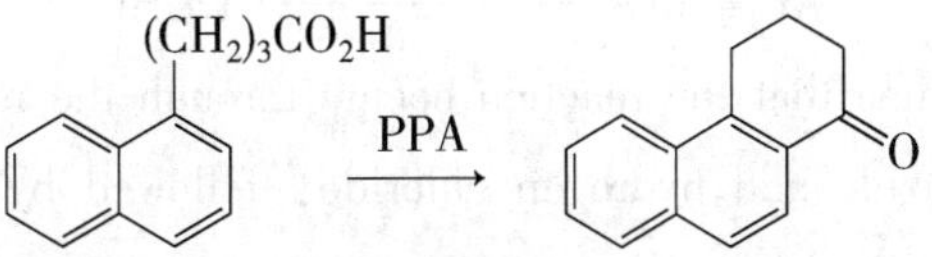

$(CH_2)_3CO_2H$ — PPA →

For example, benzene and succinic anhydride in the presence of aluminium trichloride give β-benzoylpropionic acid in 80% yield and reduction of carbonyl to methylene, conversion of the acid group to the acid chloride, and cyclization with aluminium trichloride give α-tetralone:

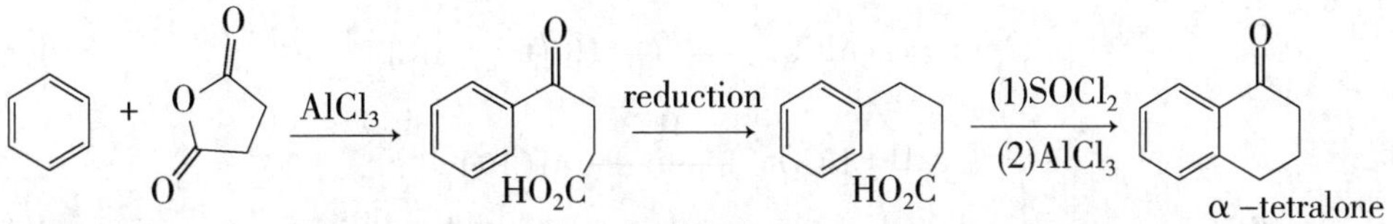

$AlCl_3$ → HO_2C — reduction → HO_2C — (1)$SOCl_2$ (2)$AlCl_3$ → α-tetralone

Many variants of this method are possible. By starting with naphthalene, naphthalene and some of its derivatives may be obtained. By using phthalic anhydride, two new aromatic rings may be built on, as in the synthesis of 1, 2-benzanthracene:

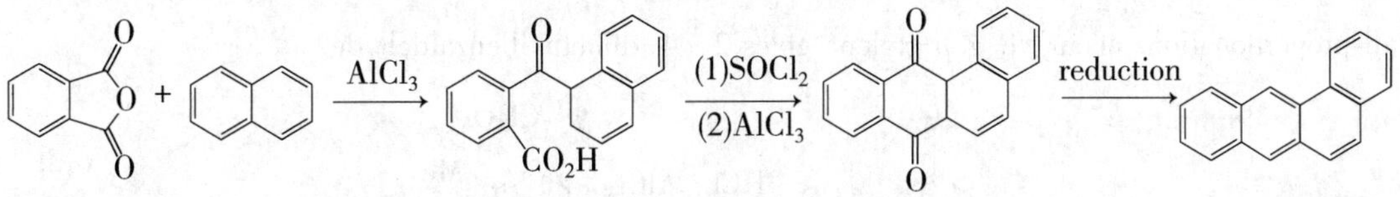

$AlCl_3$ → CO_2H — (1)$SOCl_2$ (2)$AlCl_3$ → — reduction →

Other carboxylic acid derivatives can be used in Friedel-Crafts acylation reactions, including anhydrides and esters.

A variation of the Vilsmeier-Haack reaction uses a formamide such as with benzene (or another aromatic compound) to give the aromatic aldehyde.

Carbon monoxide, hydrogen cyanide, and nitriles also react with aromatic compounds in the presence of strong acids or Friedel-Crafts catalysts to introduce formyl or acyl substituents.

Formylation with carbon monoxide:

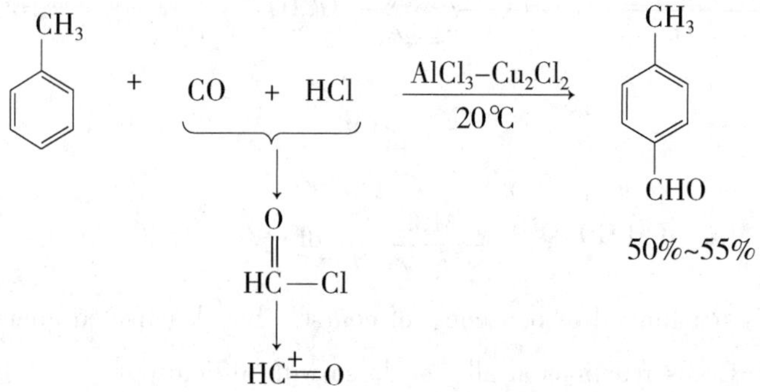

$$C_6H_5CH_3 + CO + HCl \xrightarrow[20℃]{AlCl_3-Cu_2Cl_2} p\text{-}CH_3C_6H_4CHO \quad 50\%\sim55\%$$

$$CO + HCl \longrightarrow HC(=O)-Cl \longrightarrow HC\overset{+}{\equiv}O$$

Gattermann-Koch formylation:

The formyl group, —CHO, may be introduced into aromatic compounds by treatment with carbon monoxide and hydrogen chloride in the presence of a Lewis acid:

$$Ar + CO \xrightarrow{HCl+Lewis\ acid} ArCHO$$

It was thought at one time that the reaction occurs through the formation of formyl chloride, HCOCl, from carbon monoxide and hydrogen chloride, followed by a Friedel-Crafts acylation catalyzed by the Lewis acid, but formyl chloride has never been obtained and it is now considered probable that the electrophilic species is the formyl cation, $[HC\equiv O^+ \longleftrightarrow \overset{+}{C}H=O]$, formed without the mediation of formyl chloride:

$$HCl + CO + AlCl_3 \longrightarrow HCO^+ + AlCl_4^-$$

$$ArH + HCO^+ \xrightarrow{-H^+} ArCHO$$

Formylation is unsuccessful with aromatic compounds or lower nuclear reactivity than the halobenzenes. So nitrobenzene may be used as solvent. It is also unsuccessful with amines, phenols and phenol ethers because of the formation of complexes with the Lewis acid. One drawback in the application of the reaction to polyalkylated benzenes is that rearrangements and disproportionations occur: e. g. *p*-xylene gives 2, 4-dimethylbenzaldehyde.

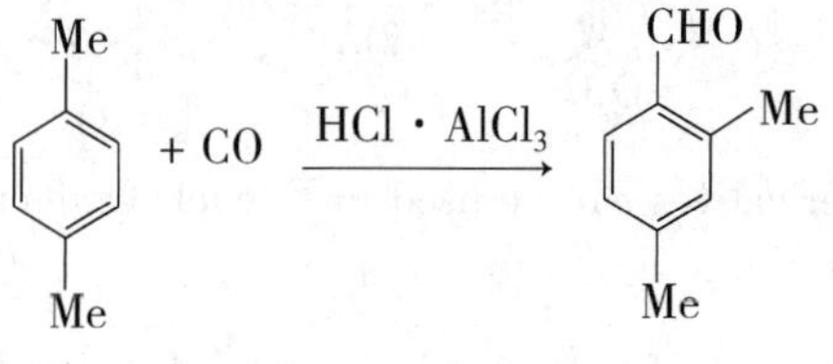

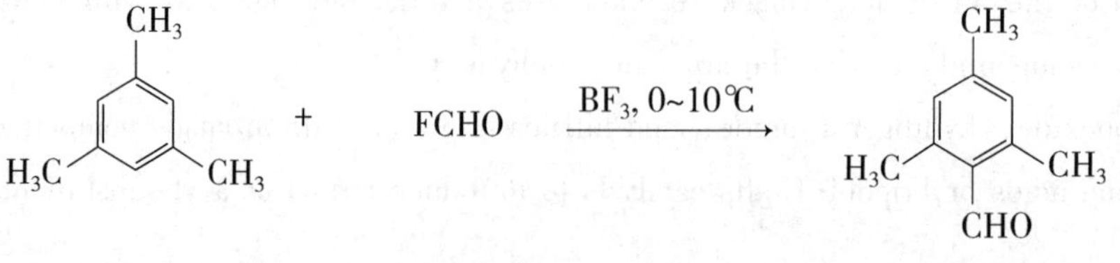

$$p\text{-}Me_2C_6H_4 + CO \xrightarrow{HCl \cdot AlCl_3} 2,4\text{-}Me_2C_6H_3CHO$$

$$1,3,5\text{-}(CH_3)_3C_6H_3 + FCHO \xrightarrow{BF_3,\ 0\sim10℃} 2,4,6\text{-}(CH_3)_3C_6H_2CHO \quad 70\%$$

Formylation with hydrogen cyanide:

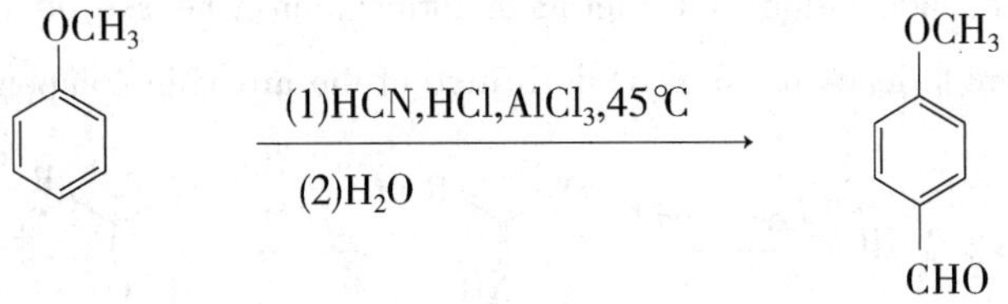

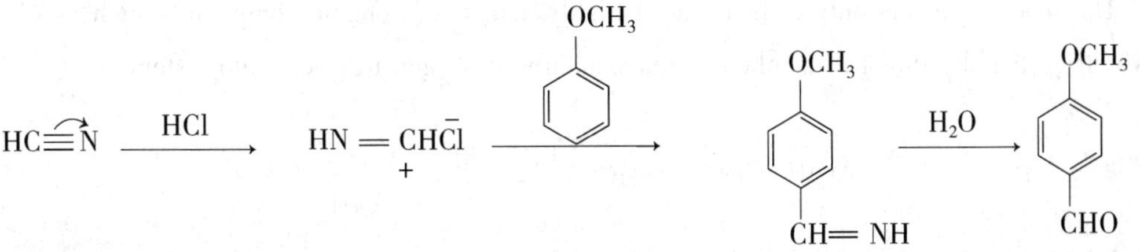

$$C_6H_5OCH_3 \xrightarrow[(2)H_2O]{(1)HCN,HCl,AlCl_3,45℃} p\text{-}OCH_3C_6H_4CHO$$

$$HC\equiv N \xrightarrow{HCl} \overset{+}{HN}=\bar{CHCl} \xrightarrow{C_6H_5OCH_3} p\text{-}OCH_3C_6H_4CH=NH \xrightarrow{H_2O} p\text{-}OCH_3C_6H_4CHO$$

Gattermann formylation:

This is an alternative to the Gattermann-Koch reaction, employing hydrogen cyanide instead of carbon monoxide. The initial product is an immonium chloride which is converted into the aldehyde with mineral acid:

$$ArH + HCN + HCl \xrightarrow{\text{Lewis acid}} ArCH=\overset{+}{N}H_2Cl^- \xrightarrow{H_3O^+} ArCHO + NH_4Cl$$

The ionic intermediate $[HC\equiv\overset{+}{N}H \longleftrightarrow \overset{+}{C}H=NH]$, analogous to the formyl cation, is thought to be the electrophilic entity. The reaction is unsuccessful with deactivated compounds such as nitrobenzene, and compounds of moderate reactivity, such as benzene and the halobenzenes, give only low yields. Yields from more reactive compounds are considerably higher, e. g. anthracene gives the 9-aldehyde in 60% yield.

Acylation with nitriles:

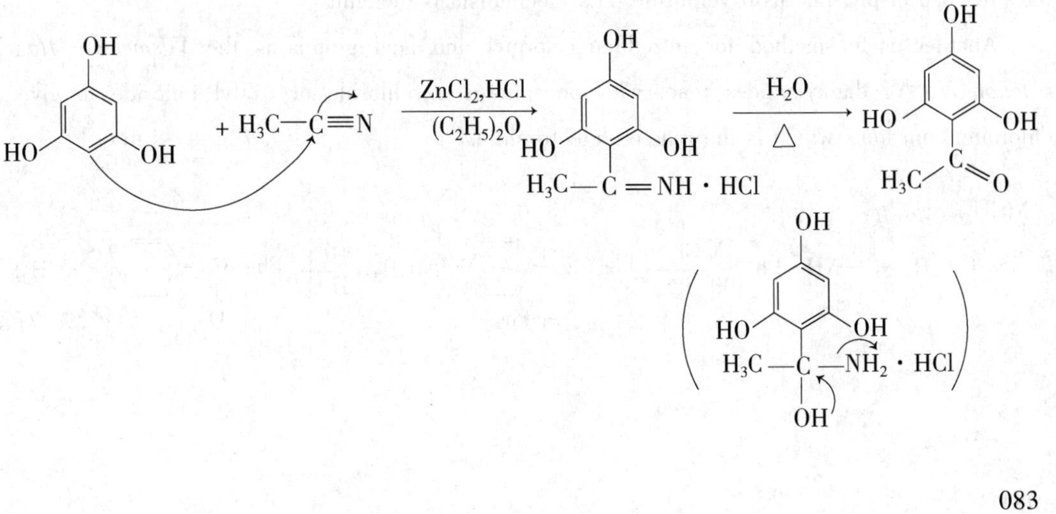

$$C_6H_3(OH)_3 + H_3C-C\equiv N \xrightarrow[(C_2H_5)_2O]{ZnCl_2,HCl} (HO)_3C_6H_2-C(CH_3)=NH\cdot HCl \xrightarrow[\triangle]{H_2O} (HO)_3C_6H_2-C(=O)CH_3$$

$$\left((HO)_3C_6H_2-C(CH_3)(OH)-NH_2\cdot HCl\right)$$

Hoesch acylation(Hoesch 酰基化)：

This reaction is an adaptation of Gattermann formylation: the use of an aliphatic nitrile in place of hydrogen cyanide leads to an acyl-derivative of the aromatic compound.

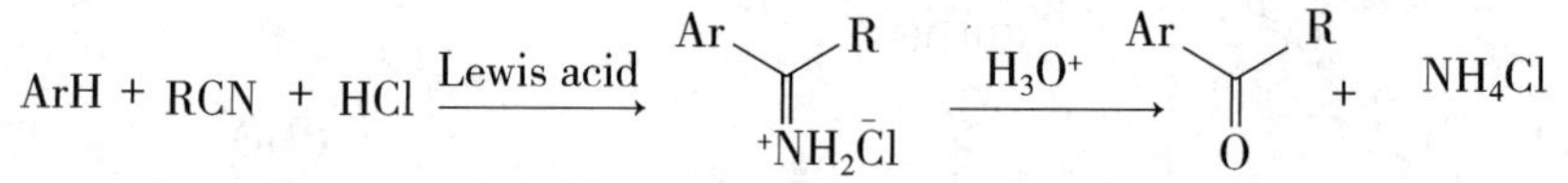

$$ArH + RCN + HCl \xrightarrow{\text{Lewis acid}} Ar(R)C{=}\overset{+}{N}H_2\bar{C}l \xrightarrow{H_3O^+} Ar(R)C{=}O + NH_4Cl$$

The reaction occurs only with the most highly activated aromatic compounds such as di-and polyhydric phenols. Monohydric phenols react mainly at oxygen to give imido-esters.

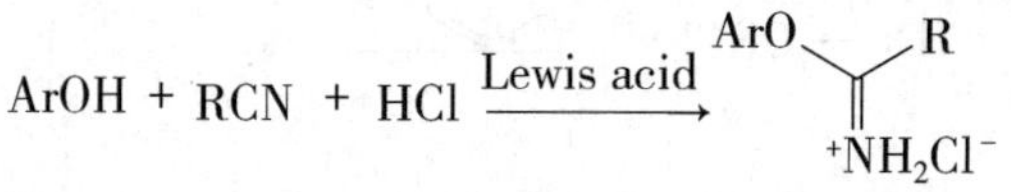

$$ArOH + RCN + HCl \xrightarrow{\text{Lewis acid}} ArO(R)C{=}\overset{+}{N}H_2Cl^-$$

But the combination of two or three hydroxyl groups meta to each other increases the reactivity of the nuclear positions ortho or para to hydroxyl that nuclear acylation occurs. For example, phloroacetophenone may be obtained in 80% yield by passing hydrogen chloride into a cooled solution of phloroglucinol (1, 3, 5-trihydroxybenzene) and acetonitrile in ether containing suspended zinc chloride and then hydrolyzing the resulting precipitate of the immonium salt by boiling in aqueous solution.

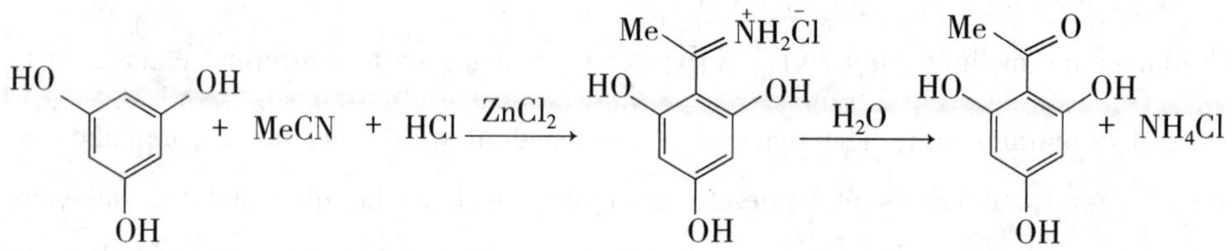

$$C_6H_3(OH)_3 + MeCN + HCl \xrightarrow{ZnCl_2} (HO)_3C_6H_2C(Me){=}\overset{+}{N}H_2\bar{C}l \xrightarrow{H_2O} (HO)_3C_6H_2C(Me){=}O + NH_4Cl$$

phloroacelophenone

Vilsmeyer formylation:

N-Formylamines, from secondary amines and formic acid, formylate aromatic compounds in the presence of phosphorus oxychloride. The mechanism is thought.

Another useful method for introducing formyl and acyl groups is the *Vilsmeier - Haack reaction*. *N*, *N* - dialkylamides react with phosphorus oxychloride or oxalyl chloride to give a chloroiminium ion, which is the reactive electrophile.

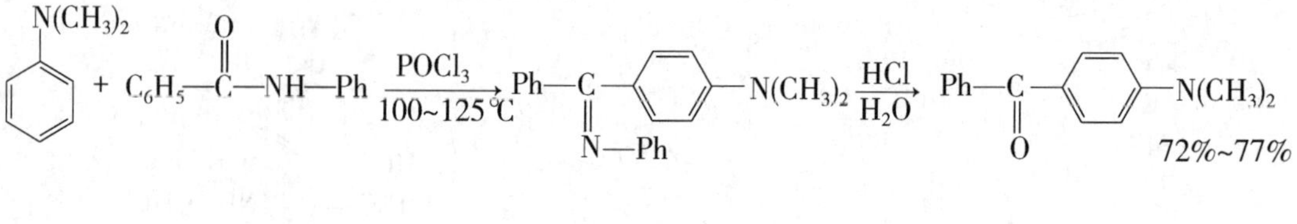

$$C_6H_5N(CH_3)_2 + C_6H_5-\overset{O}{\overset{\|}{C}}-NH-Ph \xrightarrow[100\sim125℃]{POCl_3} Ph-\underset{N-Ph}{\underset{\|}{C}}-C_6H_4-N(CH_3)_2 \xrightarrow[H_2O]{HCl} Ph-\underset{O}{\underset{\|}{C}}-C_6H_4-N(CH_3)_2 \quad 72\%\sim77\%$$

Mechanism:

$$\mathrm{Ph{-}C(=OH^+){-}NH{-}Ph} \longrightarrow \mathrm{Ph{-}C^+(OH){-}NH{-}Ph} \xrightarrow{\text{C}_6\text{H}_6} \mathrm{Ph{-}C(OH)(Ph){-}NH{-}Ph}$$

$$\xrightarrow[-H_2O]{H^+} \mathrm{Ph{-}C^+(Ph){-}NH{-}Ph} \longrightarrow \mathrm{Ph{-}C(Ph){=}N^+H{-}Ph} \xrightarrow{-H^+} \mathrm{Ph{-}C(Ph){=}N{-}Ph}$$

Many specific examples of these reactions can be found. Dichloromethyl ethers are also precursors of the formyl group via alkylation catalyzed by $SnCl_4$ or $TiCl_4$.

$$\text{mesitylene} + Cl_2CH{-}O{-}CH_3 \xrightarrow[CH_2Cl_2]{TiCl_4} \text{arenium ion (H, }ClCH{-}O{-}CH_3\text{)}$$

$$Cl_2CH{-}O{-}CH_3 \xrightarrow{-Cl^-} Cl\overset{+}{C}H{-}O{-}CH_3 \longleftrightarrow ClCH{=}\overset{+}{O}{-}CH_3$$

$$\longrightarrow \text{2,4,6-}(CH_3)_3C_6H_2{-}CH(Cl){-}O{-}CH_3 \xrightarrow[0\sim25℃]{H_2O} \text{2,4,6-}(CH_3)_3C_6H_2{-}CHO$$

There are a number of important differences between acylation and alkylation:

(ⅰ) Whereas alkylation does not require stoichiometric quantities of the Lewis acid since this is regenerated in the last stage of the reaction, acylation requires greater than equivalent quantities because the ketone which is formed complexes with the Lewis acid.

(ⅱ) Since acyl groups deactivate aromatic nuclei towards electrophilic substitution, the products of acylation are less reactive than the starting materials and the mono-acylated product is easy to isolate. This makes acylation a more useful procedure than alkylation, and alkyl-derivatives are often more satisfactorily obtained by acylation followed by reduction of carbonyl to methylene than by direct alkylation.

(ⅲ) A further advantage of acylation over alkylation is that the isomerizations and disproportionations which are characteristic of the latter process do not occur in the former. There is, however, one limitation: attempted acylation with derivatives of tertiary acids may lead to alkylation. For example, trimethylacetyl chloride reacts with benzene in the presence of aluminium trichloride to give mainly *t*-butylbenzene, liberating carbon monoxide:

$$Me_3C—COCl \xrightarrow{AlCl_3} Me_3C—\overset{+}{C}O \xrightarrow{\overset{+}{CO}} Me_3\overset{+}{C} \xrightarrow[-H^+]{PhH} Ph—CMe_3$$

The driving force for the decarbonylation no doubt resides in the relative stability of tertiary carbocations. However, more reactive aromatic compounds can react before decarbonylation occurs, e. g. anisole in the same conditions gives mainly p-CH_3O-C_6H_4-$COCMe_3$.

(ⅳ) The complex of the acylating agent and Lewis acid is evidently very bulky for ortho, para-directing monosubstituted benzenes give very little of the ortho product. For example, toluene, which gives nearly 60% of the ortho-derivative on nitration, gives hardly any o-methylacetophenone on acetylation; the *para*-isomer may be obtained in over 85% yield. When nitrobenzene is used as the solvent steric hindrance to ortho substitution is even more marked, possibly because a solvent molecule takes part in the acylating complex, e. g. whereas the acetylation of naphthalene in carbon disulfide solution gives mainly the expected 1-acetylna phthalene, reaction in nitrobenzene gives predominantly 2-acetylnaphthalene providing a convenient route to 2-naphthoic acid by oxidation with hypochlorite:

$$\text{naphthalene} + CH_3COCl \xrightarrow[-HCl]{AlCl_3/PhNO_2} \text{2-}COCH_3\text{-naphthalene} \xrightarrow{NaOCl} \text{2-}CO_2H\text{-naphthalene}$$

2-naphthoic acid

3.2 烯烃自缩合 Self-Condensation of Alkenes

$$2(CH_3)_2C{=}CH_2 \xrightarrow{50\%H_2SO_4} \underset{A}{CH_2{=}C(CH_3)CH_2C(CH_3)_3} + \underset{B}{(CH_3)_2C{=}CHC(CH_3)_3}$$

$$(CH_3)_2C{=}CH_2 \overset{H^+}{\rightleftharpoons} (CH_3)_3C^+ \overset{CH_2=C(CH_3)_2}{\rightleftharpoons} (CH_3)_3C\text{-}\underset{H}{CH}\text{-}\overset{+}{C}(CH_2\text{-}H)\text{-}CH_3 \overset{-H^+}{\rightleftharpoons} A + B$$

trimethyl-1-pentene (A) and 2, 4, 4-trimethyl-2-pentene (B).

Reaction occurs by the protonation of one molecule of the alkene to give a carbocation which adds to the methylenegroup (Markovnikov's rule) of a second molecule; the new carbocation then eliminates a proton:

$$(CH_3)_2C{=}CH_2 + H^+ \longrightarrow (CH_3)_3C^+ \xrightarrow{(CH_3)_2C=CH_2} (CH_3)_3CCH_2\overset{+}{C}(CH_3)_2 \xrightarrow{-H^+} \underset{\text{1part}}{CH_2{=}C(CH_3)CH_2C(CH_3)_3} \quad \underset{\text{4 parts}}{(CH_3)_3CCH{=}C(CH_3)_2}$$

The conditions must be carefully controlled. When dilute sulfuric acid is used, the first carbocation reacts preferentially with water to give t-butyl alcohol.

$$Me_3C^+ \xrightarrow{H_2O} Me_3C-\overset{+}{O}H_2 \xrightarrow{-H^+} Me_3C-OH$$

and if more concentrated acid is employed, the second carbocation reacts with a further molecule of isobutylene.

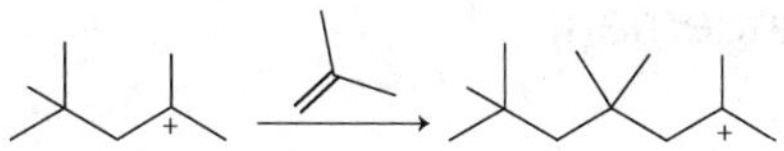

and further polymerization can occur.

Dienes undergo acid-catalyzed cyclization provided that the stereochemically favoured five-or six-membered rings are formed[6]. The example chosen is the conversion of *w*-ionone into *α*-and *β*-ionone, the second of which was used in a synthesis of vitamin A:

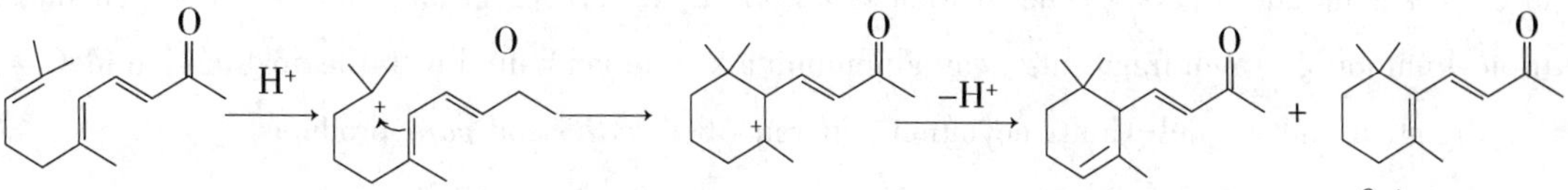

H^+　$-H^+$　+

α-ionone　　*β*-ionone

3.3 阳离子重排
Cationic Rearrangements

Carbocations can readily rearrange to more stable isomers. To be useful in synthesis, such reactions must be controlled and predictable. Cationic rearrangements can be highly selective, and carbocations can be stabilized by the migration of hydrogen, alkyl, alkenyl, or aryl groups, and, occasionally, even functional groups can migrate. A second property of carbocations that influences their synthetic utility is their propensity to rearrange to a more stable cation. Several named reactions involve cationic rearrangement.

3.3.1 Pinacol 重排
Pinacol Rearrangement

HO　Me　Me　Me　Me　OH　$\xrightarrow{H_2SO_4}$　Me　O　Me　Me　Me　67%~72%

Rearrangement of pinacol (2, 3-dimethyl-2, 3-butanediol, for which the reaction is named) to pinacolone (3, 3-dimethyl-2-butanone) is the classical example[7], but the reaction can be done with a variety of 1, 2-diols (cyclic and acyclic), utilizing both protonic and Lewis acid initiators. Rearrangement to the more stable (an oxygen-stabilized cation rather than a tertiary alkyl cation), is followed by loss of a proton to give ketone.

3.3.2 Fries 重排
Fries Rearrangement

There are also many phenol derivatives, where the oxygen atom is appended to a simple aromatic ring or to polycyclic aromatic compounds. An interesting route to such compounds exploits the propensity of aryls to rearrange under Friedel-Crafts conditions. When a phenolic ester (such as phenyl acetate) is heated with a Lewis acid such as $AlCl_3$, a rearrangement occurs to generate an ortho ketone and a para ketone in what is known as the Fries rearrangement reaction generates an acylium ion[8], which fragments to an aluminum alkoxide ion paired with the acylium ion $MeC\equiv O^+$. Migration and Friedel-Crafts acylation generates both *ortho* and *para* products.

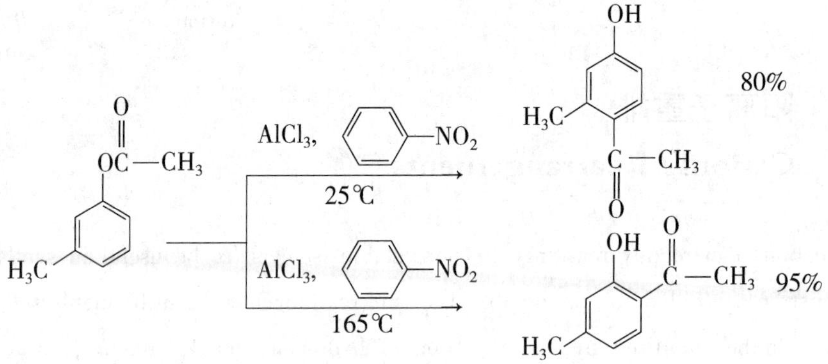

$AlCl_3$, $C_6H_5NO_2$, 25℃ → 80% (para product); $AlCl_3$, $C_6H_5NO_2$, 165℃ → 95% (ortho product)

The reaction is catalyzed by Brønsted or Lewis acids such as HF, $AlCl_3$, BF_3, $TiCl_4$ or $SnCl_4$. The acids are used in excess of the stoichiometric (定量) amount, especially the Lewis acids, since they form complexes with both the starting materials and products.

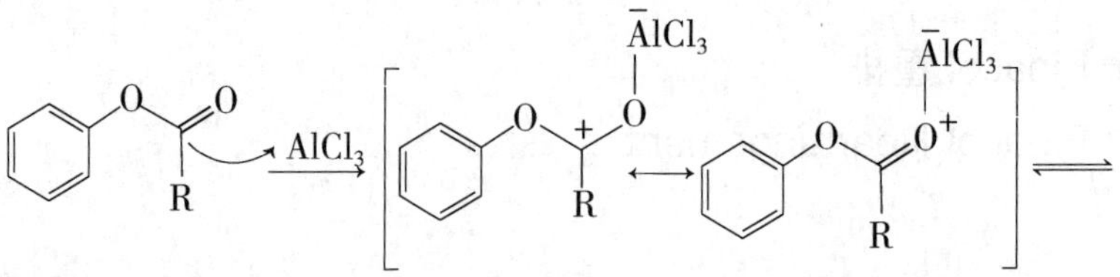

The complex can dissociate to form an acylium ion. Depending on the solvent, an ion pair can form, and the ionic species can react with each other within the solvent cage. However, reaction with a more distant molecule is also possible:

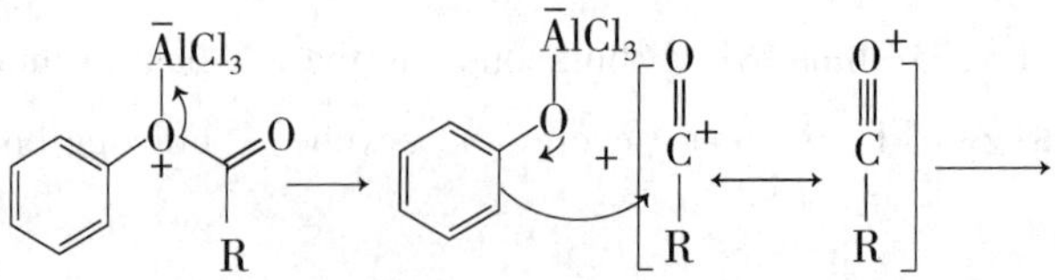

After hydrolysis, the product is liberated:

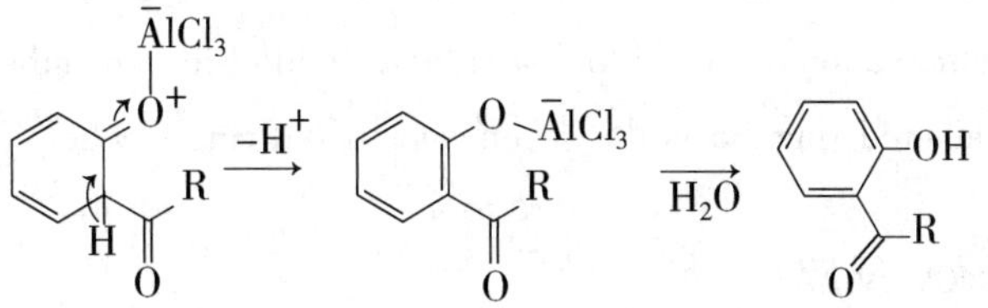

The reaction is *ortho*, *para*-selective so that, for example, the site of acylation can be regulated by the choice of temperature. Only sterically unhindered arenes are suitable substrates, since substituents will interfere with this reaction.

3.3.3 Tiffeneau-Demjanov 扩环 Tiffeneau-Demjanov Ring Expansion

Carbocations are available from precursors other than halides or sulfonate esters. Treatment of an amine with nitrous acid (HONO) generates an unstable diazonium salt, which decomposes to a transient primary cation. The initially formed cation rearranged to the larger ring cation. This reaction has been used to form ring-expanded cyclic ketones[9], a procedure known as the Tiffeneau-Demjanov reaction.

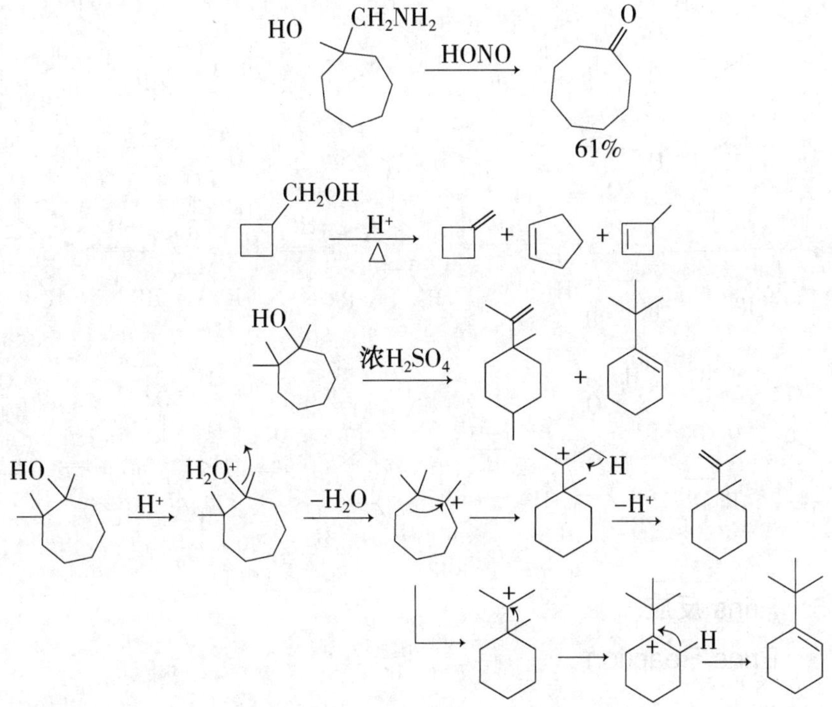

The alcohol was heated with *p*-toluenesulfonic acid in benzene to generate cation, with the positive charge adjacent to a strained four-membered ring. Opening the strained four-membered ring to the lower energy five-membered ring product occurred by a 1, 2-alkyl shift. Under the reaction conditions, elimination occurred to give a 90% yield of five-membered ring product along with 10% of four-membered ring. So is the seven-membered ring alcohol.

3.3.4 Nazarov 反应 Nazarov Reaction

The Nazarov cyclization allows the synthesis of cyclopentenones from divinyl ketones. The reaction is catalyzed by strong Lewis or Brønsted acids, and one or more equivalents of the Lewis acid are normally necessary[10]. Electron-donating and -withdrawing substituents (给电子和吸电子取代基) can polarize the conjugated system (共轭体系) in the Nazarov reaction, which facilitates the cyclization and gives better regioselectivity.

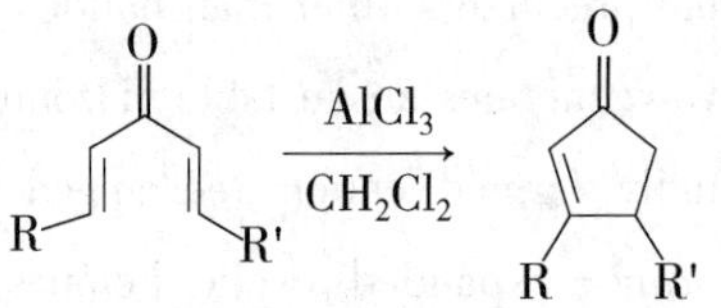

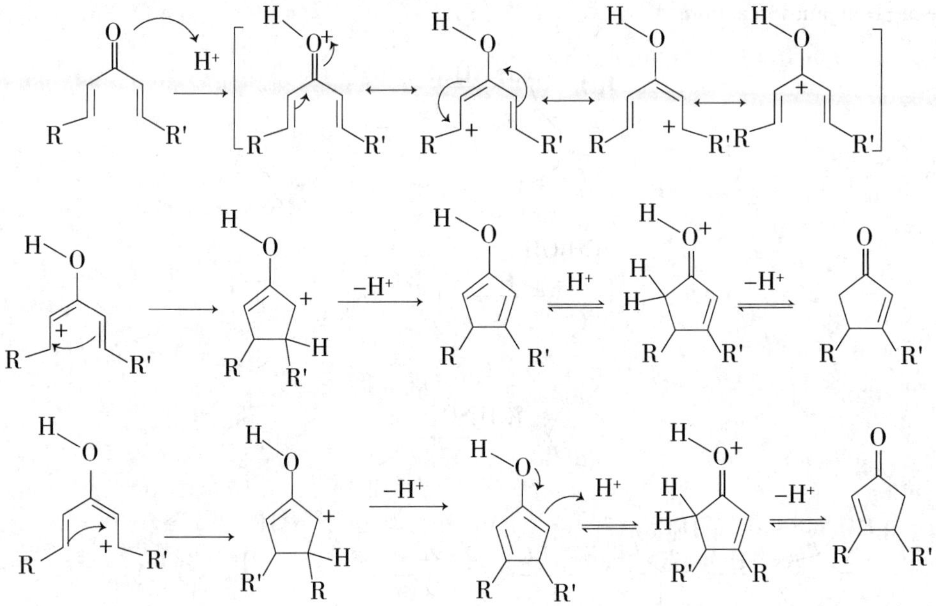

$$\text{divinyl ketone} \xrightarrow[CH_2Cl_2]{AlCl_3} \text{cyclopentenone}$$

3.3.5 Prins 反应 Prins Reaction

The Prins Reaction is the acid-catalyzed of addition aldehydes to alkenes, and gives different products depending on the reaction conditions. It can be thought of conceptually as the addition of the elements of the gem-diol carbonyl hydrate of the aldehyde across the double bond[11].

H–C(=O)–H + R–CH=CH₂ $\xrightarrow[H_2O]{H^+}$ 4-R-1,3-dioxane

3.3.6 Wagner-Meerwein 重排 Wagner-Meerwein Rearrangement

The effect of neighbouring group participation leads to Wagner-Meerwein rearrangement.

Why did the hydrogen group in **1** migrate and not the phenyl? First, migration of the phenyl would lead to a significantly less stable primary cation and the reaction would be endothermic. Second, migration of the larger phenyl group may require more energy than the smaller hydrogen. Note that the phenyl group can migrate, and a relatively low-energy phenonium ion such as **2** has been invoked in many aryl migrations. In a transition state, considerable positive charge can be present and charge dispersal (as in **2**) can facilitate the rearrangement[12,13]. A lesson from this observation is that careful analysis of a system is required to determine a priori which atom will preferentially migrate. In the absence of electronic effects, smaller alkyl groups will generally migrate before larger ones: H>Me>$CHMe_2$> CMe_3.

1 2 3

参考文献 References

1. O' Donnell M J, Weiss L. Annual Reports in Organic Synthesis - 1984. Amercian: Academic Press,

1985. 1-242.

2. Uraguchi D, Ooi T. 6. 1 C-C Bond Formation: Alkylation. Amsterdam: Elsevier, 2012. 1-36.
3. Ananya Srivastava, K. Jana C. Heterocycles via Cross Dehydrogenative Coupling. Singapore: Springer, 2019.
4. Olah G A, Krishnamurti R, et, al. Comprehensive Organic Synthesis. Oxford: Pergamon, 1991. 293-339.
5. Togo H. Advanced Free Radical Reactions for Organic Synthesis. Amsterdam: Elsevier Science, 2004. 157-170.
6. Harper S H. Rodd' s Chemistry of Carbon Compounds (Second Edition). Amsterdam: Elsevier, 1964. 147-281.
7. Rickborn B. Comprehensive Organic Synthesis. Oxford: Pergamon, 1991. 721-732.
8. Ninomiya I, Naito T. Photochemical Synthesis. London: Academic Press, 1989. 23-30.
9. Wovkulich P M. Comprehensive Organic Synthesis. Oxford: Pergamon, 1991. 843-899.
10. Török B, Schäfer C, et, al. Heterogeneous Catalysis in Sustainable Synthesis. Amsterdam: Elsevier, 2022. 491-542.
11. Snider B B. Comprehensive Organic Synthesis (Second Edition). Amsterdam: Elsevier, 2014. 148-191.
12. Hanson J R. Comprehensive Organic Synthesis. Oxford: Pergamon, 1991. 705-719.
13. Smith M B. Organic Synthesis(Fourth Edition). Boston: Academic Press, 2017.

4 通过自由基形成碳碳键

Chapter 4 Carbon-Carbon Bonds Formation: Free Radicals

Radical reactions have become an important and largely controllable synthetic tool in recent years, and should be considered as an important part of synthetic planning.

A free radical (usually just called a radical) may be defined as a species that contains one or more unpaired electrons. As with carbocations and carbanions, simple alkyl radicals are very reactive and are usually transient species. For the most part, their lifetimes are extremely short in solution, but they can be kept for relatively long periods frozen within the crystal lattices of other molecules[1,2]. This chapter will discuss methods for preparing radicals and their applications to synthesis[1].

4.1 碳自由基的形成、结构及稳定性 Formation, Structure and Stabilities of Carbon Free Radicals

4.1.1 自由基的形成 Generation of Free Radicals

Radicals can be formed in several ways. Free radicals are formed from molecules by breaking a bond so that each fragment keeps one electron. The energy necessary to break the bond is supplied in one of two ways. Many involve dissociative **homolytic cleavage** (one electron is transferred to each adjacent atom from the bond) as a key step giving two radical products. Another major route to radical intermediates involves the reaction of a radical (X·) and a neutral molecule (X-Y), producing a new radical (Y·) and a new neutral molecule (X-X). The reaction equilibrium depends on both the relative bond strength of X-Y and on the relative stabilities of X· and Y·.

There are several reactions that are used frequently to generate free radicals, both to study radical structure and reactivity and in synthetic processes. some of the most general methods are outlines here. These methods will be encountered again when we discuss special examples of free radical reactions.

4.1.1.1 热裂解 Thermal cleavage

Subjection of any organic molecule to a high enough temperature in the gas phase results in the formation of free radicals. When the molecule contains bonds with D values of 20~40kcal·mol^{-1}(80~170kJ·mol^{-1}), cleavage can occur in the liquid phase[4]. Two common examples are cleavage of diacyl peroxides to acyl radicals that decompose to alkyl radicals and cleavage of azo compounds to alkyl radicals.

The carbon-carbon bond of ethane has a bond strength of 83kcal · mol^{-1}(347.4kJ · mol^{-1}), and it fragments to methyl radicals (· CH_3) only at temperatures approaching 600℃.

$$H_3C-CH_3 \xrightarrow{600℃} 2\dot{C}H_3$$

Peroxides are a common source of radical immediate. The compounds undergo homolytic cleavage at moderate temperature to give two free radicals. e. g.

Di-tert-butyl peroxide (DTBP):

$$(H_3C)_3C-O-O-C(CH_3)_3 \xrightarrow{100\sim110℃} 2(H_3C)_3C-\dot{O}$$

$$E_d=154.9kJ \cdot mol^{-1}$$

Azobisisobutyronitrile (AIBN):

$$(H_3C)_2C(CN)-N=N-C(CN)(CH_3)_2 \xrightarrow{80\sim100℃} 2\ \dot{C}(CN)H(CH_3)_2 + N_2$$

Dibenzoyl peroxide (BPO):

$$C_6H_5-\overset{O}{\overset{\|}{C}}-O-O-\overset{O}{\overset{\|}{C}}-C_6H_5 \xrightarrow{\Delta} C_6H_5-\dot{C}O_2 \xrightarrow{-CO_2} C_6H_5\cdot$$

In principle, lower bond dissociation energies should correlate with an increased propensity for homolytic cleavage.

4.1.1.2 光化学裂解

Photochemical cleavage

The energy of light of 600~300 nm is 48~96kcal · mol^{-1}(200~400kJ · mol^{-1}), which is of the order of magnitude of covalent-bond energies. Typical examples are photochemical cleavage of alkyl halides in the presence of triethylamine, of alcohols in the presence of mercuric oxide and iodine, of alkyl 4-nitrobenzenesulfenates, of chlorine and of ketones[5].

A molecule can absorb a photon of light and the resultant photoactivated species can undergo homolytic cleavage to form radicals. Photochemical homolysis of the carbon-carbon bond of alkanes is difficult, however, for acetone, absorption of a photon leads to cleavage to the methyl radical and the acyl radical.

$$Cl_2 \xrightarrow{h\nu} 2\dot{C}l$$

$$CH_3\overset{O}{\overset{\|}{C}}CH_3 \xrightarrow{h\nu} \dot{C}H_3 + CH_3\overset{O}{\overset{\|}{C}}\cdot$$

$$CCl_3Br \xrightarrow{h\nu} \dot{C}Cl_3 + Br\cdot$$

Photolytic decomposition of *N*-hydroxypyridin-2-thione is a method that generates hydroxyl radicals. A stable dialkylphosphinyl radical has been reported and other phosphinyl radicals are known. Thiyl radicals are useful in organic synthesis. A photoreductive method has been used to generate radicals from epoxides or aziridines.

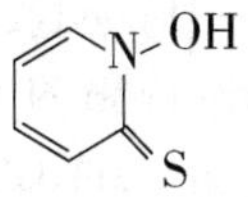

N-OH
S

4.1.1.3 金属诱导氧化还原反应 Metal-induced redox reaction

Radicals can also be formed by oxidation or reduction, including electrolytic methods. The donor may be a strongly electropositive element such as sodium, as in the acyloin reaction. Transition-metal ion may also be a donor of radical.

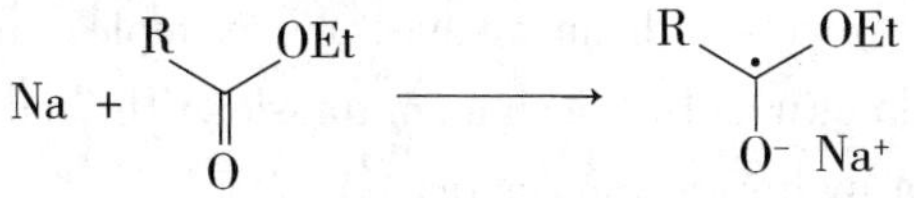

Na + RC(=O)OEt ⟶ R–Ċ(O⁻ Na⁺)–OEt

$$Co^{3+} + ArCH_3 \longrightarrow Ar\dot{C}H_2 + Co^{2+} + H^+$$

$$Fe^{2+} + H_2O_2 \longrightarrow Fe^{3+} + \dot{O}H + OH^-$$

4.1.2 自由基的结构 Structure of Radicals

A carbon radical can be viewed as a trivalent species containing a single electron in a *p* orbital. A carbanion can be viewed as a tetrahedral species containing a pair of electrons in an orbital. We have viewed a carbocation (carbenium ion) as a sp^2 hybridized, trigonal planar carbon with an empty *p* orbital. A radical, which contains one electron in an orbital, can be tetrahedral, planar or 'in between' with properties of both a carbanion and a carbocation. The two most common are pyramidal and planar[6].

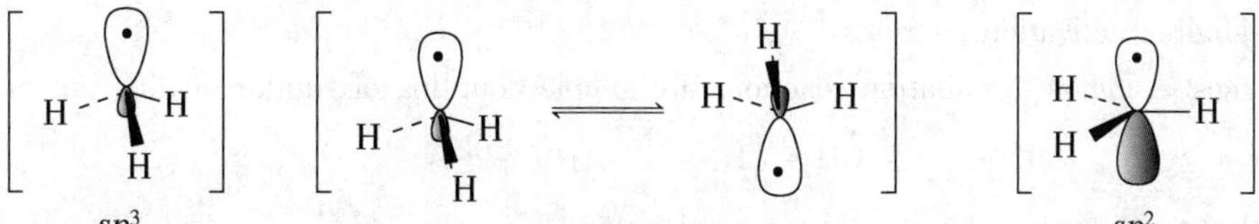

sp^3 rigid pyramidal | rapidly inverting pyramidal | sp^2 planar

4.2 自由基反应类型 Types of Free Radical Reactions

4.2.1 取代 Substitution

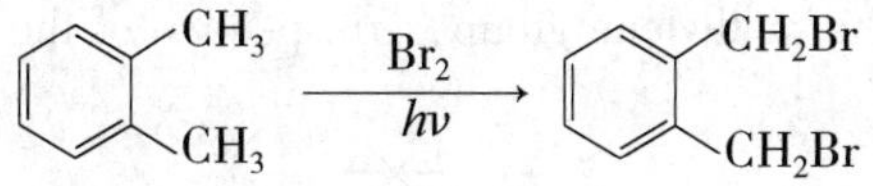

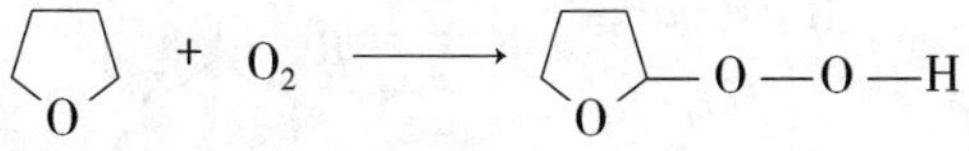

o-xylene (CH_3, CH_3) $\xrightarrow[h\nu]{Br_2}$ (CH_2Br, CH_2Br)

THF + O_2 ⟶ 2-(O—O—H)-THF

The selectivity of a free radical towards C—H bonds of different types is principally depends on both dissociation energies and polar effect. The relative order of hydrogen abstraction has been shown to be:

$CH_3—H<CH_3CH_2—H<CH_3CH_2CH_2—H< Me_2CHCH_2—H<Me_2CH—H<Me_3C—H$

4.2.2 提取或氢原子的转移 Abstraction or Hydrogen Atom Transfer

Abstraction of another atom or group, usuauy a hydrogen atom.

A bromine radical (Br·) reacts with an alkane, for example, to give HBr and a carbon radical. This type of reaction is known as hydrogen atom transfer. The reduction of a carbon radical with Bu_3SnH is an example of a hydrogen transfer reaction.

Free radicals react with saturated organic compounds by abstracting an atom, usually hydrogen, from carbon.

$$\dot{C}H_3 + C_6H_5CH_3 \longrightarrow C_6H_5\dot{C}H_2 + CH_4$$

$$CH_3CH_2=CH_2 + \cdot Br \longrightarrow \dot{C}H_2CH=CH_2 + HBr$$

$$\cdot CH_3 + BuSnH \longrightarrow CH_4 + Bu_3Sn\cdot$$

4.2.3 偶合与歧化 Combination and Disproportionation

Two free radicals can combine by dimerization. These reactions are mostly very rapid, some having negligible activation energies.

The most common termination reactions are simple coupling of similar or different radicals:

$$\dot{C}H_3 + \dot{C}H_3 \longrightarrow H_3C—CH_3$$

Another termination process is disproportionation:

$$2CH_3\dot{C}H_2 \longrightarrow H_2C=CH_2 + H_3C—CH_3$$

4.2.4 加成 Addition

The radical formed from an alkene may add to the double bond of a second equivalent of alkene, and so on. The most important of the unsaturated groups in free-radical synthesis is the C=C bond, addition to which is markedly selective. In particular, the addition to $CH_2=CHX$ occurs almost exclusively at the methylene group, irrespective of the nature of X.

R· + CH₂=CH–X ⟶ (✗) ·CH₂–CH(R)–X ; ⟶ R–CH₂–ĊH–X

The reaction generates a new radical that can propagate a chain sequence. The preferred alkenes for trapping alkyl radicals are ethene derivatives with electron-attracting groups, such as cyano, ester, or other carbonyl substituents. Radicals for addition reactions can be generated by halogen atom abstraction by stannyl radicals. The chain mechanism for alkylation of alkyl halides by reaction with a substituted alkene is outlined in section 4.3.1. There are three reactions in the propagation cycle of this chain mechanism: addition, hydrogen atom abstraction, and halogen atom transfer (see 4.3 the Reactions of Free Radicals).

The following distribution of addition products of the 1-hexyl radical show the effect of steric hindrance at the site of addition.

0.2% 99.8% $COOCH_3$　　Et 12% 88% $COOCH_3$　　t-Bu 80% 20% $COOCH_3$

4.2.5 碎裂和重排 Fragmentation and Rearrangement

Fragmentation is the reverse of radical addition. Fragmentation of radicals is often observed to be fast when the overall transformation is exothermic.

$$\dot{Y}-\overset{|}{\underset{|}{C}}-\overset{|}{\underset{|}{C}}-X \longrightarrow Y{=}C\langle + \rangle\dot{C}-X$$

Free radicals, unlike carbocationns, seldom rearrange. However, the phenyl group migrates in certain circumstances, e.g.

Ph Ph $\xrightarrow{R\dot{C}O}$ R(O)…Ph Ph $\xrightarrow[\text{Ph · migration}]{\text{rearrangement}}$ R(O)…Ph…Ph

$\xrightarrow{RCHO}$ R(O)…Ph…Ph + $R\dot{C}O$

Perhaps the best-known rearrangement is that of cyclopropylcarbinyl radicals to a butenyl radical. The rate constant for this rapid ring opening has been measured in certain functionalized cyclopropylcarbinyl radicals by picosecond radical kinetics. Substituent effects on the kinetics of ring opening in substituted cyclopropylcarbinyl radicals have been studied[7].

Rearrangements of radicals frequently occur by a series of addition-fragmentation steps. When chlorine, phenyl, acetoxy, and acyl groups are in the α-position with respect to the radical center, 1, 2-shifts have been observed. The following reactions involves radical rearrangements that proceed through addition-elimination sequences.

$_n(H_2C)$ (O) $CO_2C_2H_5$ $(CH_2)_4I$ $\xrightarrow{Bu_3Sn\cdot}$ $(CH_2)_n$ ($\dot{O}$) $CO_2C_2H_5$ $\longrightarrow$ $(CH_2)_n$ (O) $CO_2C_2H_5$ $\xrightarrow{Bu_3SnH}$ $(CH_2)_n$ (O) $CO_2C_2H_5$

4.3 自由基的反应 Reactions of Free Radicals

4.3.1 加成反应 Addition Reactions

Alarge group of synthetically useful radical-catalyzed reactions is based on the addition of aliphatic carbon radicals to C=C bonds. The reaction of bromoform with 1-butene is illustrative:

$$PhC(=O)O\text{-}OC(=O)Ph \xrightarrow{heat} PhC(=O)O\cdot$$

$$PhC(=O)O\cdot + CHBr_3 \xrightarrow{heat} PhC(=O)OH + \cdot CBr_3$$

propagation:

$$CH_3CH_2CH=CH_2 + \cdot CBr_3 \longrightarrow CH_3CH_2\dot{C}HCH_2CBr_3$$

$$CH_3CH_2\dot{C}HCH_2CBr_3 + CHBr_3 \longrightarrow CH_3CH_2CH_2CH_2CBr_3 + \cdot CBr_3$$

One of the most useful sources of free radicals in preparative chemistry is the reaction of halides with stannyl radicals. The fundamentals of the reaction can be illustrated by the AIBN induced reaction of tributyltin hydride and **RX** (X = Cl, Br, I, SPh, or SePh), which generates radical. Stannanes undergo hydrogen abstraction reactions and the stannyl radical can then abstract halogen from the alkyl group to alkyl radical. For example, net addition of an alkyl group to a reactive double bond can follow halogen abstraction by a stannyl radical[8].

Reaction is usually initiated with AIBN, either thermally or photochemically. The radical $\cdot CMe_2CN$ readily abstracts a hydrogen atom from the weak Sn—H bond in the tributylstannane, and the chain consists of the following steps (X=Cl, Br, I, SPh, or SePh):

Initiation

$$AIBN \xrightarrow[\text{or } h\nu]{\triangle} H_3C\dot{C}(CN)CH_3 \xrightarrow{H—Sn(Bu\text{-}n)_3} H_3C\text{-}CH(CN)CH_3 + \cdot Sn(Bu\text{-}n)_3$$

Propagation

$$R—X + \cdot Sn(Bu\text{-}n)_3 \longrightarrow R\cdot + X—Sn(Bu\text{-}n)_3 \quad (1)$$

$$R\cdot + CH_2=CHY \longrightarrow RCH_2\dot{C}HY \quad (2)$$

$$RCH_2\dot{C}HY + H—Sn(Bu\text{-}n)_3 \longrightarrow RCH_2CH_2Y + \cdot Sn(Bu\text{-}n)_3 \quad (3)$$

There are also unwanted reactions:

$$R\cdot + HSn(Bu\text{-}n)_3 \longrightarrow RH + \cdot Sn(Bu\text{-}n)_3 \quad (4)$$

(5)

There may be little selectivity between the various C—H bonds and the yields of individual products are then low. it is therefore in general better to use an attracting radical which abstracts a group or an atom other than hydrogen in preference to a hydrogen atom and to locate that group or atom in the reactant in the required position. Such a radical is tributylstannyl, $Bu_3Sn\cdot$, and the atoms and PhSe. In contrast, Since the Sn—H bond is weak, it does not abstract hydrogen from C—H. This method, while of limited application in intramolecular analogue.

$$R\text{—}Br \xrightarrow[h\nu]{H\text{-}Sn(Bu\text{-}n)_3} R\cdot \xrightarrow{CH_2{=}CHCN} RCH_2\dot{C}HCN \longrightarrow RCH_2CH_2CN + \cdot Sn(Bu\text{-}n)_3$$

When a carbon radical ($R\cdot$) is generated in the presence of tributyltin hydride (n-Bu_3SnH), a hydrogen atom is transferred to the radical to give RH and a new radical, n-$Bu_3Sn\cdot$. The tin radical usually undergoes rapid coupling to another tin radical to give n-$Bu_3Sn\ Sn$-n-Bu_3, which effectively terminates the chain radical process. The carbon radical is reduced ($R\cdot \rightarrow RH$) as a result of the hydrogen atom transfer, and the tin dimer can be removed from the reaction. Again, hydrogen atom transfer is simply a variation of the radical reaction known as atom transfer.

Therefore, stannanes can prevent olefin polymerization. If the initiator is BPO, the product is completely different.

$$n\ CH_2{=}CHCN \xrightarrow[\Delta]{BPO} (\text{polymer})_n$$

$(C_4H_9)_3SnH$, $CH_2{=}CHCN, h\nu$, Et_2O — 72%

$(C_4H_9)_3SnH$, AIBN, benzene — 70%

$(C_4H_9)_3SnH$, AIBN

To improve the reaction selectivity, using allyl tributylstanne as a radical trap can affect the allylation of alkyl radicals.

88%

4.3.2 取代反应(还原) Substitution Reactions (Reduction)

When a radical is generated in the presence of a hydrogen transfer agent, it is possible to reduce various functional groups. Since an X group is replaced with H, this is a form of the substitution reaction, but it is separated into a different section because of its synthetic importance and because it is formally a reduction. When cyclization is slow relative to hydrogen transfer, the dominant reaction is usually hydrogen atom transfer to give reduced products. Photolysis of the tin dimer generates the radical intermediate without a hydrogen atom transfer agent present (other than the substrate and the solvent). The best example shows that an alkyl halide is reduced by treatment with tributyltin hydride, illustrated here by the reduction of iodide in the presence of AIBN and Bu_3SnH to give a 98% yield of product. AIBN is a radical initiator, and leads to loss of iodide to give the radical. The intermediate radical reacted with Bu_3SnH via hydrogen transfer to give product and $Bu_3Sn\cdot$, which generated Bu_3SnI. This reaction is a very effective method for the controlled reduction of halides[9].

A very useful radical-based reaction has been developed that can be applied to alcohols. conversion of an alcohol to a thionocarbonate followed by treatment with tributyltin hydride under radical conditions gives cleavage to the C—O bond to give the reduction product. This transformation called the Barton deoxygenation.

Another example illustrating the utility of this reagent in cyclization reactions is Curran and Chang' s attempt to cyclize **I**, which gave only **II** on treatment with tin hydride/AIBN and no **III**. Cyclization with dibutyltin dimer and photochemical induction gave a radical intermediate that was quenched by reaction with the iodine radical byproduct to give **III**. The iodine could be reduced by reaction with tin hydride.

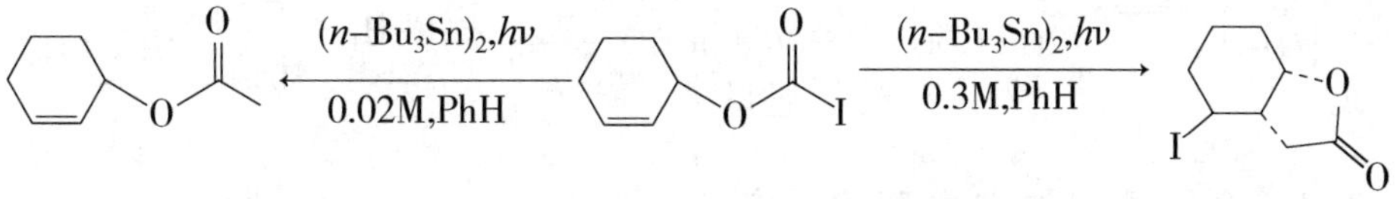

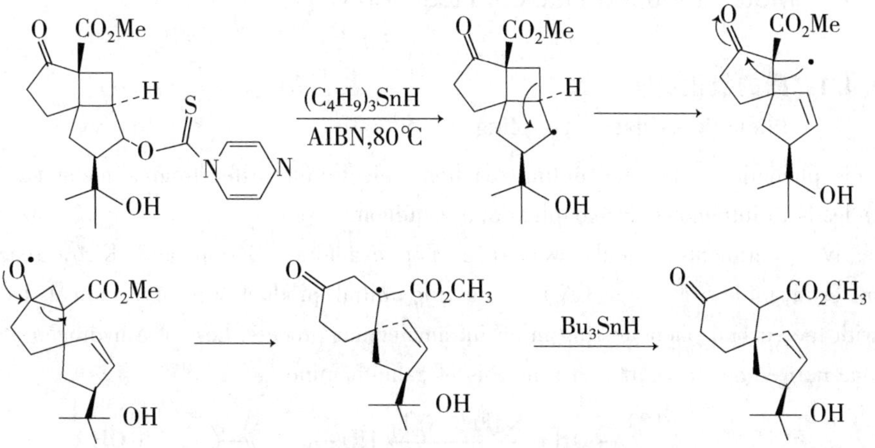

4.3.3 Kolbe 电合成 Kolbe Electrolytic Synthesis

Electrolysis of carboxylate ions leads to decarboxylation, and combination of the resulting radicals to give the coupling product R-R. This coupling reaction is called the Kolbe reaction or the Kolbe electrosynthesis. It is used to prepare symmetrical R—R, where R is straight chained, since little or no yield is obtained when there is branching. Much larger and more complex molecules can be synthesized by Kolbe electrosynthesis. The reaction is very efficient when two identical acids are coupled together. The reaction is not successful for R=aryl. Many functional groups may be present, though many others inhibit the reaction. Unsymmetrical R—R′ have been made by coupling mixtures of acid salts. The Kolbe reaction has been done using solid-supported bases[10,11].

$$2CH_3(CH_2)_{14}COOK \xrightarrow{e^-} \underset{88\%}{CH_3(CH_2)_{28}CH_3} + 2CO_2$$

A free - radical mechanism is involved, via formation of RCOO·, loss of CO_2, and formation of R·.

$$2RCO_2^- \xrightarrow{-2\bar{e}} 2RC\dot{O}_2 \xrightarrow{-CO_2} 2R\cdot \longrightarrow R—R$$

The Kolbe synthesis of two different acids (RCOOH + R'COOH) generally leads to a mixture of products arising from statistical coupling (R—R, R—R', R'—R'). For example, using an excess of one acid can lead to the desired mixed-coupling product (R—R'), as in the reaction of the unsaturated acid with an excess of heptanoic acid to give coupling adduct in 80% yield.

$$n\text{-}C_{18}H_{17}\diagup\!\!=\!\!\diagdown(CH_2)_7CO\bar{O} + CH_3(CH_2)_5COO^- \xrightarrow{-e^-} n\text{-}C_{18}H_{17}\diagup\!\!=\!\!\diagdown(CH_2)_{12}CH_3 \quad 80\%$$

过量

4.3.4 金属诱导的自由基的反应 Metal-Induced Radical Reactions

4.3.4.1 酚的氧化偶联 Phenolic oxidative coupling

In this phenolic oxidative coupling reaction, electron transfer from a metal salt to a bis (phenol) leads to intramolecular coupling and a quinone product[12].

In early experiments, yields were poor. For example, Barton and Kirby reacted with potassium ferricyanide [$K_3(Fe(CN)_6]$] and the initial product was aryl radical. This radical reacted with the second phenolic ring in an intramolecular process. Loss of a hydrogen atom led to the quinone narwedine, as part of a synthesis of galanthamine[13].

$$C_6H_5\text{-}OH \xrightarrow{K_3[Fe(CN)_6]} HO\text{-}C_6H_4\text{-}C_6H_4\text{-}OH$$

Mechanism:

$$C_6H_5\text{-}OH \xrightarrow{K_3[Fe(CN)_6]} C_6H_5\text{-}O\cdot$$

$$2\ C_6H_5\text{-}O\cdot \longleftrightarrow \cdot(H)C_6H_4\text{=}O \xrightarrow{\text{dimerization}} O\text{=}C_6H_4(H)\text{-}(H)C_6H_4\text{=}O$$

$$\xrightarrow{\text{rearrangement}} HO\text{-}C_6H_4\text{-}C_6H_4\text{-}OH$$

If substituents at the *ortho* or *para* site of benzene ring, migration of hydrogen atom of quinone molecules is difficult.

$$p\text{-}CH_3C_6H_4OH \xrightarrow{K_3[Fe(CN)_6]} \text{(dienone intermediate: } H_3C, CH_3, OH, O) \longrightarrow \text{(tricyclic ketone: } H_3C, CH_3, O, O)$$

$$\text{(HO, } H_3CO, H_3CO, OCH_3, OH \text{ substrate)} \xrightarrow{[Fe(DMF)_3Cl_2][FeCl_3]} \text{(spiro dienone: HO, } H_3CO, H_3CO, OCH_3, O) \quad 67\%$$

Recognizing the potential symmetry greatly shortened the synthesis. Usnic acid(松香酸) is an example of a molecule that possesses potential symmetry. The synthesis of usnic acid is shown in the following scheme, with a phenolic oxidative coupling of a single phenol derivative to give the biphenyl intermediate via coupling of the initial oxidation product. The phenolic oxidative coupling reaction is a radical process by which two phenol moieties are joined. The two phenolic moieties reacted via an acid-catalyzed Michael reaction to give a dimer, which cyclized via the phenol hydroxyl group. Treatment with concentrated sulfuric acid completed the two-step synthesis of usnic acid.

$K_3[Fe(CN)_6]$

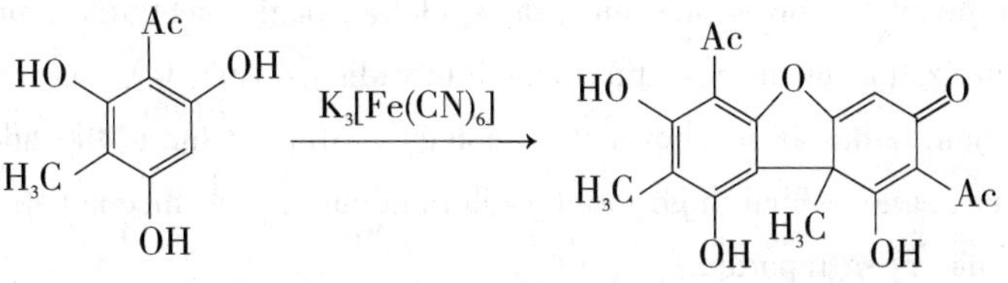

isnic acid

Mechanism:

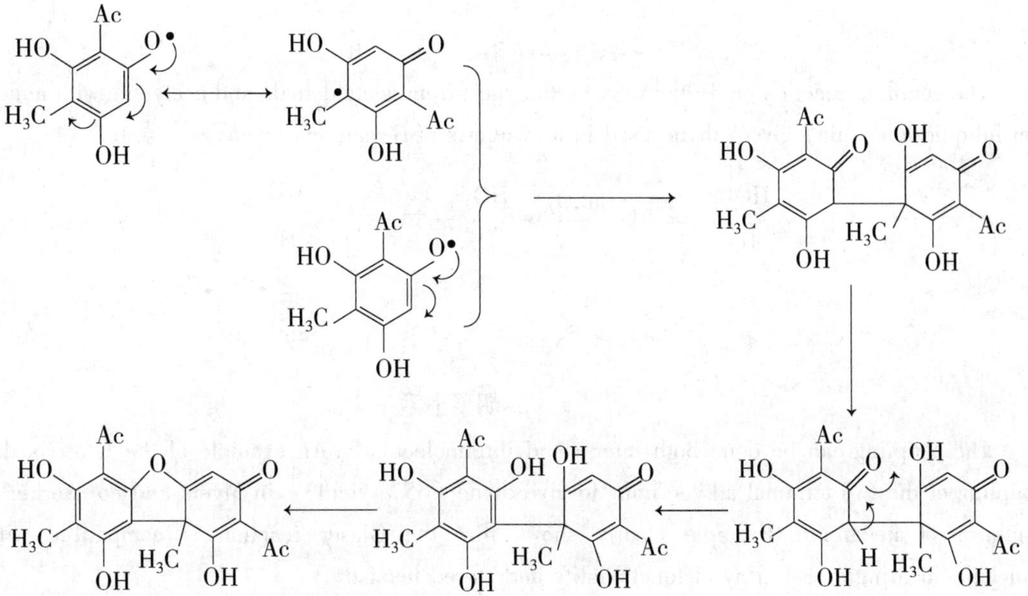

4.3.4.2 铜催化的偶联反应 Copper-catalyzed coupling reactions

(ⅰ) Glaser 反应

Glaser reaction

Terminal alkynes can be coupled by heating with stoichiometric amounts of cupric salts in pyridine or a similar base. This reaction, which produces symmetrical diynes in high yields, is called the Eglinton reaction[14].

Another common procedure is the use of catalytic amounts of cuprous salts in the presence of ammonia or ammonium chloride (this method is called the Glaser reaction). Atmospheric oxygen or some other oxidizing agent, such as permanganate or hydrogen peroxide, is required in the latter procedure.

In the Glaser reaction, a terminal alkyne such as phenylacetylene reacts with basic cupric chloride ($CuCl_2$) and subsequent air oxidation to give a diyne (1, 4-diphenyl-1, 3-butadiyne) in 90% yield.

$$2Ph-\equiv-H \xrightarrow[(2)air]{(1)CuCl_2,NH_4OH} Ph-\equiv-\equiv-Ph$$

The reaction probably involves one-electron oxidation of the acetylide anion by copper (II) ion followed by dimerization of the resulting acetylide radicals. Cu (II) has been proposed as an oxidant in the reaction. It has been shown that molecular oxygen forms the adducts with Cu(I) supported by tertiary amines which might be the intermediates, and mechanistic considerations for this variation have also been reported.

$$R-\equiv-H \xrightarrow[(\bar{O}H)]{-H^+} R-\equiv C^- \xrightarrow[oxidation]{Cu^{2+}} R-\equiv\dot{C} + Cu^+$$

$$2R-\equiv\dot{C} \longrightarrow R-\equiv-\equiv-R$$

The coupling reaction on 3-hydroxyy-1-butyne (from acetaldehyde and acetylene with amide ion inliquid ammonia) give a diyne used in a synthesis of β-carotene(β-胡萝卜素).

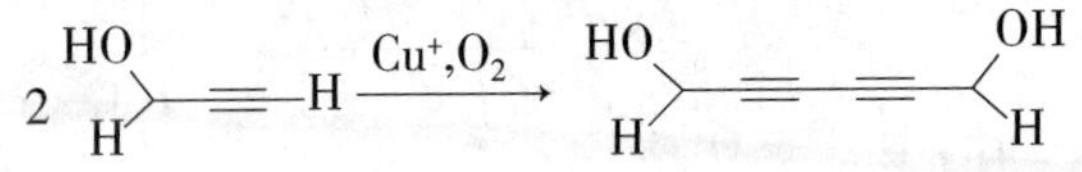

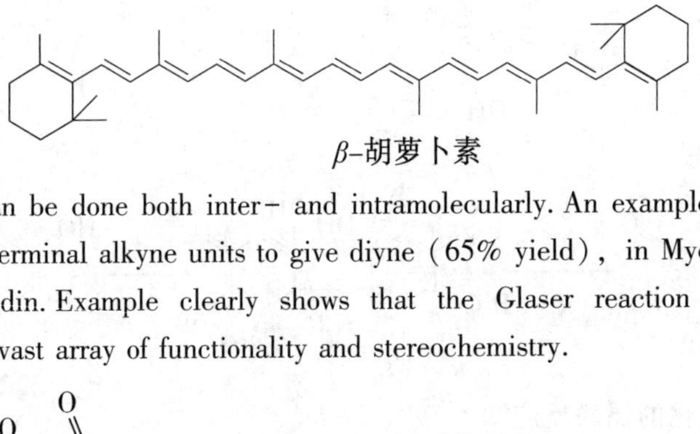

$$2\ CH_3CH(OH)-\equiv-H \xrightarrow{Cu^+,O_2} CH_3CH(OH)-\equiv-\equiv-CH(OH)CH_3$$

β-胡萝卜素

The coupling can be done both inter- and intramolecularly. An example of the latter is the coupling of the two terminal alkyne units to give diyne (65% yield), in Myers and co-worker' s synthesis of kedarcidin. Example clearly shows that the Glaser reaction is compatible with molecules bearing a vast array of functionality and stereochemistry.

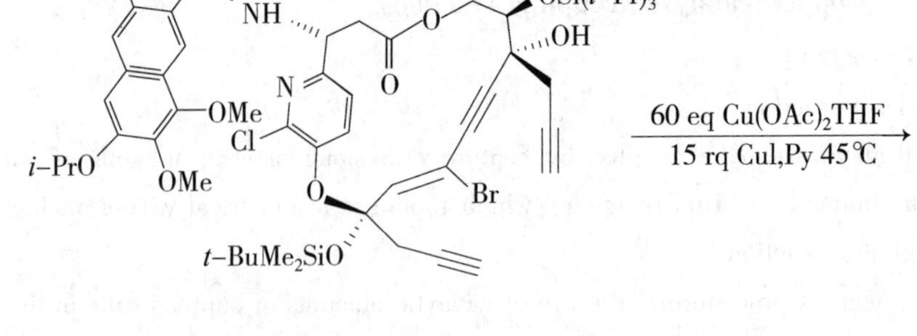

$$\xrightarrow[15\ rq\ CuI,Py\ 45℃]{60\ eq\ Cu(OAc)_2THF}$$

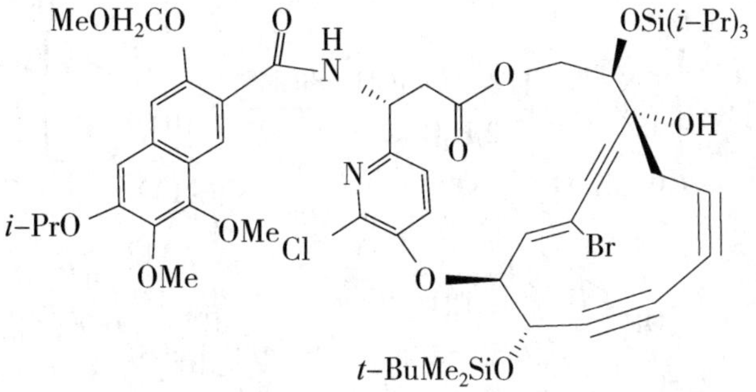

MeOH$_2$CO O H N OSi(i-Pr)$_3$ O OH O N i-PrO OMe Cl Br OMe O t-BuMe$_2$SiO

（ⅱ）Ullmann 反应

Ullmann reaction

The Ullmann reaction is very similar in that aryl halides are coupled to form biaryls (such as biphenyl) by heating with copper. Once again, the copper generates an aryl radical, which reacts with Cu^+ to form an arylcopper(Ⅰ) species. Subsequent reaction of the aryl copper with iodobenzene leads to a coupling reaction that gives biphenyl, which is related to the organocuprate coupling reactions[15].

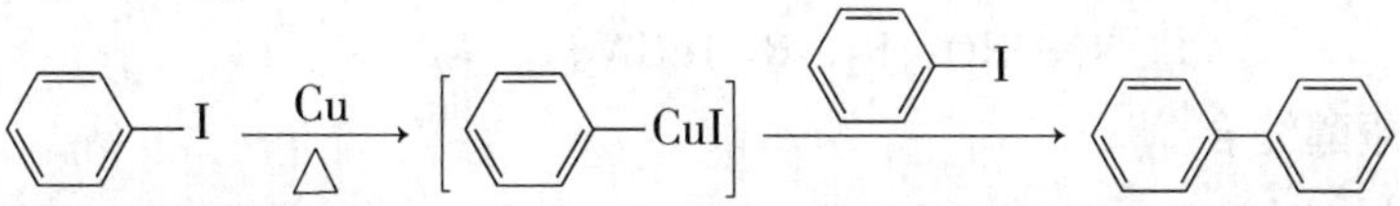

$$PhI \xrightarrow[\Delta]{Cu} [PhCuI] \xrightarrow{PhI} Ph-Ph$$

4.3.4.3 Pinacol 偶联

Pinacol coupling

Alkali metals react with a ketone such as 3-pentanone to produce a radical anion via electron transfe. Other reagents developed that give pinacol coupling, include magnesium metals, samarium (Ⅱ), titanium (Ⅳ) and lanthanide. If a monovalent metal such as sodium or potassium is used in ethanol, the carbonyl is reduced to the alcohol. If a bivalent metal such as magnesium is used, however, the ketyl is stabilized, and reduction is slow[16].

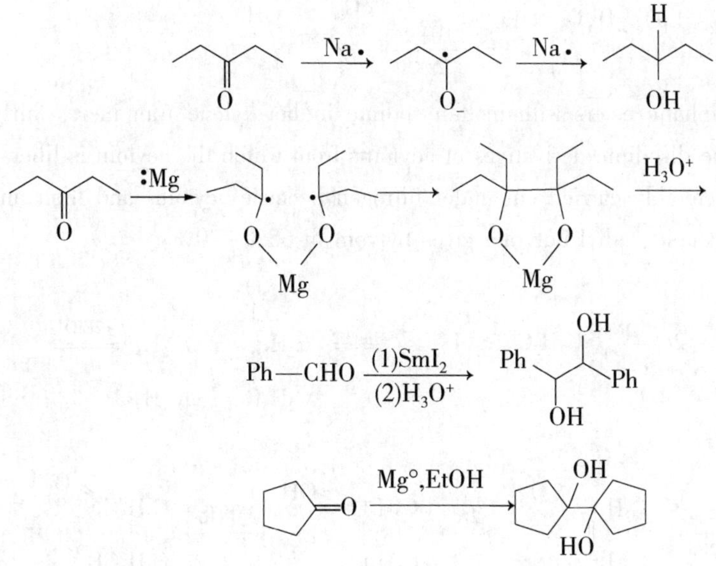

Na· Na· H O O OH

:Mg H$_3$O$^+$ O O O O HO OH Mg Mg

$$Ph-CHO \xrightarrow[(2)H_3O^+]{(1)SmI_2} Ph-CH(OH)-CH(OH)-Ph$$

Mg°,EtOH OH O HO

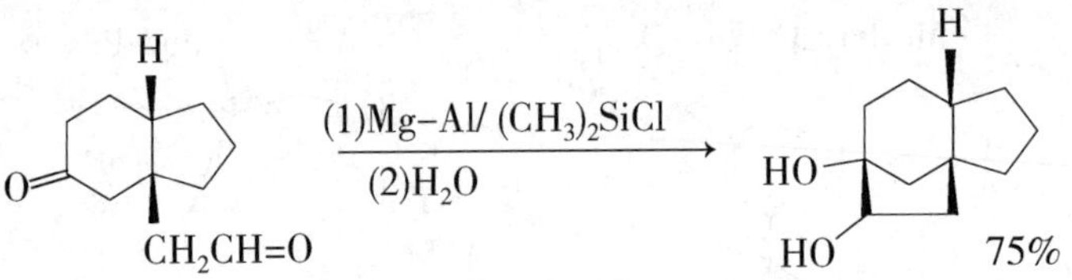

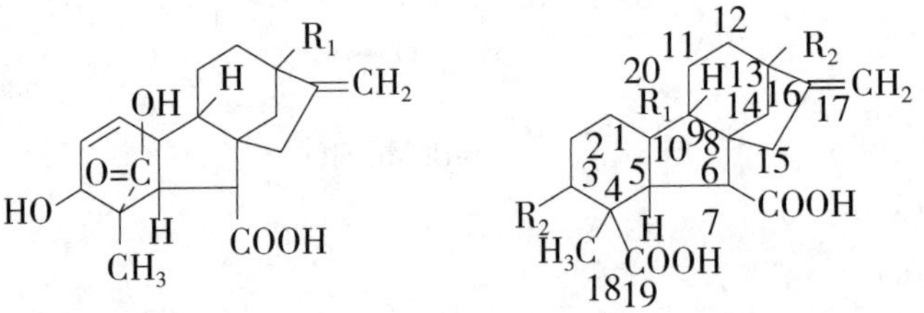

$$\xrightarrow[(2)H_2O]{(1)Mg-Al/(CH_3)_2SiCl}$$ 75%

C19赤霉素 C20赤霉素

R_1=H:GA_7 R_1=CH_3，R_2，R_3=H:GA_{18}

R_1=H:GA_3 R_1=CH_3，R_2=OH，R_3=H:GA_{19}

R_1=CHO，R_2，R_3=H:GA_{24}

4.3.4.4 酮醇缩合 Acyloin condensation

The reaction of esters with sodium metal (Na°) in refluxing xylene, which gave a good yield of 2-hydroxycycloheptanone after hydrolysis. Bouveault and Loquin reported the condensation of a α, β-diester to a α-hydroxy ketone, the acyloin condensation in 1905[17].

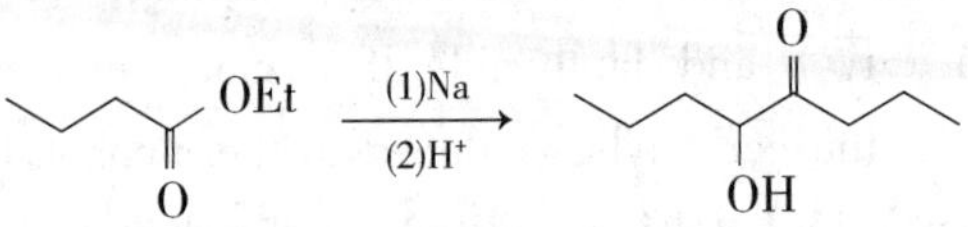

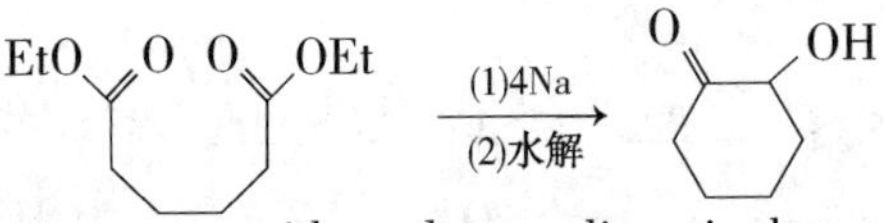

$$\xrightarrow[(2)H^+]{(1)Na}$$

$$\xrightarrow[(2)水解]{(1)4Na}$$

The treatment of alphatic esters with molten sodium in hot xylene (an inert, fairly high-boiling solvent) gives the disodium derivatives of acyloins from which the acyloin is liberated with acid. The reaction is preferably carried out under nitrogen because acyloins and their anions are readily oxidized. For example, ethyl butyrate gives bytyoin in 65%~70%:

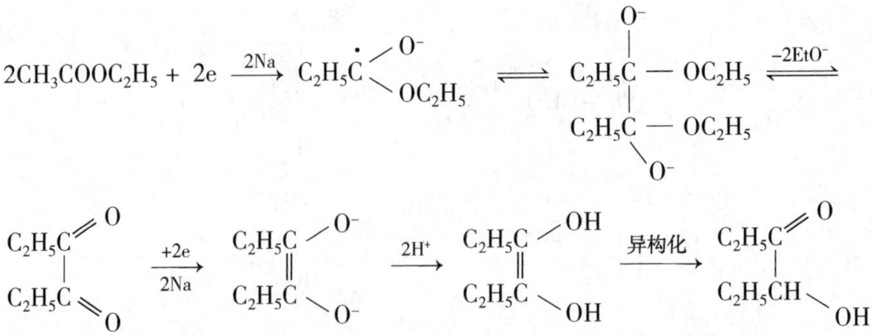

$$2CH_3COOC_2H_5 + 2e \xrightarrow{2Na} C_2H_5\dot{C}(O^-)(OC_2H_5) \rightleftharpoons C_2H_5C(O^-)(OC_2H_5)-C(O^-)(OC_2H_5)C_2H_5 \xrightleftharpoons{-2EtO^-}$$

$$C_2H_5C(=O)-C(=O)C_2H_5 \xrightarrow[2Na]{+2e} C_2H_5C(O^-)=C(O^-)C_2H_5 \xrightarrow{2H^+} C_2H_5C(OH)=C(OH)C_2H_5 \xrightarrow{异构化} C_2H_5C(=O)-CH(OH)C_2H_5$$

Reaction is initiated by electron transfer to the carbonyl group of ether, the resulting radicals dimerize, alkoide groups are eliminated, and further electron transfers give the disodium derivative of the acyloin, e. g.

$$\text{EtOOC-CH}_2\text{CH}_2\text{-COOEt} \xrightarrow[-2C_2H_5OSi(CH_3)_3]{4Na,4Me_3SiCl} \text{1,2-bis(OSiMe}_3\text{)cyclobutene} \xrightarrow[-2CH_3OSiMe_3]{MeOH} \text{2-hydroxycyclobutanone (71–86\%)}$$

It is even possible to make four-membered cyclic acylic acyloins in this way. A modification of the acyloin condensation adds chlorotrimethylsilane to trap the alkoxide intermediate as a bis (silyl enol ether). This modification has become the standard version of the acyloin condensation to obtain the acyloin in a good yield.

$$\text{CO}_2\text{CH}_3 / \text{H}_3\text{CO}_2\text{C diester} \xrightarrow{Na,甲苯} \text{cyclobutene-diolate (O}^-\text{, O}^-\text{)} \xrightarrow{TMSCl} \text{bis(OSiMe}_3\text{)cyclobutene} \xrightarrow{H_3O^+} \text{cyclobutanone (OH, O)}$$

4.3.4.5 硼烷的偶联

Coupling of boranes

Alkylboranes can be coupled by treatment with silver nitrate and base. Since alkylboranes are easily prepared from alkenes, this is essentially a way of coupling and reducing alkenes[18].

$$R_3B \xrightarrow[\text{or OH-电解}]{AgNO_3,OH^-} R\text{—}R$$

$$(CH_3CH_2CH_2CH_2CH_2CH_2)_3B \xrightarrow[KOH,EtOH]{AgNO_3} n\text{-}C_{12}H_{26}\ \ 70\%$$

$$R_3B + OH^- \longrightarrow R_3\bar{B}OH \xrightarrow{-e} R_3\dot{B}OH \longrightarrow R\cdot + R_2BOH \quad 2R\cdot \longrightarrow R\text{—}R$$

4.3.5 Hunsdiecker 反应

Hunsdiecker Reaction

Reaction of a silver salt of a carboxylic acid with bromine is called the Hunsdiecker reaction and is a method of decreasing the length of a carbon chain by one unit. The reaction is of wide scope, giving good results for n-alkyl R from 2 to 18 carbons and for many branched R too, producing primary, secondary, and tertiary bromides. Many functional groups may be present as long as they are not α substituted.

The silver(Ⅰ) salts of carboxylic acids react with halogens to give unstable intermediates which readily decarboxylate thermally to yield alkyl halides[19].

$$RCO_2Me \xrightarrow[(Hg^{2+} \text{or } Pb^{4+})]{AgNO_3} RCO_2Ag \xrightarrow[CCl_4]{Br_2} RBr + AgBr + CO_2$$

$$t\text{-}BuCH_2COOAg \xrightarrow[62\%]{Br_2} t\text{-}BuCH_2Br$$

$$RCOOH \xrightarrow[\text{MeCN, r.t. 1-27h}]{\text{5 mol\% Ag(Phen)}_2\text{OTf, 1.5eq. } t\text{-BuOCl}} R\text{—}Cl \quad \text{R:alkyl,benzyl}$$

The reaction is believed to involve homolysis of the C—C bond and a radical chain mechanism.

$$RCOOAg + Br_2 \longrightarrow RCOOBr + AgBr$$

$$RCOOBr \longrightarrow R\dot{C}O_2 + \dot{B}r$$

$$RCOO\cdot \xrightarrow{-CO_2} R\cdot \xrightarrow[\text{or } Br\cdot]{RCOOBr} RCO_2\cdot + RBr$$

4. 3. 6 氧化
Oxidation

C—H bonds in a wide of variety of environments are oxidized on standing in air to hydroperoxide groups. Reactions is apparently initiated by the appearance of stray radicals produced, for example, by sunlight photolysis, and thereafter a chain process operates:

$$RH + R'\cdot \longrightarrow R\cdot + R'H$$

$$R\cdot + O_2 \longrightarrow R\text{—}O\text{—}O\cdot$$

$$R\text{—}O\text{—}O\cdot + RH \longrightarrow R\text{—}O\text{—}OH$$

In the alkane series the order of reactivity is, tertiary>secondary>primary C—H, as usual in free-radical reactions.

Ethers are particularly prone to autoxidation, e. g. ether, tetrahydrofuran gives the α-hydroperoxide.

$$EtO\text{—}CH_2CH_3 \xrightarrow{O_2} EtO\text{—}CH(OOH)CH_3 \qquad \text{THF} \longrightarrow \text{2-(O—OH)-THF}$$

Since hydroperoxides can explode on heating it is essential to remove them from ethers [e. g. by reduction with aqueous iron(II) sulfate] before using the ethers as solvents for reactions which reqire heat.

Aldehydes also autoxidizes readily but the initial product, a peroxyacid, reacts with more of the aldehyde to give the carboxylic acid. For example, benzaldehyde gives benzoic acid on standing in air.

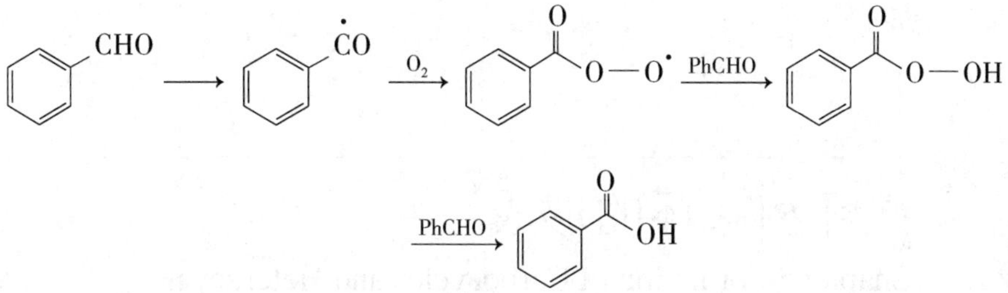

Hydroperoxides are not themselves of much synthetic value but they are employed as intermediates in certain reactions. For example, cumene (异丙苯) is converted industrially into phenol and acetone via cumene hydroperoxide.

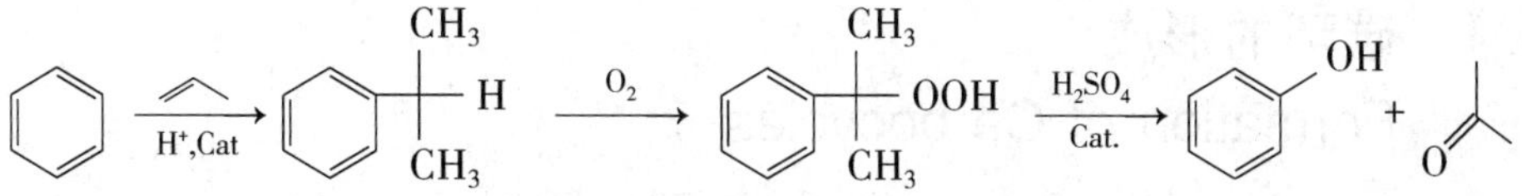

参考文献
References

1. Deepthi A, et, al. *J. Tetrahedron Letters*. 2018, 59 (29): 2767-2777.
2. James O O, Maity S. Hydrocarbon Biorefinery. Amsterdam: Elsevier, 2022. 297-325.
3. Brimberry M A, et, al. *J. Journal of Inorganic Biochemistry*, 2022, 226: 111636.
4. Hammer S G, Heinrich M R. Comprehensive Organic Synthesis (Second Edition). Amsterdam: Elsevier, 2014. 495-516.
5. Griesbeck A G, Franke M. Comprehensive Organic Synthesis (Second Edition). Amsterdam: Elsevier, 2014. 129-158.
6. Iwahashi H. Studies in Natural Products Chemistry. Amsterdam: Elsevier, 2021. 1-22.
7. Guin A, Deswal S, et, al. Comprehensive Aryne Synthetic Chemistry. Amsterdam: Elsevier, 2022. 223-266.
8. Loertscher B M, Castle S L. Comprehensive Organic Synthesis (Second Edition). Amsterdam: Elsevier, 2014. 742-809.
9. Xu T. Comprehensive Organometallic Chemistry IV. Oxford: Elsevier, 2022. 332-346.
10. Schäfer H J. Comprehensive Organic Synthesis. Oxford: Pergamon, 1991. 633-658.
11. Nişancı B. Nontraditional Activation Methods in Green and Sustainable Applications. Amsterdam: Elsevier, 2021. 329-347.
12. Quideau S, Deffieux D, et, al. Comprehensive Organic Synthesis (Second Edition). Amsterdam: Elsevier, 2014. 656-740.
13. Whiting D A. Comprehensive Organic Synthesis. Oxford: Pergamon, 1991. 659-703.
14. Bakker A, Gao H Y, et, al. Encyclopedia of Interfacial Chemistry. Oxford: Elsevier, 2018. 272-284.
15. Hietschold M. Encyclopedia of Interfacial Chemistry. Oxford: Elsevier, 2018. 499-508.
16. Margetić D, Štrukil V. Mechanochemical Organic Synthesis. Boston: Elsevier, 2016. 55-139.
17. Togo H. Advanced Free Radical Reactions for Organic Synthesis. Amsterdam: Elsevier Science, 2004. 39-56.
18. Hamdaoui M, Varkhedkar R, et al. Synthetic Inorganic Chemistry. Oxford: Elsevier, 2021. 343-389.
19. Crich D. Comprehensive Organic Synthesis. Oxford: Pergamon, 1991. 717-734.

5 碳环和杂环的形成

Chapter 5 Formation of Carbocycles and Heterocycles

5.1 碳环的形成 Formation of Carbocycles

Cyclic compounds play an important role in organic synthesis. The desired compound is not always commercially available, however, and must often be prepared by cyclization reactions from acyclic precursors[1]. This is particularly true for large ring (macrocyclic) compounds and polycyclic molecules. In the latter case, a cyclic molecule acts as a template and the other rings are built onto the template. This section will discuss the salient features of ring-forming reactions commonly encountered in synthesis[2].

5.1.1 通过卡宾形成三元环 Formation of 3-Membered Rings via Carbenes

5.1.1.1 卡宾的结构和制备 Preparation and structure of carbenes

(ⅰ) Preparation of carbenes

Photolysis of ketene (烯酮的光解作用):

Photolysis of ketene with light of wavelength 300~370nm produces the reactive intermediate carbene, which is capable of a variety of insertion and addition reactions. Photolysis of other alkyl and aryl ketenes can generate alkyl and aryl carbenes[3].

$$CH_2{=}C{=}O \xrightarrow[\text{or}\,\Delta]{h\nu} \ddot{C}H_2 + CO\uparrow$$

Photolysis of diazoalkane (重氮烷烃的光解作用):

Another important precursor for the preparation of carbene and related molecule is diazoalkane. Since diazomethane can detonate on contact with ground glass, its preparation requires the use of specialized glassware. Diazomethane is also toxic.

$$RCHN_2 \xrightarrow{h\nu} RCH: + N_2$$

The simplest diazoalkane is diazomethane, represented by several resonance hybrid structures, which are zwitterionic, although it can also be written as a nitrene[4].

$$[H_2C{=}N^+{=}N^- \longleftrightarrow H_2C^+{-}N{=}N^- \longleftrightarrow H_2\bar{C}{-}\overset{+}{N}{\equiv}N \longleftrightarrow H_2\bar{C}{-}N{=}\overset{+}{N}]$$

generates carbenes is a rearrangement of α－diazocarbonyl compounds (called the **Wolff rearrangement**), in which a diazocarbonyl compound loses nitrogen (N_2) to give an acyl carbene.

Halocarbenes（卤代烷烃）：

Polyhalomethanes can be converted to halocarbenes via photolysis, thermolysis, or treatment with base. Haloforms react with potassium *tert*－butoxide to form dihalocarbenes, which add smoothly to olefins giving 1, 1－dihalocyclopropanes. The reaction does not appear to be complicated by insertion and is therefore of great synthetic use.

Chloroform is a particularly useful carbene precursor. Reaction of chloroform with hydroxide begins with an acid-base reaction to generate the anion ($Cl_3C:^-$) and water. Loss of chloride ion generates dichlorocarbene ($Cl_2C:$). Potassium *tert*-butoxide is the base most commonly used to generate dichlorocarbene.

$$CHCl_3 \xrightarrow{B^-} BH+\bar{C}Cl_3 \longrightarrow :CCl_2+Cl^-+BH$$

$$R_2CBr_2+R'Li \longrightarrow R_2CBrLi+LiBr \longrightarrow R_2C:+LiBr$$

Dihalocarbenes from Phenyl(trihalomethyl)mercury Compounds［来自苯基(三卤甲基)汞化合物的二卤代卡宾］：

$$C_6H_5HgCBr_3 \longrightarrow :CBr_2+C_6H_5HgBr$$

It should be pointed out that the mono-, di-, and tribromo derivatives of the reagent all react considerably more rapidly than the trichloro reagent（三氯试剂）. For example, the tribromo compound reacts with cyclohexene in about 2 hours, while the trichloro compound requires 36 to 48 hours.

（ⅱ）Structure of carbenes

A carbene is a divalent carbon species linked to two adjacent groups by covalent bonds, possessing two nonbonded electrons and six valence electrons. The H—C—H angle of the triplet state, as determined from the ESR spectrum is 125°～140°. The H—C—H angle of the singlet state is found to be 102° by electronic spectroscopy. Although generally considered to reactive intermediates, stable carbenes are known, particularly persistent triplet diarylcarbenes and heteroatom-substituted singlet carbenes（三重态二芳基卡宾和杂原子取代的单线态卡宾）.

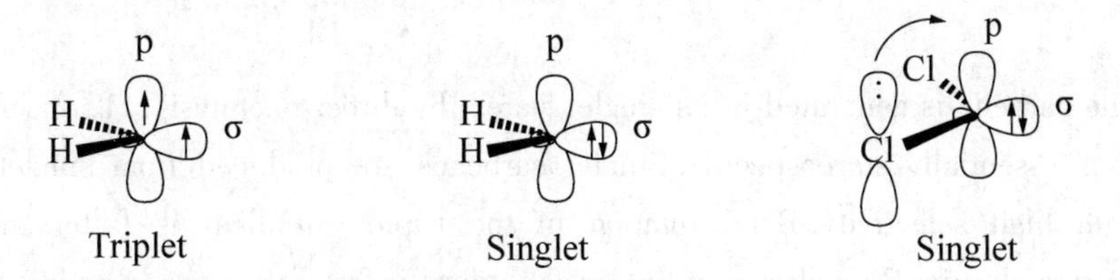

Features：

(a) Due to electron repulsion, there is an energy cost (8～10 kcal/mol) in pairing both electron in the *s* orbital.

(b) The electrons will remain unpaired (triplet) if energy difference between the *s* and *p* orbitals is small; or else, the electrons will be paired in the *s* orbital (singlet).

(c) Carbenes substituted with *p*-donor atoms (N, O, halogen), whose *p* orbital is raised high enough in energy to make the pairing of the electrons in the *s* orbital energetically favorable, are often in the singlet state.

5.1.1.2 卡宾的反应

Reactions of carbenes

(ⅰ) Addition of carbene to alkene and alkyne

The most useful reaction of carbenes is with alkenes to form cyclopropane derivatives. Polyhalomethanes can be converted to halocarbenes via photolysis, thermolysis, or treatment with base. A typical reaction is that of cyclohexene and chloroform with potassium *tert*-butoxide to give dichlorocyclopropanes.

$$\text{cyclohexene} + :CH_2 \longrightarrow \text{bicyclo[4.1.0]heptane}$$

$$\text{cyclohexene} \xrightarrow[t\text{-BuOK}]{CHCl_3} \text{7,7-dichlorobicyclo[4.1.0]heptane (Cl, Cl)}$$

$$\text{cyclopentene} + CHCl_3 \xrightarrow{KOC(CH_3)_3} \text{dibromobicyclo[3.1.0]hexane (}H_2O\text{, Br, Br, }Ag^+\text{)} \xrightarrow[H_2O]{AgNO_3} \text{2-bromocyclohex-2-enol (OH, Br)}$$

Addition of carbene to alkyne:

$$R-C\equiv C-R + :CH_2 \longrightarrow R\text{–(bicyclobutane)–}R$$

$$R-C\equiv C-R + :CR_2 \longrightarrow \text{cyclopropene}=CR_2$$

$$H_3C-CH=\underset{C_2H_5}{C}-C\equiv CH \xrightarrow[t\text{-BuOK}]{CHCl_3} H_3C-HC-\overset{C_2H_5}{C}(CCl_2)-C\equiv CH$$

When the carbene is generated in its singlet state, by direct photolysis, its insertion into the double bond is essentially stereospecific. Singlet carbenes are produced from singlet diradicals, and close with high selectivity. Bond rotation in the triplet diradical is faster than $T_1 \rightarrow S_1$ conversion or ring closure. Ring closure of the singlet radical is faster than bond rotation. For example, *cis*-2-butene and dicarbomethoxymethylene give almost entirely the *cis*-dimethyl-cyclopropane.

Me Me / H H $\xrightarrow[\text{singlet state}]{:C(CO_2Me)_2}$ Me Me / H H / MeO_2C CO_2Me

However, when the carbene is generated by photolysis in the presence of a triplet sensitizer, it is formed in its triplet state and stereospecificity is lost; for example, *cis*-2-butene gives largely *trans*-dimethylcyclopropane. This is because there is a time lag between formation of the two C—C bonds during which spin-inversion occurs and this allows rotation about the original C—C bond. Both triplet diradicals are slowly converted to singlet diradicals, which give the *cis*-cyclopropane and the trans-cyclopropane.

Me Me / H H $\xrightarrow{:C(CO_2Me)_2}$ Me ↑ Me / H H / MeO_2C CH_2O_2Me $\xrightarrow{\text{spin-inversion}}$ Me Me / H H / MeO_2C CO_2Me

triplet stat ⇅

Me ↑ H / H Me / MeO_2C CH_2O_2Me $\xrightarrow{\text{spin-inversion}}$ Me H / H Me / MeO_2C CO_2Me

(ⅱ) 重氮甲烷的反应

Reactions of diazomethane

When diazoalkanes are treated with transition metals in the presence of an alkene, particularly copperor rhodium derivatives, cyclopropanation occurs although the reactive intermediate may not be a free carbene. Diazoalkanes are converted to the corresponding carbene photolytically, and in the presence of copper derivatives such as Cu(acac)$_2$.

$+ N_2CHCOOC_2H_5 \xrightarrow{Cu}$ $-COOC_2H_5$

O / $COOCH_3$ / N_2 $\xrightarrow{CuSO_4}$ O / $COOCH_3$ / CH_3 / H

Cl_3C / CH_3 / CH_3 $+ N_2$=$COOEt$ $\xrightarrow{\text{Chiral catalyst}}$ CH_3 / H H / Cl_3C / CH_3 / COOEt $\xrightarrow[EtOH]{KOH}$ CH_3 / H H / Cl_2C= / CH_3 / COOEt

Pyrethrin(除虫菊酯)

Reaction with aromatic derivatives leads to ring expansion to cycloheptatriene derivatives. Both of these reactions (addition to an alkene or arene insertion) involve generation of an intermediate carbene and addition to a π bond.

$+ :CH_2 \longrightarrow \longrightarrow$

$+ \ CH_2N_2 \xrightarrow[25℃]{CH_3OH}$ 63% + 15%

(ⅲ) 卡宾与锌铜偶联剂加成

Carbene addition by the zinc-copper couple

Photolytically generated carbene, as mentioned above, undergoes a variety of undiscriminated addition and insertion reactions and is therefore of limited synthetic utility. The discovery of the generation of carbenes by the zinc-copper couple, however, makes carbene addition to double bonds synthetically useful. The most commonly used carbenoid is generated by reaction of diiodomethane and a Zn/Cu couple. When the carbenoid formed in this manner adds to alkenes, it is called the Simmons-Smith reaction.

$\xrightarrow[Zn]{CH_2I_2}$ 65%

OH $\xrightarrow[Zn \ (or \ Et_2Zn)]{CH_2I_2}$ OH

$CH_2{=}CHCOOC_2H_5 + CH_2I_2 \xrightarrow{Zn{-}Cu}$ ▷—$COOCH_3$ 50%

HO + $CH_2I_2 \xrightarrow{Zn{-}Cu}$ HO 66%

$+ \ CH_3CHI_2 + (C_2H_5)_2Zn \longrightarrow$ H, CH_3 + CH_3

O + $CH_2I_2 \xrightarrow{Zn{-}Cu}$ O 50%

There is a pronounced neighboring-group effect when a hydrogen-bonding oxygen is a substituent. The oxygen may " coordinate with the reagent, increasing the rate and control the stereochemistry of the addition. "

Electron-rich alkenes such as vinyl ethers and silyl enol ethers usually react faster than other alkenes:

$$\text{CH}_3\text{COCH=CH}_2 + \text{ClSi(CH}_3)_3 \xrightarrow{\text{Et}_3\text{N}} \text{CH}_2\text{=C(OSi(CH}_3)_3)\text{CH=CH}_2 \xrightarrow[\text{Zn-Cu}]{\text{CH}_2\text{I}_2} (\text{CH}_3)_3\text{SiO-C}_3\text{H}_4\text{-CH=CH}_2 \quad 75\%$$

The iodomethylzinc iodide complex is believed to function by electrophilic addition to the double bond in a three-center transition state giving essentially cis addition. Use of the system for the preparation of norcarane is given in the procedure. The zinc must be activated or the yield of cyclopropane product is poor.

$$CH_2I_2 + \text{Zn-Cu} \rightleftharpoons (ICH_2)_2Zn \cdot ZnI_2 \xrightarrow{\text{C=C}} [\text{C=C} \cdots CH_2 \cdots ZnI / I] \longrightarrow \text{C}_2CH_2$$

5.1.2 四元环形成 Formation of 4-Membered Rings

One is the [2+2]-cycloaddition that occurs between two alkenes, or between an alkene and a carbonyl or alkyne. Cyclobutanes and cyclobutylenes are prepared by this reaction, and β-lactams and β-lactones have been prepared[4,5].

According to Woodward-Hoffmann rules, photochemical [2+2] cycloaddition (and other 4*n* cycloadditions) that form cyclobutane derivatives is a symmetry-allowed process.

$$\| + \| \xrightarrow{h\nu} \square$$

$$\| + ||| \xrightarrow{h\nu} \text{cyclobutene}$$

$$2 \text{ cyclopentene} \xrightarrow[\text{CH}_3\text{COCH}_3]{h\nu} \text{tricyclic dimer}$$

$$\text{3-methylcyclohex-2-enone} + \| \xrightarrow[\text{CH}_2\text{Cl}_2]{h\nu, -70℃} \text{bicyclic ketone (H, CH}_3)$$

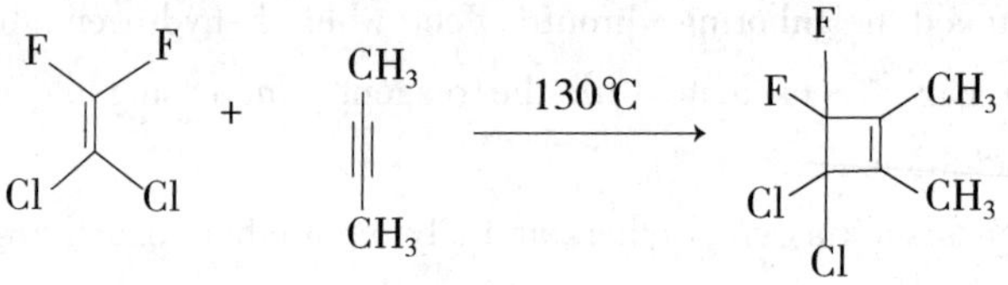

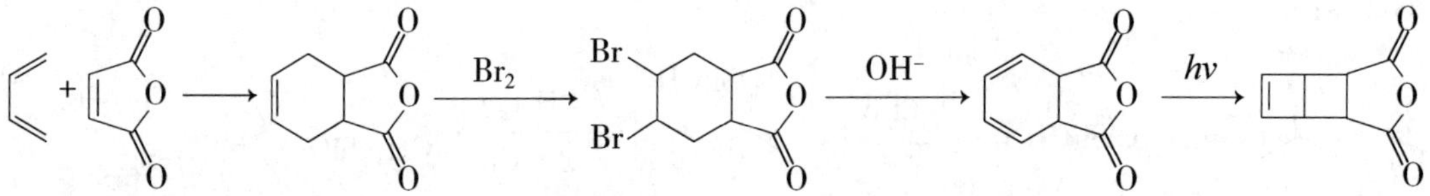

Ketenes ($R_2C=C=O$)(烯酮) react at the C=C unit via thermal [2+2]-cycloaddition reactions to give cyclobutanone derivatives. In a typical reaction, dimethylketene and ethene were heated to give cyclobutanone in a reaction controlled by the HOMO alkene-LUMO ketene interaction.

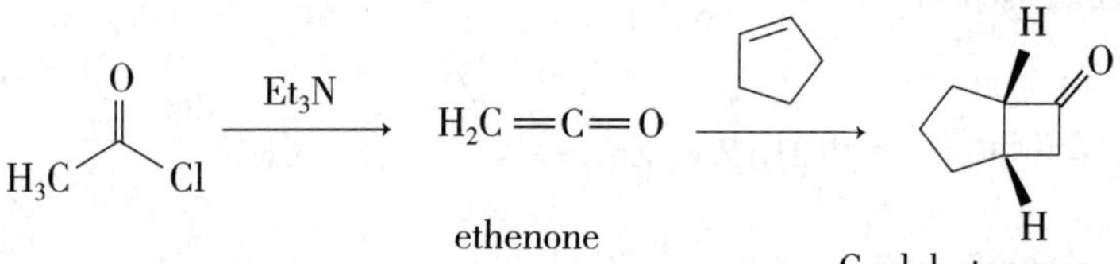

ethenone

Cyclobutanone

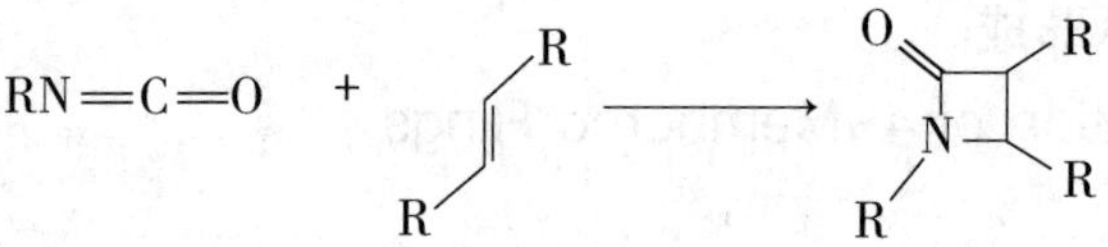

5.1.3 六元环形成 Formation of 6-Membered Rings

5.1.3.1 Diels-Alder reaction

The reaction between butadiene and ethylene is the simplest example of a general procedure for forming 6-membered rings, developed by Diels and Alder, in which a conjugated diene reacts with a compound containing a C=C or C≡C bond (a dienophile). The cycloaddition of alkenes and dienes is a very useful method for forming substituted cyclohexenes[6].

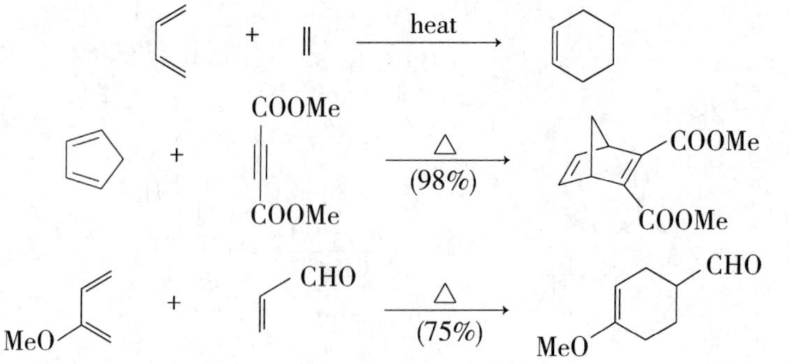

This thermal reaction of a diene and an alkene (or other two π-electron system, generically known as a dienophile) is now called the Diels-Alder reaction (D-A 反应). Diels and Alder were awarded the Nobel Prize for this reaction in 1950. The reaction is explained in terms of molecular orbital theory (分子轨道理论), and is recognized as a [$4n + 2n$]-cycloaddition reaction.

Diels-Alder reactions have large negative entropies of activation, reflecting not only the loss of translational entropy when the two molecules come together but also the high degree of ordering which corresponds to the mutual orientation of four atomic centers. However, the activation enthalpy is often small, so that rates of reaction are large even at moderate temperatures. The reverse reaction can often be brought about at high temperatures.

A brief account these reactions in terms of frontier-orbital analysis was given earlier. A more detailed treatment of the interaction between the highest-occupied molecular orbital (HOMO) of one reactant and the lowest-occupied molecular orbital (LUMO) of the other will be given here.

$HOMO_{diene}$ ≡

$LUMO_{alkene}$

≡

A [4+2]-reaction was shown to be thermally allowed. The reaction proceeds thermally, presumably by donation of electrons from the HOMO diene to the LUMO alkene.

(ⅰ) Reactivity in the Diels-Alder reaction

The Dienophile:

Diels - Alder reactions of cyclopentadiene with different alkenes proceed at different rates. Generally, simple alkenes and electron-rich alkenes (bearing an electron-releasing group) will react with butadiene or cyclopentadiene only under very vigorous conditions[7].

Relative rate:

+ tetracyanoethylene (Dienophile) —20℃→ D.A.cycloadduct

(a) (NC)₂C=C(CN)₂ > CH₂=C(CN)₂ > maleic anhydride > NC–CH=CH–CN > CH₂=CH–CN

$t_{1/2}$: 0.002sec. · 2sec. · 20sec. · 20mins. · 28h

(b) CH₂=CH–COCl > CH₂=CH–COPh > CH₂=CH–COMe > CH₂=CH–COOEt > CH₂=CH–CN

The Diene:

In the for reaction to occur, the diene must be capable of achieving the *s*-cis conformation which the formation of a 6-membered ring requires. This is always possible for acyclic dienes but it can result in wide variations in reactivity. For example, cis-butadienes are less reactive than their tran-isomers because the substituents in the former suffers steric crowding in the *s*-cis conformation which raises the activation energy.

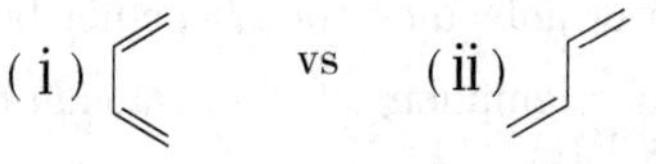

(i) s-*cis* vs (ii) s-*trans*

Cyclic dienes react only if they are of cis type; for example:

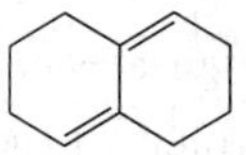

fails to react.

In the following reactions the reaction rate of (i) is 10^3 times faster than (ii), because that this poor reactivity was attributed to the energy barrier required for Dienophile (亲双烯体) to assume a cisoid conformation.

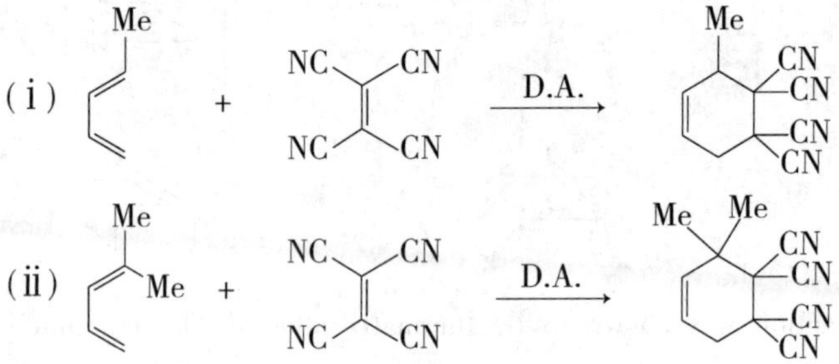

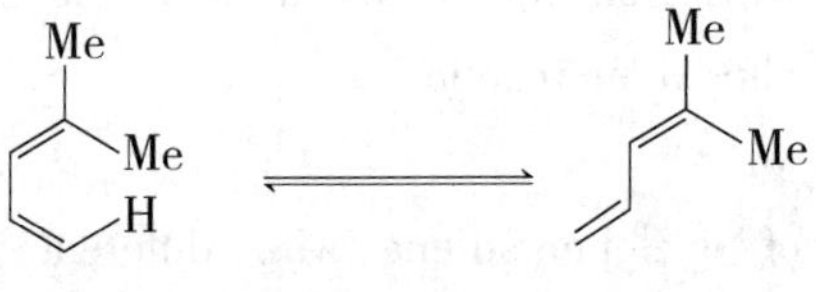

(i) Me + NC CN / NC CN —D.A.→ Me, CN, CN, CN, CN

(ii) Me Me + NC CN / NC CN —D.A.→ Me Me, CN, CN, CN, CN

Me Me H ⇌ Me Me

If the diene contains an electron - withdrawing substituent, the dienophile requires an electron-releasing one for ready reaction.

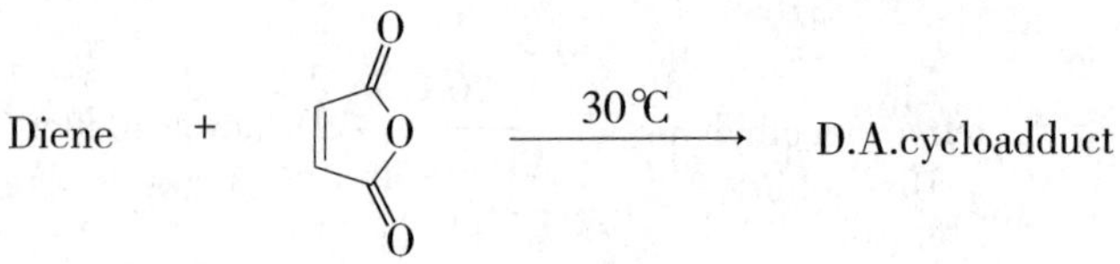

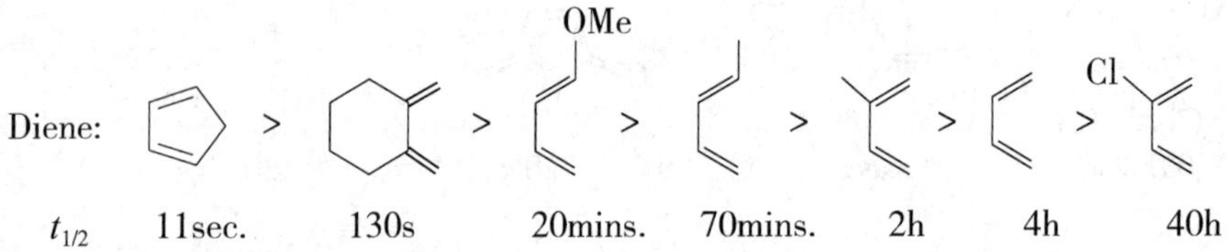

Diene + O O O —30℃→ D.A.cycloadduct

Diene:	(cyclopentadiene)	>	(dimethylenecyclohexane)	>	OMe	>		>		>		>	Cl
$t_{1/2}$	11sec.		130s		20mins.		70mins.		2h		4h		40h

Diels－Alder cycloaddition reactions most commonly involve electron－rich dienes and electrondeficient dienophiles（usually alkenes）. According to the Frontier Molecular Orbital（FMO）model（分子前线轨道理论），the reaction is driven by the interaction of the most available occupied and unoccupied molecular orbitals. Specifically，electron transfer from the highest-occupied molecular orbital（HOMO）on the diene to the lowest-unoccupied molecular orbital（LUMO）on the dienophile stabilizes the transition state. This implies that the smaller the HOMO-LUMO gap，the faster will be the reaction. Therefore，it is reasonable to expect that the energy difference between HOMO of the diene and LUMO of the dienophile will relate to the activation energy for the reaction.

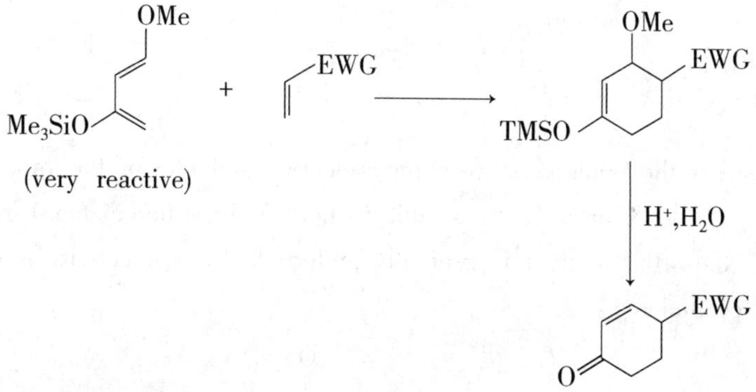

（ⅱ）D-A 反应的选择性

Selectivity in the Diels-Alder reaction

Cis-Trans Selectivity（顺反选择性）：

The cycloaddition is stereospecific（立体环加成）. The diene and dienophile configurations are retained in the adduct（the *cis* principle）.

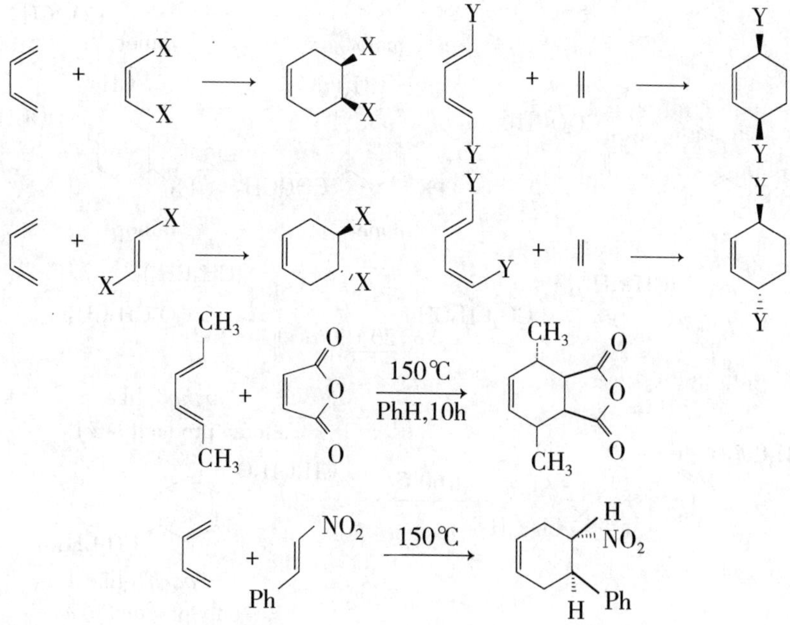

Regioselectivity of Cycloaddition Reactions（环加成反应的区域选择性）:

When both diene and dienophile contain substituents, more than one product can be formed. If the *diene* and the *dienophile* are both unsymmetrical the Diels–Alder addition may occur in four ways:

Type A

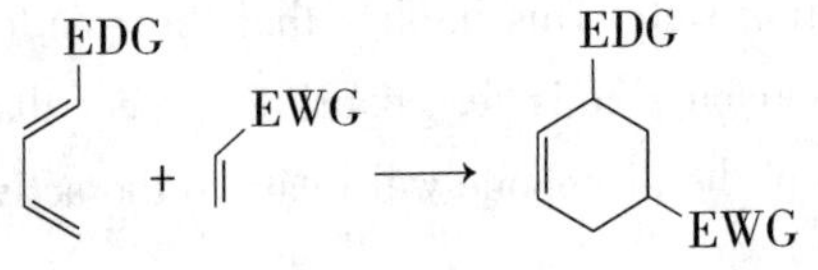

EDG + EWG ⟶ EDG, EWG

Type B

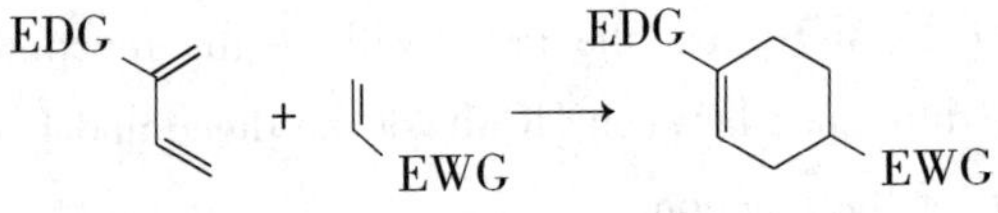

EDG + EWG ⟶ EDG, EWG

Type C

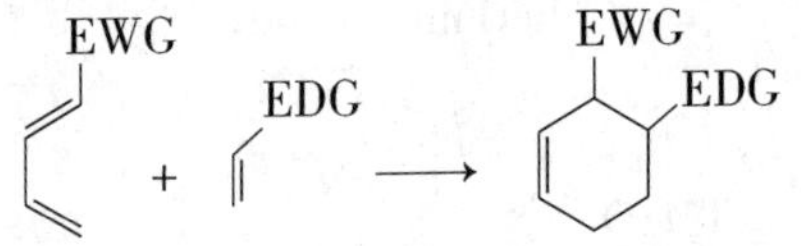

EWG + EDG ⟶ EWG, EDG

Type D

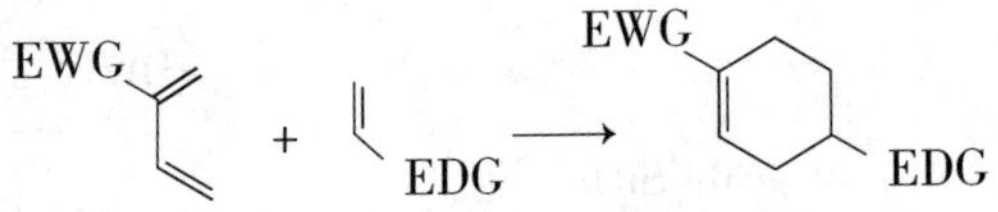

EWG + EDG ⟶ EWG, EDG

However, usually the reaction is very regioselective and one of the products is produced dominantly. When a C ═ X or a C ≡ X unit (where X is a heteroatom) reacts with a 1-substituted diene, the ortho product is generally preferred. Ortho selectivity is explained by the size of the orbital coefficients.

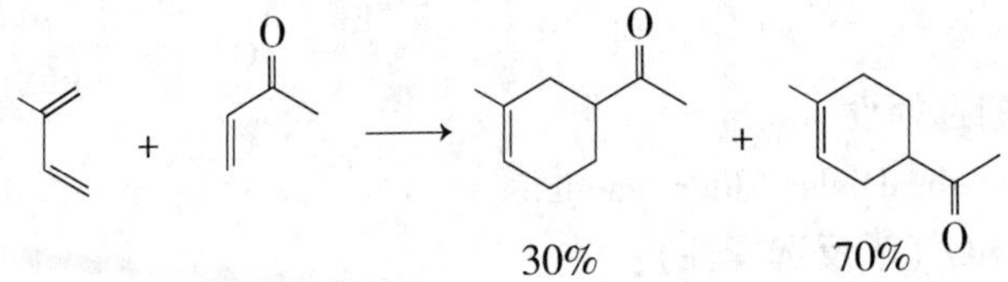

30% 70%

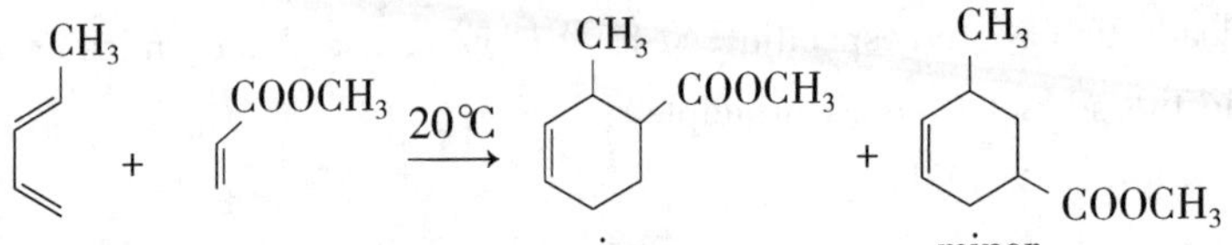

CH_3 + $COOCH_3$ —20℃→ CH_3, $COOCH_3$ (major) + CH_3, $COOCH_3$ (minor)

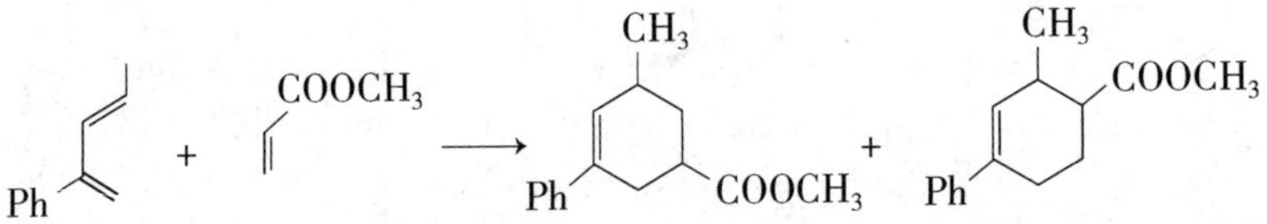

Ph + $COOCH_3$ ⟶ CH_3, Ph, $COOCH_3$ (major) + CH_3, $COOCH_3$, Ph (minor)

$N(CH_2CH_3)_2$ + $CO_2CH_2CH_3$ —20℃→ $N(CH_2CH_3)_2$, $CO_2CH_2CH_3$

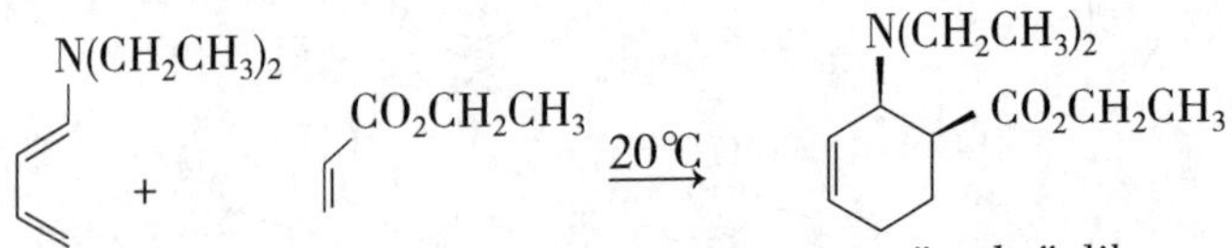

"*ortho*"-like

only product(94%)

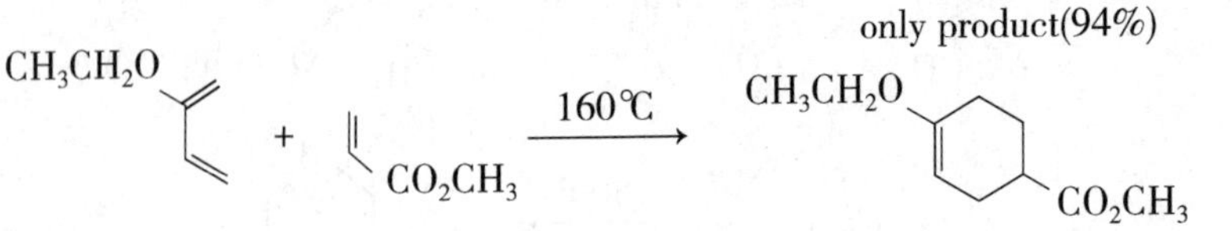

CH_3CH_2O + CO_2CH_3 —160℃→ CH_3CH_2O, CO_2CH_3

"*para*"-like

only product(50%)

This is can be explained in terms of frontier orbital theory. It is believed that 1,2-products predominate in these cases because bonding at the transition state is not only more effective when the HOMO of one reactant and the LUMO of the other are more closely matched in energy but also when the sizes of the orbitals at reacting termini are the more closely matched, e. g.

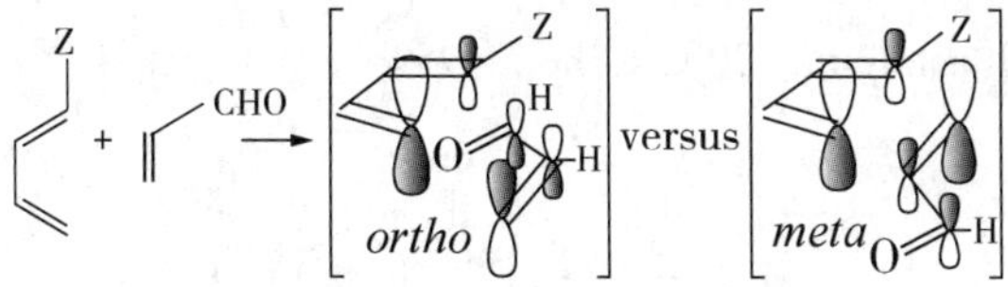

Z + CHO → [*ortho*] versus [*meta*]

(ⅲ) 空间异构体中的内型选择性

Endo selectivity

In the reaction, the observed stereochemistry demands that the benzoquinone is under the diene at the time of reaction, which is the normal reaction of dienophiles bearing a substituent with a π-bond and is called an **endo mode** of addition (the group of interest is under or down, relative to the diene, rather than out or up), and the endo orientation is preferred, which is the so-called **endo rule** or **Alder endo rule**.

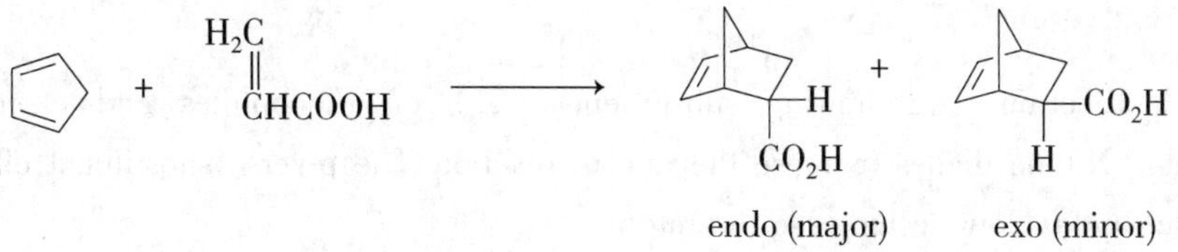

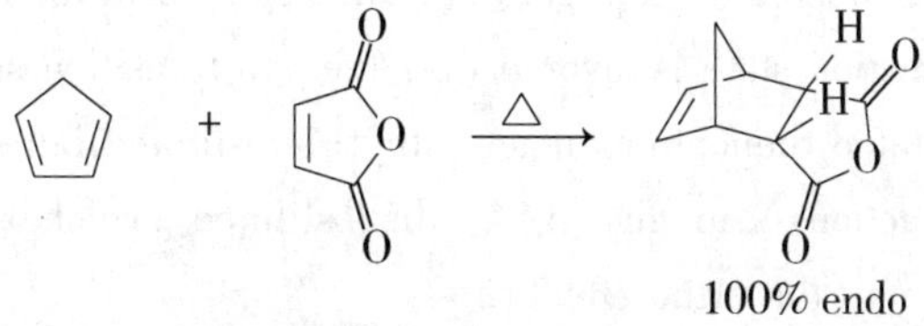

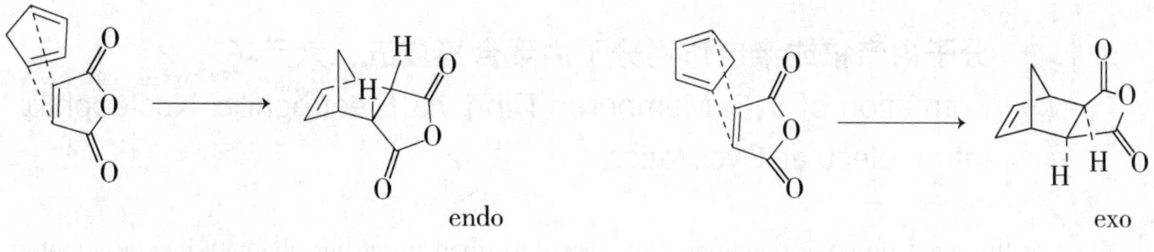

+ $H_2C{=}CHCOOH$ ⟶ endo (major) + exo (minor)

+ △ ⟶ 100% endo

endo

exo

(ⅳ) Lewis 酸催化剂

Catalysis by Lewis acids

Lewis acids are, in general, excellent catalysts for Diels - Alder reaction. Their catalytic activity lies in the ability of Lewis acid to complex a heteroatom component of the diene or dienophile. In general, the Lewis acid decreases the reaction time and increases product yield and selectivity, e. g.

$CO_2C_2H_5$ + $CO_2C_2H_5$ ⟶ $CO_2C_2H_5$ + $CO_2C_2H_5$

No catalyst 120℃, 6h 70% 30%

$AlCl_3$ catalyst 20℃, 3h 95% 5%

O + ⟶ O CH_3 CH_3 H + O CH_3 CH_3 H

No catalyst 150℃, 142h, 20% *para/ortho* 1.9

$AlCl_3$ catalyst 25℃, 17h, 97% *para/ortho* 36

Summary:

X + Z ⟶ X Z

(ⅰ) Electron - withdrawing substituents (Z) on dienophiles and electron releasing substituents (X) on dienes increase the rate of reaction. The reverse substituent effects (inverse electron demand) likewise increase the rate.

(ⅱ) The diene and dienophile configurations are retained in the adduct (the *cis* principle).

(ⅲ) The endo transition state is favored over the exo transition state (the *endo* rule).

(ⅳ) The Z-substituted dienophiles react with 1-substituted butadienes (in normal electron demand Diels - Alder reactions) to give 3, 4 - disubstituted cyclohexenes, independent of the nature of diene substituents (the ortho effect).

5.1.4 分子内亲核或亲电作用分子内环合形成五、六元环 Formation of 5,6-Membered Ring via Electrophilic/Nucleophilic Intramolecular Cyclization

Many of the bond-forming reactions that were described in earlier chapters maybe adapted to produce cyclic compounds, as the following examples show: Dieckmann condensation, aldol condensation, Michael addition, Friedel - Crafts reaction, pericyclic reaction, acyloin condensation, etc.

5.1.4.1 Dieckmann 缩合 Dieckmann condensation

The base - catalyzed intramolecular condensation of a diester. The Dieckmann condensation works well to produce 5 - or 6 - membered cyclic β - keto esters, and is usually affected with sodium alkoxide in alcoholic solvent.

If this geometry to be attained, ring formation is possible (favored), and we make the predication that the reaction will succeed. If the proper geometry cannot be attained, ring formation is difficult (disfavored) and competitive processes often dominate. most of the cyclization reactions for small- and medium-sized rings encountered[8].

CH_3 O COCH$_3$ $CH_2CH_2COOCH_3$ $\xrightarrow{NaOCH_3}$ CH_3 O^- $COOCH_3$

$\longrightarrow$ CH_3 O $COOCH_3$ 97%

The yields are good if the product has an enolizable proton; otherwise, the reverse reaction (cleavage with ring scission) can compete.

H_3C ($CH_2CO_2C_2H_5$)($CO_2C_2H_5$)($CH_2CH_2CH_2CO_2C_2H_5$) $\longrightarrow$ $C_2H_5O_2C$ O CH_3 $CH_2CO_2C_2H_5$ (major) + H_3C $C_2H_5O_2C$ O $CO_2C_2H_5$ (minor)

H_3C ($CH_2CO_2C_2H_5$)($CO_2C_2H_5$)($CH_2CH_2CH_2CO_2C_2H_5$) $\xrightarrow{NaOC_2H_5}$ O $CO_2C_2H_5$ $CO_2C_2H_5$ (major) + $C_2H_5O_2C$ O $CH_2CO_2C_2H_5$ (minor)

The general reaction can be described as:

CO_2R $(CH_2)_n$ CO_2R $\xrightarrow{RO^-}$ CO_2R CH $(CH_2)_{n-1}$ C=O

the ring number n is 4,5 or 6; if $n>6$ in a dilute solution;

$C_2H_5OOC(CH_2)_4COOC_2H_5 \xrightarrow{EtONa}$ O $COOC_2H_5$ $\xrightarrow[-CO_2]{Decarboxylation}$ O

$CH_3OOC(CH_2)_7COOCH_3 \xrightarrow[dimethylbenzene]{NaH}$ O $COOCH_3$ 48%

In short chains, an advantage in terms of entropy is offset by an increase in enthalpy due to extremely large strain energies. Ziegler first used this principle of a ring-shaped transition state to generate large membered rings by is known as the **high dilution method.**

$$(CH_3)_2C(CH_2COOC_2H_5)_2 + \begin{matrix} COOC_2H_5 \\ | \\ COOC_2H_5 \end{matrix} \xrightarrow{NaOEt} \text{(cyclopentane-1,2-dione with } 3,5\text{-di}(C_2H_5OOC)\text{, } 4,4\text{-di}(H_3C)\text{)}$$

the intermediate of camphor(樟脑)

$$C_2H_5OOC(CH_2)_{14}COOC_2H_5 \xrightarrow[\text{二甲苯}]{(CH_3)_3COK} (CH_2)_{13}\text{-ring: } CH(CO_2C_2H_5)-C{=}O \xrightarrow[\triangle]{H_3O^+} (CH_2)_{13}\text{-ring: } CH_2-C{=}O \quad 48\%$$

Thorpe-Ziegler 反应：

$$NC(CH_2)_5CN \xrightarrow[NaOEt]{Na} \text{2-cyanocyclohexanimine} \xrightarrow[(2)H^+]{(1)OH^-,H_2O} \text{2-oxocyclohexanecarboxylic acid} \xrightarrow[\triangle]{-CO_2} \text{cyclohexanone}$$

5.1.4.2 酰偶姻缩合

Acyloin condensation

$$\text{cyclohexane-1,2-}(CO_2C_2H_5)_2 \xrightarrow[(2)H_3O^+]{(1)Na,dimethylbenzene} \text{2-hydroxycyclohexanone} \quad 57\%$$

$$CH_3O_2C(CH_2)_{16}CO_2CH_3 \xrightarrow[(2)H_3O^+]{(1)Na,dimethylbenzene,N_2} (CH_2)_{16}\text{-ring: } CH(OH)-C{=}O \quad 96\%$$

The bimolecular reductive coupling of carboxylic esters by reaction with metallic sodium in an inert solvent under reflux gives an α-hydroxyketone, which is known as an acyloin. With longer alkyl chains, higher boiling solvents can be used, he intramolecular version of this reaction has been used extensively to close rings of different sizes, e. g. paracyclophanes（对位环芳）or catenanes(索烃).

The Benzoin condensation produces similar products, although with aromatic substituents and under different conditions. When the acyloin condensation is carried out in the presence of chlorotrimethylsilane, the enediolate intermediate is trapped as the bis-silyl derivative. This can be isolated and subsequently is hydrolysed under acidic condition to the acyloin, which gives a better overall yield.

5.1.4.3 分子内羟醛缩合 Intramolecular aldol condensation

Intramolecular aldol condensations are more favorable than intermolecular aldol condensations. Cyclization occurs if the α−carbon atom and the second carbonyl carbon atom can bond to form a five− or six−membered ring. If two or more reactions can yield these rings, it is necessary to consider which process is favored. The various possible enolates exist in low concentration under equilibrium conditions. Thus, the enolate that is the better nucleophile attacks the more reactive carbonyl carbon atom and dominates the product formed. In general, for example, intramolecular aldol condensations where the enolate attacks the carbonyl carbon atom of an aldehyde are favored over addition to the carbonyl carbon atom of a ketone[9].

KOH, H_2O, 100℃ — 73%

CHO, CHO, OH^-, H_2O, O, CH, CHO, O^-, CHO, H_2O, OH, CHO, $-H_2O$, CHO

R_2NH, O, O, OH, O, $-H_2O$, △, O

$CH_2CH_2COCH_3$, O, $NaOCH_3$, $CH_2CH_2COCH_3$, O, O, O, 73%

5.1.4.4 分子内烷基化反应 Intramolecular alkylation

Annulation via intramolecular S_N2 process:

$Cl—CH_2—CH_2—CH_2—CN$ $\xrightarrow{NaNH_2}$ cyclopropyl−CN 50%

O, $\bar{C}H_2$, $H_2C—\overset{H_2}{C}—CH_2—\overset{H_2}{C}—Br$ $\xrightarrow[\text{benzene } \triangle]{\text{sodium neopentyl oxide}}$ O

cyclohexanone + Br(CH₂)₄Br $\xrightarrow[C_6H_6, 5h\ \triangle]{t\text{-BuOK}/t\text{-BuOH}}$ O 70%

Malonic ester(丙二酸酯) may also be used for the synthesis of the three- and four-membered alicyclic compounds from dibromides (see Chapter 2).

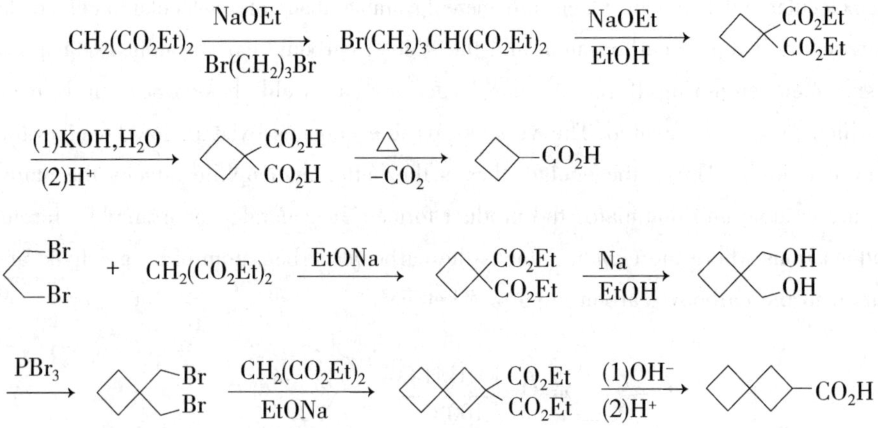

$$CH_2(CO_2Et)_2 \xrightarrow[Br(CH_2)_3Br]{NaOEt} Br(CH_2)_3CH(CO_2Et)_2 \xrightarrow[EtOH]{NaOEt}$$

$$\xrightarrow[(2)H^+]{(1)KOH,H_2O} \qquad \xrightarrow[-CO_2]{\triangle}$$

$$+ \; CH_2(CO_2Et)_2 \xrightarrow{EtONa} \qquad \xrightarrow[EtOH]{Na}$$

$$\xrightarrow{PBr_3} \qquad \xrightarrow[EtONa]{CH_2(CO_2Et)_2} \qquad \xrightarrow[(2)H^+]{(1)OH^-}$$

5.1.4.5 Robinson 环化反应 Robinson annulation

The Robinson annulation is a useful reaction for the formation of six-membered rings in polycyclic compounds, such as steroids. It combines two reactions: the Michael addition and the aldol condensation. The first step in the process is the Michael addition to an α, β-unsaturated ketone, such as methyl vinyl ketone (ethyl vinyl ketone is shown above). The newly formed enolate intermediate must first tautomerize for the conversion to continue: The subsequent cyclization via aldol addition is followed by a condensation to form a six-membered ring enone[10] (see Chapter 2).

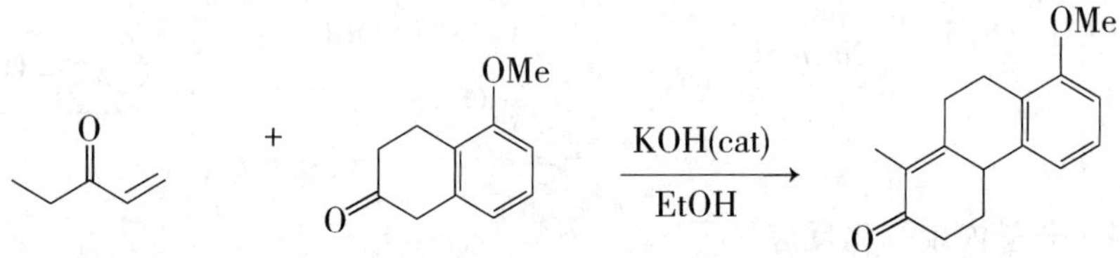

$$\xrightarrow[EtOH]{KOH(cat)}$$

5.1.4.6 分子内傅克反应 Intramolecular Friedel-Crafts reaction

Formation of five and six-membered fused systems has been achieved via acid promoted intramolecular Friedel-Crafts reaction.

$$\xrightarrow{H^+}$$

57%

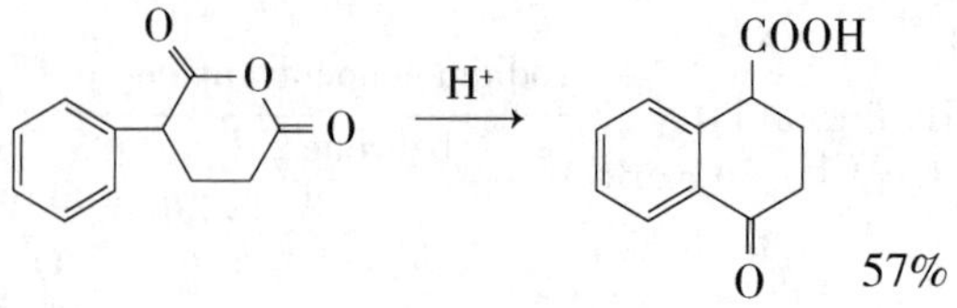

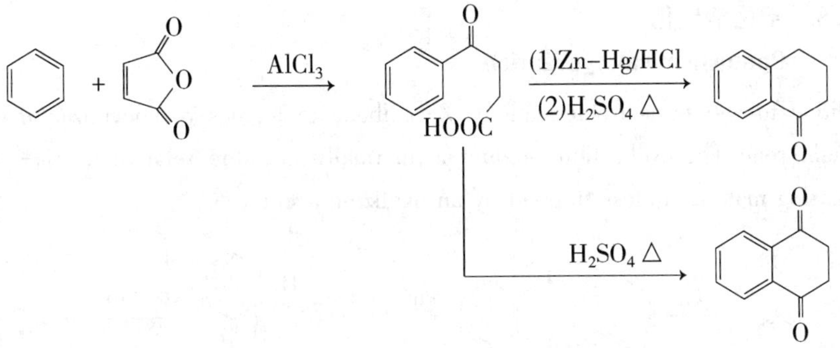

7-membered may be formed in the reaction in a dilute solution:

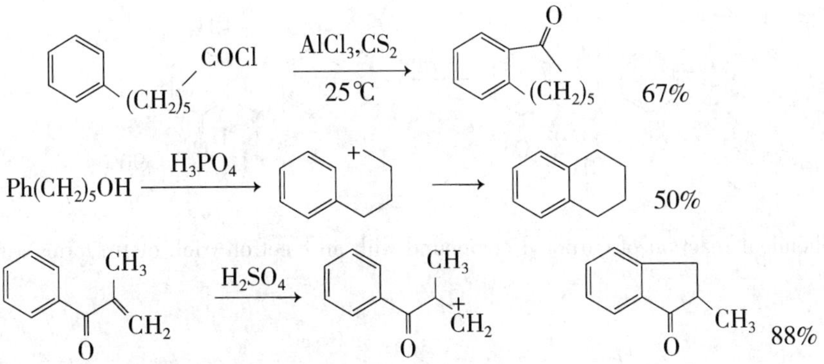

5.1.4.7 通过1,3-偶极环加成形成五元环

Formation of 5-membered ring via 1,3-dipolar cycloaddition

The 1,3-dipolar cycloaddition, also known as the Huisgen cycloaddition or Huisgen reaction, is an organic chemical reaction belonging to the larger class of cycloadditions. It is the reaction between a 1,3-dipole and a dipolarophile, most of which are substituted alkenes, to form a five-membered ring.

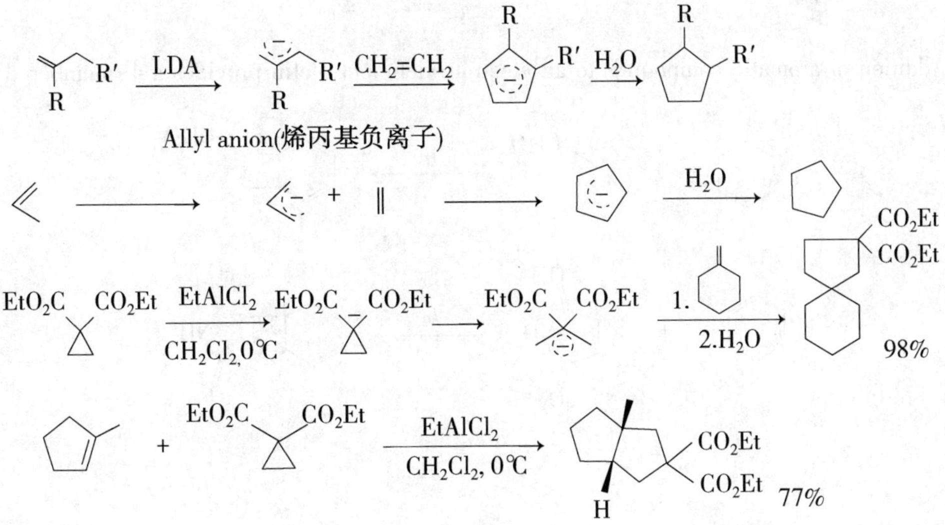

Allyl anion(烯丙基负离子)

5.1.4.8 光化学环化

Photochemical cyclization

In addition to cis-trans isomerization, Z-stilbene undergoes photocyclization to 4a, 4b-dihydrophenanthrene. The cyclization product is thermally unstable relative to cis-stilbene and reverts to starting material unless trapped by an oxidizing agent.

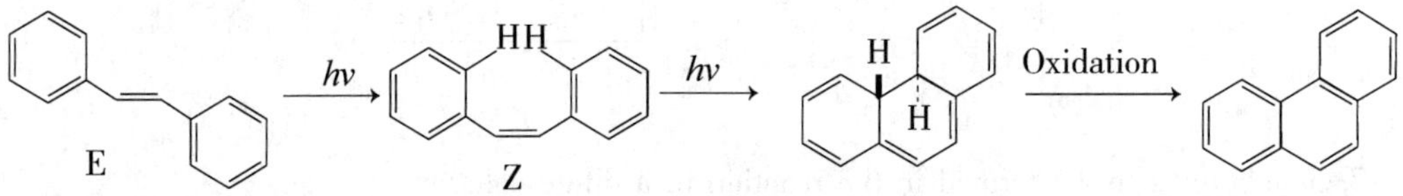

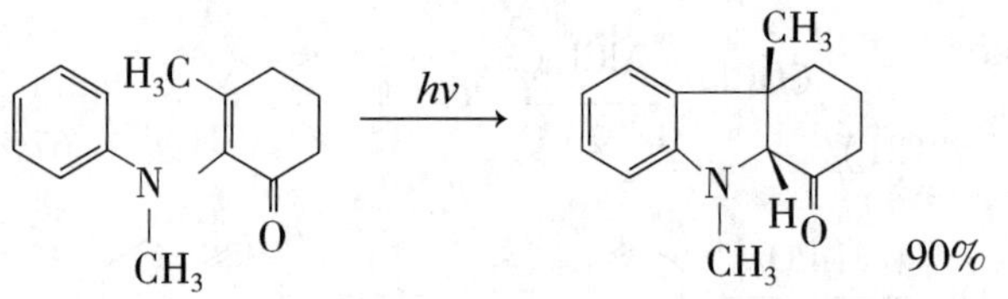

E HH hv Z hv H H Oxidation

H_3C N CH_3 O hv CH_3 N H O CH_3 90%

Photochemical reaction of carbonyl compound with an electron-rich olefin forms a trimethylene oxide[11].

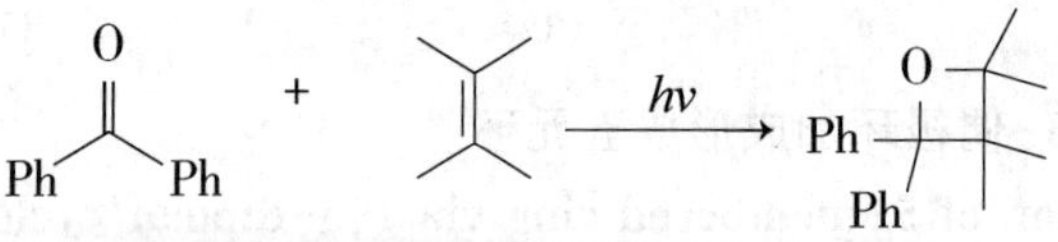

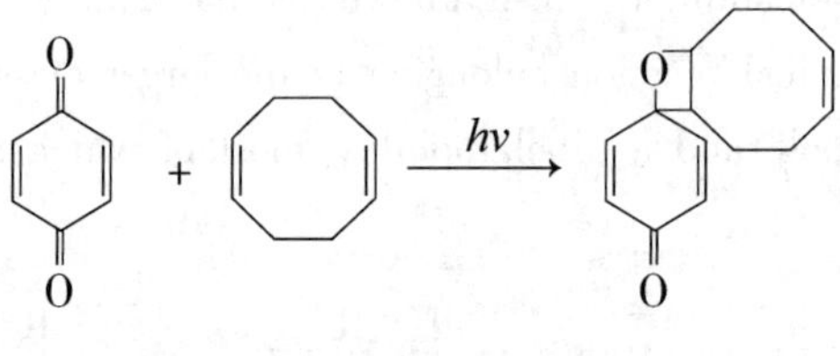

O Ph Ph + hv O Ph Ph

O O + hv O O

Addition of aromatic compounds to an electron-deficient olefin provides a 4-membered ring.

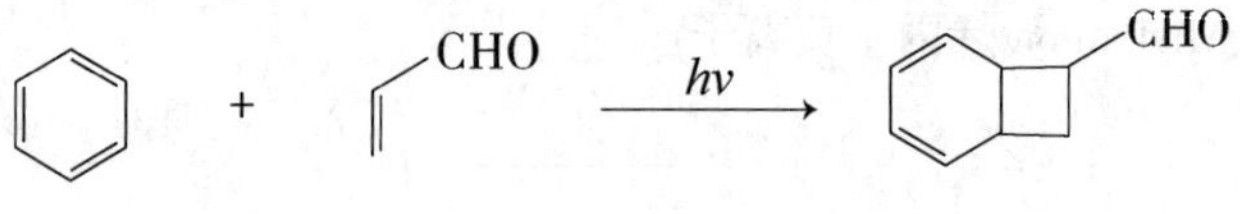

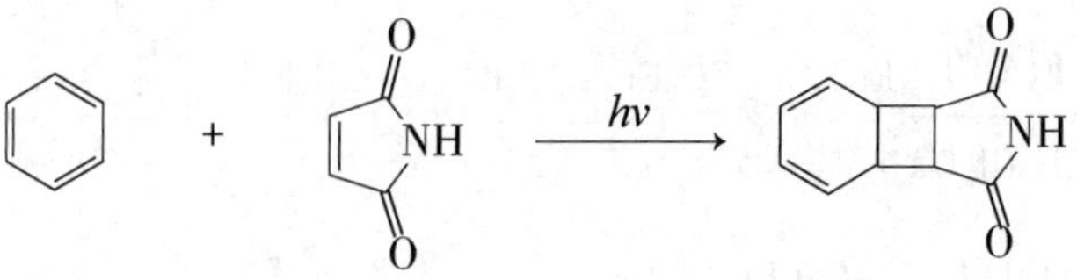

+ CHO hv CHO

+ O NH O hv O NH O

5.2 杂环的形成 Formation of Heterocycles

5.2.1 五元杂环 5-membered Heterocycles

Pyrrole itself may be extracted from coal char and bone oil by distillation, so that its syntheses are relatively unimportant. Substituted pyrroles are usually made by combining two aliphatic fragments in one two ways[12].

5.2.1.1 1,4-二羰基化合物 1,4-dicarbonyl compound (Paal-Knorr reaction)

A 1,4-dicarbonyl compound is treated with ammonia or a primary amine. Successive reactions of the nucleophilic nitrogen at the carbonyl groups are followed by dehydration, which occurs readily because the product is aromatic:

H H + $\ddot{Z}H_2$ ⇌ H H $^-$O ZHO ⇌ H H HO OH Z $\xrightarrow{-2H_2O}$ H H Z

$\xrightarrow[110\sim115℃ NH_3]{(NH_4)_2CO_3}$ N H

$\xrightarrow[ZnCl_2,\ H^+]{(CH_3CO)_2O}$ O

1,4-dicarbonyl compounds react with phosphorus *tri*- or *penta*-sulfide, the former usually being more satisfactory.

$\xrightarrow{P_2S_5}$ S

5.2.1.2 Condensation of enolates to α-amino aldehyde or ketone (Knorr reaction)

This is the most important route to pyrroles. The principle was outlined above: an α-amino-ketone is reacted with a ketone containing an activated α-methylene group in acetic acid solution.

R^3 = EWG

While R^3 is a group such as ester that promotes rapid acid-catalyzed enolization. The α-amino-ketone is usually prepared from a β-keto-ester and an alkyl nitrile, followed by reduction, usually in situ, with sodium dithionite or zinc and acetic acid.

5.2.1.3 Condensation of enolate to α-halo aldehyde or ketone in the presence of amine (Hantzsch reaction)

α, β-keto-ester is treated with an α-chloro-ketone in the presence of ammonia:

5.2.1.4 Hinsberg's procedure

An α-dicarbonyl compound reacted with a thioether in which the sulfur atom is adjacent to two activated methylene groups, e. g.

5.2.1.5 1,3-偶极环加成反应 1,3-Dipolar cycloaddition reactions

Dipolar cycloaddition reactions are useful both for syntheses of heterocyclic compounds and for carbon-carbon bond formation. 1,3-dipoleshave π electron systems that are isoelectronic with allyl or propargyl anions, consisting of two filled and one empty orbital. Each molecule has at least one charge-separated resonance structure with opposite charges in a 1,3-relationship, and it is this structural feature that leads to the name 1,3-dipolar cycloadditions for this class of reactions.

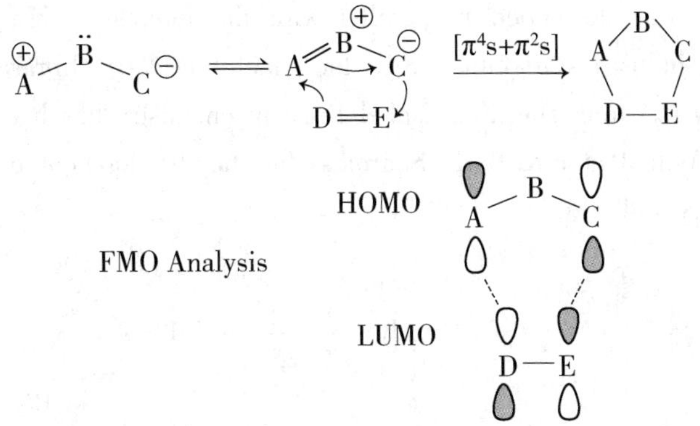

FMO Analysis

HOMO

LUMO

The Huisgen cycloaddition is the reaction of a dipolarophile with a 1,3-dipolar compound that leads to 5-membered (hetero) cycles. Examples of dipolarophiles are alkenes and alkynes and molecules that possess related heteroatom functional groups (such as carbonyls and nitriles). 1,3-Dipolar compounds contain one or more heteroatoms and can be described as having at least one mesomeric structure that represents a charged dipole.

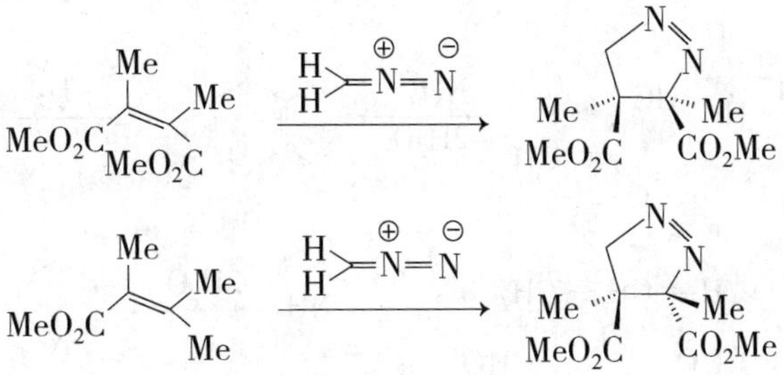

Huisgen azide-alkyne cycloaddition (Huisgen 叠氮炔烃的环加成反应)

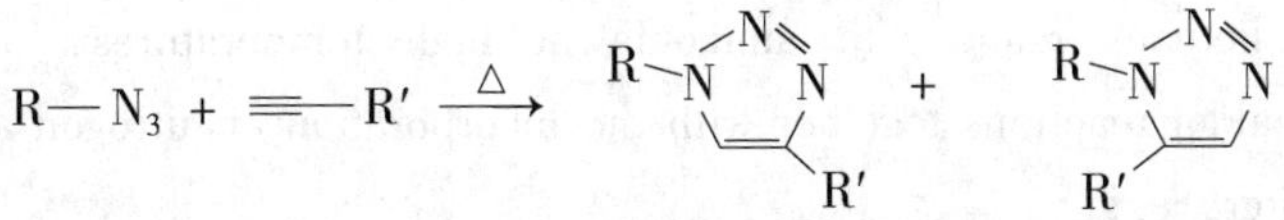

Unfortunately, the thermal Huisgen 1,3-dipolar cycloaddition of alkynes to azides requires elevated temperatures and often produces mixtures of the two regioisomers when using asymmetric alkynes. In this respect, the classic 1,3-dipolar cycloaddition fails as a true click reaction. A copper-catalyzed variant that follows a different mechanism can be conducted under aqueous conditions, even at room temperature. Additionally, whereas the classic Huisgen 1,3-dipolar cycloaddition often gives mixtures of regioisomers, the copper-catalyzed reaction allows the synthesis of the 1,4-disubstituted regioisomers specifically. By contrast, a later developed ruthenium-catalyzed reaction gives the opposite regioselectivity with the formation of 1,5-disubstituted triazoles. Thus, these catalyzed reactions comply fully with the definition of click chemistry and have put a focus on azide-alkyne cycloaddition as a prototype **Click Reaction**. "Click Chemistry" is a term that was introduced by K. B. Sharpless in 2001 to describe reactions that are high yielding, wide in scope, create only byproducts that can be removed without chromatography, are stereospecific, simple to perform, and can be conducted in easily removable or benign solvents. This concept was developed in parallel with the interest within the pharmaceutical, materials, and other industries in capabilities for generating large libraries of compounds for screening in discovery research. The 2022 Nobel Prize in chemistry has been awarded to Carolyn R. Bertozzi, Morten Meldal and K. Barry Sharpless for the development of click chemistry and bioorthogonal chemistry.

$$R-N=\overset{\oplus}{N}=\overset{\ominus}{N} + HC\equiv C-R' \xrightarrow[0\text{–}25℃]{Cu^{I}(cat.)} \text{1-R-4-R'-1,2,3-triazole}$$

5. 2. 2 六元杂环 6-membered Heterocycles

5. 2. 2. 1 1,5-dicarbonyl compound

$$H_3C-CO-CH(CH_3)-CH(CH_3)-CH(CH_3)-CO-CH_3 \xrightarrow[-2H_2O]{NH_3} \text{dihydropyridine (N-H)} \xrightarrow{-H_2} \text{pyridine}$$

$$\underset{CHO}{CH}=CH-\underset{CHO}{CH_2} + NH_3 \xrightarrow{-2H_2O} \text{pyridine}$$

5. 2. 2. 2 From the aldehydes or ketones and ammonia

Aldehydes or ketones react with ammonia at high temperatures, under pressure, by ammonia-catalyzed aldol reactions together with the incorporation of nitrogen atom of ammonia by Michael-type addition, e. g.

$$4CH_3CHO + NH_3 \longrightarrow \text{5-}C_2H_5\text{-2-}H_3C\text{-pyridine}$$

$$H_3C-CHO + H_3C-CHO + NH_3 \longrightarrow \left[CH_3CH=CHCHO \xrightarrow{NH_3} CH_3C(NH_2)=CHCHO \right] \xrightarrow{CH_3CHO} \text{5-}C_2H_5\text{-2-}H_3C\text{-piperidine (N-H)}$$

$$\longrightarrow \text{dihydropyridine (}C_2H_5\text{, }H_3C\text{)} \xrightarrow{CH_3CHO} \text{(=CHCH}_3\text{, }H_3C\text{)} \longrightarrow \text{5-}C_2H_5\text{-2-}H_3C\text{-pyridine} \quad 53\%$$

The products are usually mixtures which require separation by chromatography or distillation.

5.2.2.3 The Hantzsch synthesis

Two molecules of β-Ketoester and one of aldehyde react in the presence of ammonia to give a dihydropyridine which is then hydrogenated, usually with nitric acid. The aldehyde reacts with one molecule of the β-Ketoester, under the influende of ammonia or an added base, and ammonia itself reacts with the second molecule of the β-Ketoester the two resulting units are then joined by a Michael-type addition followed by ring closure.[13]

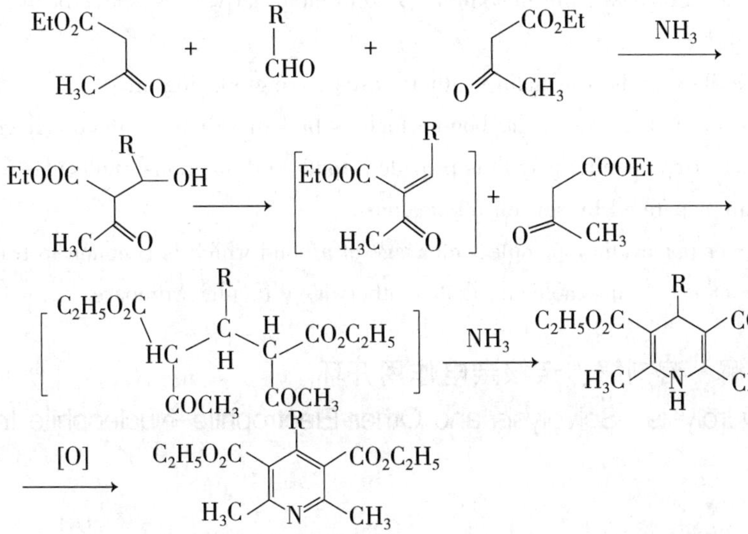

EtO_2C H_3C O + R CHO + CO_2Et O CH_3 $\xrightarrow{NH_3}$

R EtOOC OH H_3C O ⟶ [EtOOC R H_3C O] + COOEt O CH_3 ⟶

[$C_2H_5O_2C$ R C H HC H C $CO_2C_2H_5$ $COCH_3$ $COCH_3$] $\xrightarrow{NH_3}$ R $C_2H_5O_2C$ $CO_2C_2H_5$ H_3C N H CH_3

$\xrightarrow{[O]}$ R $C_2H_5O_2C$ $CO_2C_2H_5$ H_3C N CH_3

For example:

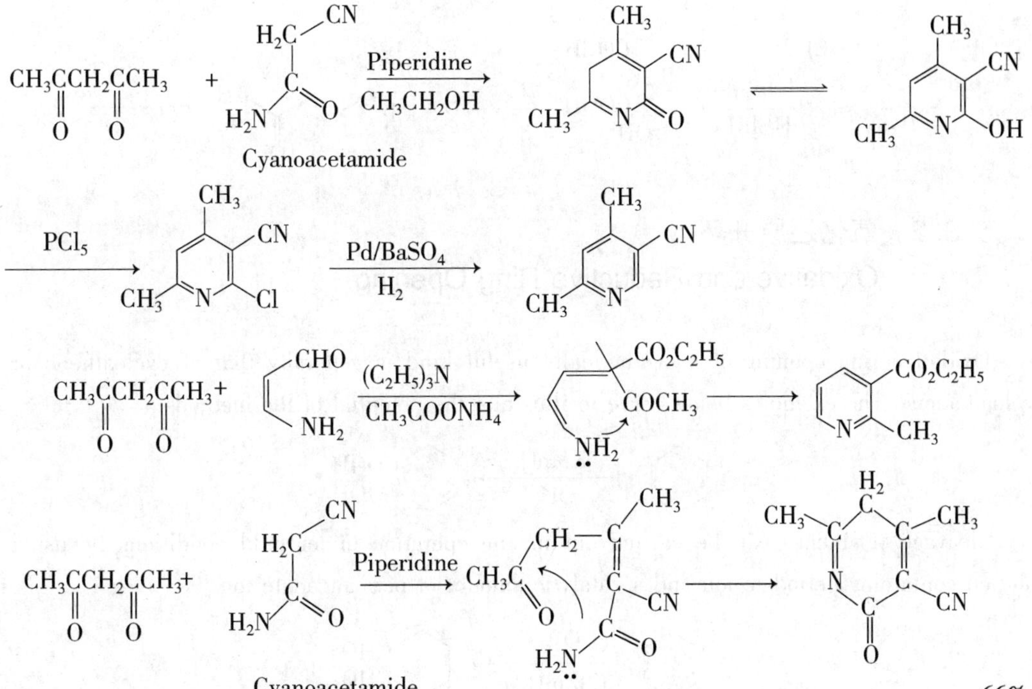

$CH_3CCH_2CCH_3$ (O O) + H_2C CN C H_2N O (Cyanoacetamide) $\xrightarrow[CH_3CH_2OH]{Piperidine}$ CH_3 CN CH_3 N O ⇌ CH_3 CN CH_3 N OH

$\xrightarrow{PCl_5}$ CH_3 CN CH_3 N Cl $\xrightarrow[H_2]{Pd/BaSO_4}$ CH_3 CN CH_3 N

$CH_3CCH_2CCH_3$ (O O) + CHO NH_2 $\xrightarrow[CH_3COONH_4]{(C_2H_5)_3N}$ $CO_2C_2H_5$ $COCH_3$ NH_2 ⟶ $CO_2C_2H_5$ N CH_3

$CH_3CCH_2CCH_3$ (O O) + H_2C CN C H_2N O (Cyanoacetamide) $\xrightarrow{Piperidine}$ CH_3C O CH_2 C CH_3 C CN C O H_2N ⟶ CH_3 C H_2 C CH_3 C N C C CN O

66%

5.3 开环反应 Ring Opening

The value of ring opening as a synthetic procedure is not as obvious as that of ring closure: indeed, we have discussed synthesis so far only in terms of bond formation and examples of bond cleavage (e. g. the decarboxylation of malonic or β-keto-acid derivatives) have been incidental to the main theme[14].

Apart from the above, the two main synthetic uses of ring opening are:

(ⅰ) the atoms at either end of the bond which is broken will bear functional group in the ring-opened product; ring opening may thus provide a route to difunctional molecules in which the functional groups are separated by several other atoms.

(ⅱ) in a bi- or polycyclic molecule, cleavage of a bond which is common to two rings may lead to a medium- or large-ring molecule that is otherwise difficult to prepare.

5.3.1 水解、溶剂解、亲核亲电作用开环 Hydrolysis, Solvolysis and Other Electrophile-Nucleophile Interaction

(1)KOH,CH_3OH; (2)CH_3I → (1)$Ba(OH)_2$; (2)H^+ → COOH, CH_3, O

3

$COCH_3$ → heat, HCl,H_2O → $COCH_3$, COOH, OH → Cl, CH_3, O 75%

5.3.2 氧化还原开环 Oxidative and Reductive Ring Opening

Oxidative ring opening of a synthetically useful kind is generally that of cycloalkene or a cycloalkanone. The examples below serve to illustrate the potential of the methods.

$KMnO_4$, OH^- → COOH, COOH

Cleavage of alkenes can be carried out in one operation under mild conditions by using a solution containing periodate ion and a catalytic amount of permanganate ion.

(1)O_3; (2)Zn,HAc → CHO, CHO

Reductive ring opening is of less general value, although hydrogenolysis of some sulfur-containing compounds provides a notable exception, e. g.

$$(H_3C)_3C\text{-thiophene-}COOH \xrightarrow{H_2,Ni} (H_3C)_3C\text{-tetrahydrothiophene-}COOH \longrightarrow (H_3C)_3C\text{—}(CH)_4\text{—}COOH \quad 70\%$$

$$\xrightarrow[Et_2O]{LiAlH_4} \quad 99\%$$

5.3.3 热解开环 Pyrolysis

Pericyclic ring was opened by pyrolysis. Many of these involve the cleavage of a bicyclic Diels Alder adduct which has itself been formed from a cyclic diene and a dienophile. The cleavage reaction becomes effectively irreversible if one of the cleavage products is volatile[15].

Electrocyclic ring opening is another important pericyclic reaction and is the exact opposite of the electrocyclic ring closure. The thermal ring opening of a cyclohexadiene is thus disrotatory and that of a cyclobutene is conrotatory.

5.3.4 重排反应 Rearrangement

The rearrangement is generally stereospecific, although the configuration of the product is not always predictable, depending as it does on the conformation in the transition state.

An acid-induced rearrangement of oximes to give amides via **Beckmann Rearrangement,** which is the important immediate for synthesis of nylon. In the reaction process, an electropositive nitrogen is formed that initiates an alkyl migration[16]. Oximes generally have a high barrier to inversion, and accordingly this reaction is envisioned to proceed by protonation of the oxime hydroxyl, followed by migration of the alkyl substituent "*trans*" to nitrogen. The N—O bond is simultaneously cleaved with the expulsion of water, so that formation of a free nitrene is avoided.

$$(CH_2)_n\begin{matrix}CH_2\\ \\ CH_2\end{matrix}C=N-OH \longrightarrow (CH_2)_n\begin{matrix}CH_2-C=O\\ |\\ NH\\ CH_2\end{matrix}$$

$$\text{cyclohexanone oxime} \xrightarrow{H_2SO_4} \text{caprolactam (NH)} \xrightarrow{-H_2O} \left[NH(CH_2)\overset{O}{\overset{\|}{C}} \right]_n \longrightarrow \text{nylon}$$

$$\begin{matrix}R\\ \\ R'\end{matrix}C=N-OH \xrightarrow{H^+} \begin{matrix}R\\ \\ R'\end{matrix}C=N-\overset{+}{O}H_2 \xrightarrow{-H_2O} R-C^+=N-R' \xrightarrow[-H^+]{H_2O} \begin{matrix}HO\\ \\ R'\end{matrix}C=N\begin{matrix}R'\\ \\ \end{matrix} \rightleftharpoons \begin{matrix}O\\ \\ R'\end{matrix}C-N\begin{matrix}R'\\ \\ H\end{matrix}$$

5. 3. 5 三元环、四元环裂解 Cleavage Due to Tension of 3- or 4-Membered Ring

$$\triangle + H_2 \xrightarrow[80℃]{Ni} \text{propane} \qquad \square + H_2 \xrightarrow[200℃]{Ni} \text{butane}$$

$$\text{cyclopentane} + H_2 \xrightarrow[>300℃]{Pt} \text{pentane}$$

Cyclopropane is prone to break because the ring tension of small ring is larger than thatof five-membered ringor six-membered ring.

参考文献
References

1. Diederich F. *J. Chem. Educ.* 1990, **67** (10): 813.
2. Hilvert D. *J. Pure Appl. Chem.* 1992, 64 (8): 1103.
3. Hanson J R. Comprehensive Organic Synthesis. Oxford: Pergamon, 1991. 705-719.
4. Empel C, et, al. Unlocking novel reaction pathways of diazoalkanes with visible light. *J. Chemical Communications*. 2022, 58(17): 2788-2798.
5. Anaya J, Sánchez R M. Progress in Heterocyclic Chemistry. Amsterdam: Elsevier, 2021. 53-91.
6. Santos C M M, Silva A M S. Progress in Heterocyclic Chemistry. Amsterdam: Elsevier, 2021. 501-563.
7. Rebiere F, et, al. Asymmetric Diels - Alder reaction catalysed by some chiral Lewis acids. *J. Tetrahedron: Asymmetry*. 1990, 1(3): 199-214.
8. Song X, Wang W. Comprehensive Organic Synthesis (Second Edition). Amsterdam: Elsevier, 2014. 86-118.
9. Purich D L. Enzyme Kinetics: Catalysis & Control. Boston: Elsevier, 2010. 685-728.
10. Wang D, et al. *J. Journal of Physical Organic Chemistry*. 2017, 30(1): 62-66.
11. Mandal D K. Stereochemistry and Organic Reactions. New York: Academic Press, 2021. 529-558.
12. Lalezari I. Comprehensive Heterocyclic Chemistry. Oxford: Pergamon, 1984. 333-363.
13. Zheng C, You S L. Comprehensive Chirality. Amsterdam: Elsevier, 2012. 586-607.
14. Thirumalaikumar M. Ring Opening Reactions of Epoxides. *J. Organic Preparations and Procedures International*. 2022, 54(1): 1-39.
15. McNab H. Comprehensive Organic Functional Group Transformations. Oxford: Elsevier Science, 1995. 771-791.
16. Chandrasekhar S. Comprehensive Organic Synthesis (Second Edition). Amsterdam: Elsevier, 2014. 770-800.

第二部分 分子骨架修饰

Part Ⅱ Molecular Skeleton Modification

6 官能团互换
Chapter 6 Functional Group Interconversion

Functional group interconversions (FGIs) may be regarded as largely tactical in nature, and in a sense are the ' glue' which holds the strategy together. For example, a functional group arising from a strategy level transformation involving C—C bond formation may have to be converted into another functional group before the next strategy-level, C—C bond-forming transformation is possible. FGIs leave the carbon skeleton unchanged (almost always), and usually involve exchange of heteroatoms (i. e. not C or H).

In most syntheses, the main focus is usually on construction of the molecule using carbon-carbon bond forming reactions. Most of the actual chemical reactions in a synthesis, however, are those that incorporate or change functional groups. Such reactions are known as functional group exchange reactions, and this chapter will review major reaction types involved in functional group exchanges.

6.1 官能团的导入与互换 Introduction and Transformation of Functional Groups

Important aspects of synthesis are the introduction of functional groups into a molecule and the interconversion of functional groups. We shall show that in some instances it is relatively easy to functionalize certain positions whereas in other situations functionalization is impossible and the desired product can only be obtained by a series of interconversions of functional groups.

In this chapter, we shall attempt to bring together, in outline only, a variety of reactions which successful synthetic chemists will require to have at their command. Further details on the reactions mentioned in this chapter will be found in standard works on organic chemistry and Sykes describes the mechanisms of many of the reactions[1].

Functional groups are so called because they impart specific types of reactivity to organic molecules. In general, the characteristic reactions functional groups are observed, irrespective of the precise molecular environment in which the functional group is situated. It should be obvious that the synthesis of a complicated molecule containing several functional groups depends on the chemoselectivity of the individual reaction steps. Reagents must be chosen which react only at the desired functional group or groups, and if necessary other functionality in the molecule must first be protected in order to prevent unwanted side reactions. The use of such protective groups in describes later in this chapter.

6.1.1 烷烃的官能化 Functionalization of Alkanes

The unreactivity of alkanes towards electrophilic and nucleophilic reagents will be familiar to the reader. Alkanes are, however, reactive in radical reactions, particularly halogenation. Such reactions are nevertheless of limited synthetic use because of the difficulties encountered in attempts to control them.

Because of the higher reactivity of Cl · than Br ·, chlorination tends to be less selective than bromination and, indeed, 2-bromo-2-methypropane is almost exclusively formed when 2-methylpropane reacts with bromine at 300℃ whereas chlorination results in a 2 : 1 mixture of 1-chloro- and 2-chloro-2-methylpropane. Bromides are essential functional groups, easy to form, easy to remove as a leaving group[2,3].

Adamantane $\xrightarrow[\text{r.t.}\sim 105℃]{Br_2/h\nu}$ 1-bromoadamantane (Br) 85%

$$2\ \text{cyclohexane} + Br_2 + HgO \xrightarrow[\text{reflux}]{CCl_4} 2\ \text{bromocyclohexane} + HgBr_2 + H_2O\ 80\%$$

Br-octahydronaphthalene 80% $\xleftarrow{\text{NBS(1mol)}}$ octahydronaphthalene $\xrightarrow{\text{NBS(2mol)}}$ tetrahydronaphthalene 50%

CH_2Br-naphthalene $\xleftarrow{\text{NBS(1mol)}}$ CH_3-naphthalene $\xrightarrow{\text{NBS(2mol)}}$ $CHBr_2$-naphthalene

$$-\underset{H}{\underset{|}{\overset{|}{C}}}-\overset{O}{\overset{\|}{C}}- \xrightarrow[\text{acid}]{X_2} -\underset{Br}{\underset{|}{\overset{|}{C}}}-\overset{O}{\overset{\|}{C}}-$$

On the other hand, rearrangements are encountered in the intermediate radicals with less frequency than in the corresponding carbocation. Thus, only 1-chloro-2,2-dimethylpropane results when 2,2-dimethylpropane is chlorinated:

$$H_3C-\underset{CH_3}{\underset{|}{\overset{CH_3}{\overset{|}{C}}}}-CH_3 + Cl_2 \xrightarrow{h\nu} H_3C-\underset{CH_3}{\underset{|}{\overset{CH_3}{\overset{|}{C}}}}-CH_2Cl$$

6.1.2 烯烃的官能化 Functionalization of Alkenes

Unlike alkanes, alkenes contain two sites at which functionalization can be carried out with a high degree of regio- and stereospecificity. These are (a) at the C═C double bond and (b) at the carbon adjacent to the double bond——the allylic position[4].

The chemistry of alkenes is largely concerned with reactions of electrophiles with the double bond. The mechanism of these reactions and the stereochemistry will not be discussed in detail in this chapter. It is, however, necessary to recall that addition of electrophiles to unsymmetrical alkenes proceeds through the more stable carbocation, resulting in the product in which the more positive moiety of reagents has become attached to the less substituted alkene carbon. Scheme 6-1 summarizes addition reactions involving propene. Strong acids, e. g. HCl, HBr, HI, H_2SO_4 and CF_3COOH, add to alkenes directly, but weaker acids, e. g. CH_3CO_2H and H_2O, require catalysis by a stronger acid (e. g. H_2SO_4). An alternative to the last of these, viz. acid-catalyzed addition of water, is provided by oxomercuration, using mercury (II) acetate (Hg^{II} being a Lewis acid and thus an electrophile) followed by reaction with sodium borohydride and hydrolysis. This method avoids the use of a strong protic acid[5].

In all of these cases, and also for the hypohalous acids, Markovnikov addition is observed. The more positive (electrophilic) end of the dipolar molecule become attached to the less substituted carbon.

$$CH_3CH{=}CH_2 + V{-}W \longrightarrow CH_3\overset{+}{C}H{-}CH_2W$$

$$\downarrow$$

$$CH_3CHVCH_2W$$

$$CH_3CH(OH)CH_2Br \xleftarrow{HOBr} CH_3CH{=}CH_2 \xrightarrow{ICl} CH_3CHClCH_2I$$

In the case of addition of water through hydroboration, although the anti - Markovnikov product is eventually formed, the addition step itself (of a borane) actually follows Markovnikov's rule.

In the case of addition of HBr, however, Markovnikov's addition is observed only if the alkene is rigorously purified so that peroxide impurities are excluded. Otherwise, anti - Markovnikov's addition occurs. This is because, in the presence of peroxide, a radical mechanism is followed; the attacking radical (Br·) becomes attached to the less substituted carbon (less hindered; conditions of kinetic control). The fact that the more stable of the possible radical intermediates is usually produced (secondary usually more stable the primary) is merely a bonus.

The intermediate in reactions involving halogens and hypohalous acids is a halonium ion(I), reaction of which with a nucleophile leads to a trans addition product. In the case of addition of hypohalous acid, the trans halogeno-alcohol formed can be converted, by treatment with base, into an oxirane (epoxide):

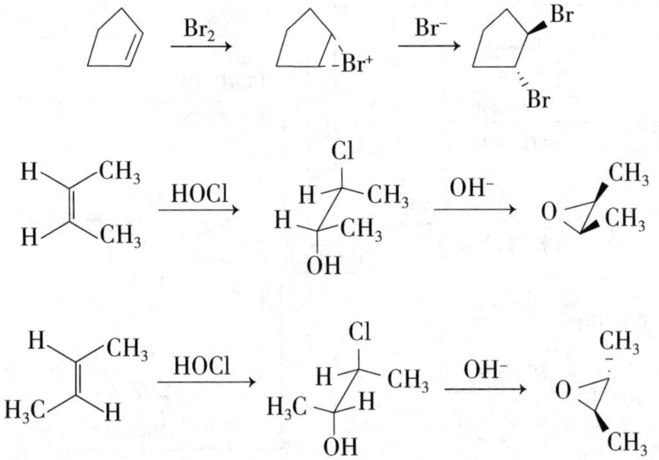

Br_2 → Br^+ → Br^- → Br, Br

$$H(CH_3)C=C(CH_3)H \xrightarrow{HOCl} \text{H, Cl, } CH_3 / \text{H, OH, } CH_3 \xrightarrow{OH^-} \text{O, } CH_3, CH_3$$

$$H(CH_3)C=C(H)H_3C \xrightarrow{HOCl} \text{H, Cl, } CH_3 / H_3C\text{, OH, H} \xrightarrow{OH^-} \text{O, } CH_3, CH_3$$

An alternative means by which alkenes may be functionalized is reaction at the allylic position. Carbon-halogen bonds adjacent to the carbon-carbon double bond, the allylic hydrogens, are susceptible to oxidation and to halogenation. Although the majority of these halogenation reactions are free-radical process, ionic reactions can also take place.

The most commonly used reagent for bromination is *N*-bromosuccinimide (NBS) and, since the related involves an intermediate allyl radical, a mixture of bromides can be expected:

$$RCH_2CHCH_2 \xrightarrow[(PhCO_2)_2]{NBS} RCH{=}CH\dot{C}H_2 \longleftrightarrow R\dot{C}HCH{=}CH_2$$

$$\downarrow NBS$$

$$RCH{=}CHCH_2Br + RCHBrCH{=}CH_2$$

(Z-and E-isomers)

$$C_6H_5CH{=}CHCH_3 \xrightarrow{NBS} C_6H_5CH{=}CHCH_2Br$$

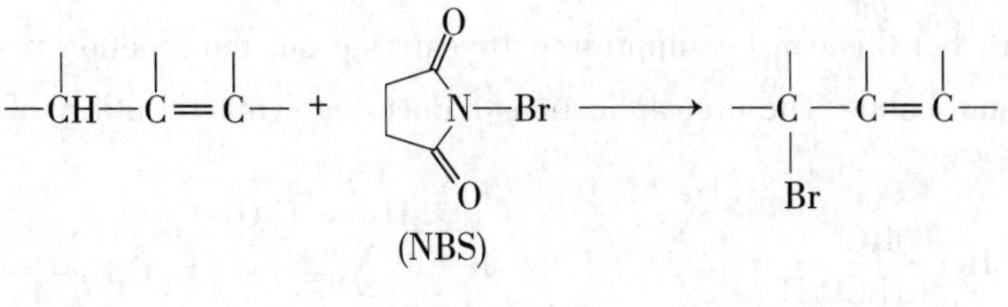

$$-CH-C{=}C- + \text{N-Br (NBS)} \longrightarrow -\underset{Br}{C}-C{=}C-$$

(NBS)

However, in simple case such as cyclohexane, a good yield of the bromoalkene is obtained. The introduction of oxygenated functional groups at allylic positions will be discussed later.

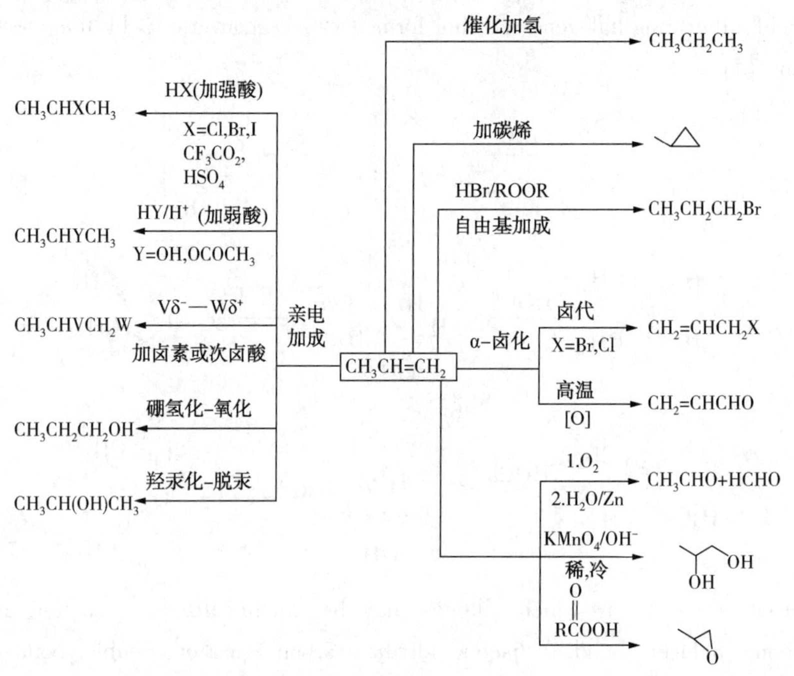

Scheme 6-1 Reactions of Alkenes

6.1.3 炔烃的官能化 Functionalization of Alkynes

Most of the chemistry of alkynes is concerned with their reactivity towards electrophiles. As in the case of alkenes, reactions with halogens, hydrogen halides and acids are synthetically useful. Hydrogenation of alkynes is also of considerable significance. In addition, an alklyne is a weak acid and the anion derived from it is of importance in carbon-carbon bond-forming reactions[6].

Reaction of bromine with an alkyne is a trans addition and addition of lithium bromide to the reaction mixture increases the yield of the product. Reaction with hydrogen halide is of greater complexity often following a cis stereochemistry. However, when the triple bond is not conjugated with an aromatic ring, the trans addition predominates. Also, the addition of solvent may be competing reaction, but this can be suppressed by carrying out the reaction in the presence of a quaternary ammonium halide. These complications reduce the synthetic utility of the reaction[7]:

$$C_2H_5C\equiv CC_2H_5 \xrightarrow[CH_3CO_2H]{HCl} \underset{40\%\sim72\%}{(C_2H_5)(H)C{=}C(Cl)(C_2H_5)} + \underset{<1\%}{(C_2H_5)(C_2H_5)C{=}C(Cl)(H)} + \left[C_2H_5{-}\underset{H}{C}{=}C(OCOCH_3)(C_2H_5)\right]$$

$$\xrightarrow{H_2O} \underset{28\%\sim60\%}{C_3H_7COC_2H_5}$$

$$C_2H_5C\equiv CC_2H_5 \xrightarrow[CH_3CO_2H]{HCl,\ (CH_3)_4N^+Cl^-} \underset{91\%\sim97\%}{(C_2H_5)(Cl)C{=}C(H)(C_2H_5)} + \underset{<0.2\%}{(C_2H_5)(Cl)C{=}C(C_2H_5)(H)} + \underset{3\%\sim9\%}{C_3H_7COC_2H_5}$$

$$PhC\equiv CCH_3 \xrightarrow[CH_3CO_2H]{HCl} \underset{67\%\sim78\%}{(Ph)(Cl)C{=}C(CH_3)(H)} + \underset{11\%\sim19\%}{(Ph)(Cl)C{=}C(H)(CH_3)} + \underset{10\%\sim13\%}{PhCOC_2H_5}$$

Addition of water and of carboxylic acids to alkynes is catalyzed by mercuric oxide. In the former case the product from a terminal alkyne is a methyl ketone and in the later it is an enol ester:

$$\text{1-ethynylcyclohexanol} \xrightarrow[HgO]{H_2O,H^+} \text{1-(1-hydroxyvinyl)cyclohexanol} \longrightarrow \text{1-acetylcyclohexanol}\ \ 91\%$$

The commonly used synthetic procedures are shown in Scheme 6-2.

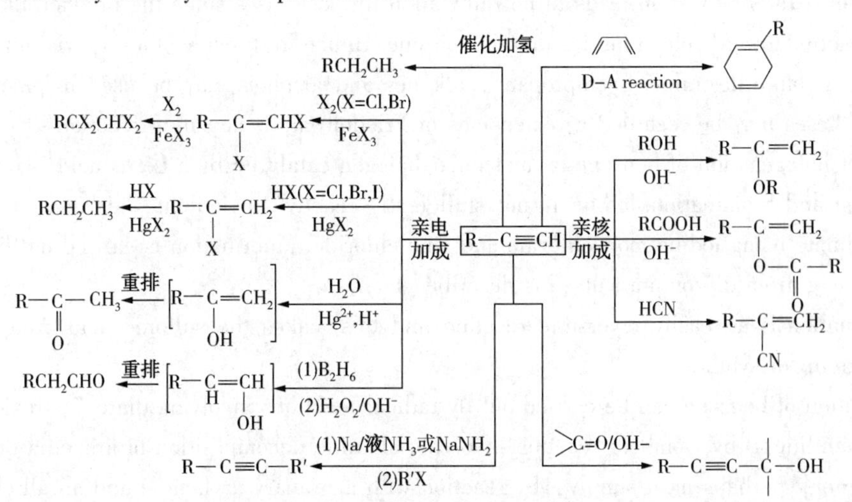

Scheme 6-2 Reactions of Alkynes

6.1.4 芳烃的官能化 Functionalization of Aromatic Compounds

6.1.4.1 环上的取代反应 **Substitution at a ring position**

The Friedel-Crafts reaction is a very important method for introducing alkyl substituents on an aromatic ring by generation of a carbocation or related electrophilic species. The characteristic reaction of benzene is an electrophilic addition-elimination reaction, the overall effect of which is substitution. This is the most widely used procedure for introduction of functional groups on to the benzene ring. Scheme 6-3 outlines some of the more important reactions[8].

Scheme 6-3 Reactions of Benzene

The usual method of generating these electrophiles involves Friedel-Crafts alkylation between an alkyl halide and a Lewis acid. The most common Friedel-Crafts catalyst for preparative work is $AlCl_3$, but other Lewis acids such as SbF_5, $TiCl_4$, $SnCl_4$, and BF_3 can also promote reaction. Alternative routes to alkylating species include reaction of alcohols or alkenes with strong acids.

Friedel-Crafts alkylation leads to polyalkylation in most cases since the product alkylbenzene is more reactive towards electrophiles than is benzene. Hence in direct synthesis, via acylation and reduction, is often desirable. Cyclopropane, alkenes and alcohols may be used in place of alkyl halide syntheses may be regarded as extensions of Friedel-Crafts acylation reaction[9].

Direct halogenation of benzene by molecular halogen catalyzed by a Lewis acid is restricted to chlorination and bromination. Iodine is not sufficiently reactive to iodinate benzene, but toluene can be iodinate using iodine monochloride and zinc chloride. Fluorination is carried out by indirect methods, e. g. from diazonium salts, as described later.

Sulfonation is an easily reversible reaction and this makes the sulfonic acid group a useful blocking group in synthesis.

Acylation of benzene can be carried out by radical reactions involving diaroyl peroxides or *N*-nitrosoacetanilides, by Gomberg reaction involving alkaline decomposition of arenediazonium salts in benzene or, perhaps most simply, by reaction with a primary arylamine and an alkyl nitrite.

6.1.4.2 侧链的反应
Reaction in the side chain

Alkylbenzene can be functionalized either in the side chain or in the ring. The former will be discussed shortly. The side chain is susceptible to attack by radicals and also to oxidation at the position adjacent to the ring (the benzylic position) the oxidation of a methyl group involves three levels of oxidation: $-CH_2OH$, $-CHO$ and $-CO_2H$.

The benzylic position is also susceptible to autooxidation and the commercially valuable synthesis of phenol and acetone from cumene makes use of this.

Halogenation at benzylic positions normally proceeds by a free radical mechanism and, in the absence of other reactive functional groups, is normally carried out using molecular chlorine or bromide. Chlorination may also be carried out using *t*-butyl hypochlorite or sulfuryl chloride, and bromination using *N*-bromosuccinimide. In all case, the reaction is stepwise and the steps become slower with increase halogen substitution. It is, therefore, feasible to prepare benzyl chloride, α, α-dichlorotoluene and α, α, α-trichlorotoluene by varying he reaction conditions.

$$PhCH_3 \longrightarrow PhCH_2Cl \longrightarrow PhCHCl_2 \longrightarrow PhCCl_3$$

6.1.4.3 取代苯衍生物的官能化

Functionalization of substituted benzene derivatives

Substituted benzene derivatives undergo electrophilic and radical substitution reactions analogous to those described previously for benzene. Electrophilic substitution is generally highly regioselective, depending on the substituents already present in the ring. These substituents also affect the rate of substitution to such an extent that certain reactions (e.g. alkylation of nitrobenzene) cannot be carried out, and others not possible with benzene can take place (e.g. reactions of sodium phenoxide with diazonium salts)[10].

Two points are worth noting at this stage. Firstly, when more than one substitution is already on the benzene ring, the most strongly electron-donating group controls the position of further substitution. Secondly, in order to mimerize possible substitution at nitrogen, aromatic amines are usually converted into acetanilides before substitution is carried out. This also serves to reduce the reactivity of the ring towards electrophilic substitution. Below are some examples which may help the reader to understand the application of the rules:

$$C_6H_5OH \xrightarrow{\text{fuming } HNO_3} 2,4,6\text{-}(O_2N)_3C_6H_2OH$$

Mononitration takes place with dilute nitric acid, indicating that phenol is much more reactive than benzene. The hydroxyl group is *o*-/*p*-directing.

$$C_6H_5NH_2 \xrightarrow[H_2O]{Br_2} 2,4,6\text{-}Br_3C_6H_2NH_2$$

No Lewis acid catalyst is required and the reaction cannot be stopped at the mono- or the dibromo stage. The amino group, as a powerful (mesomeric) electron donor, is *o*-/*p*-directing and controls the orientation of addition rather than the weakly electron-donating bromine. Monobromination can be affected by way of acetanilide:

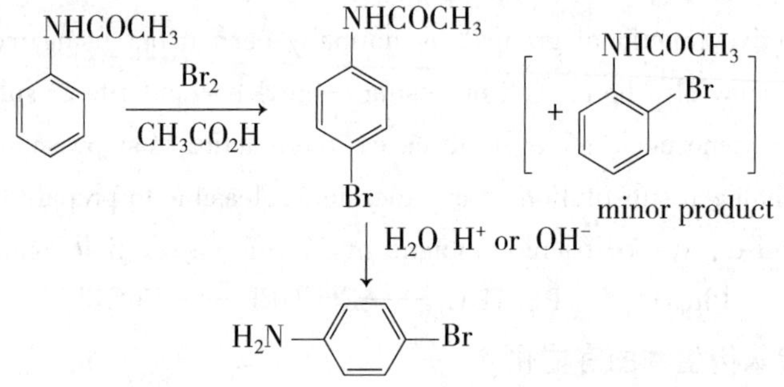

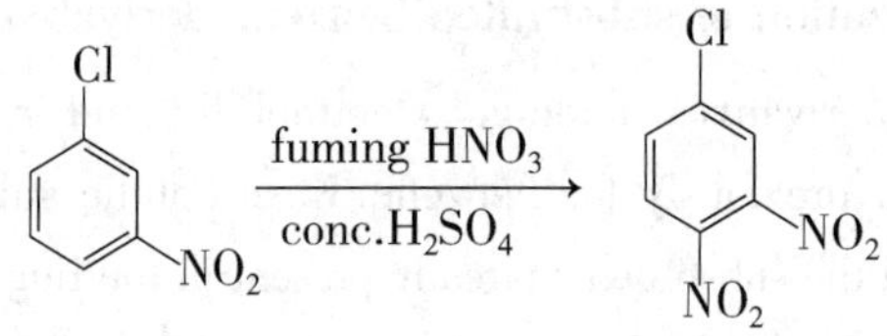

Much more vigorous conditions are required for this reaction since both substitutions retard nitration. The orientation is governed by the *o*-/*p*-directing chlorine.

The directional effects in radical substitution reactions are much less pronounced and it is normal to expect all three isomers from, for example, phenylation of a monosubstituted benzene:

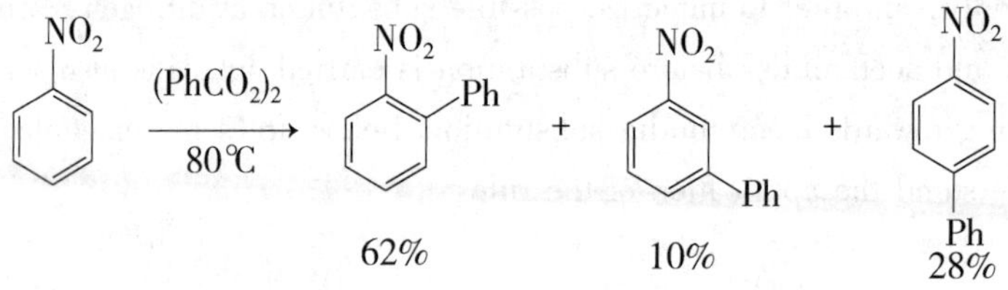

Nucleophilic substitution is accelerated by electron-withdrawing substituents, e. g. NO_2, however, a leaving group such as halogen is also required and the reaction is not considered at this point.

6.1.5 芳香杂环化合物的官能化 Functionalization of Heteroaromatic Compounds

In the space available, it is only possile to deal in ouline with some of the more important reactions of simple heterocyclic compounds.

Pyridine is a weak base which processes a considerable degree of aromatic character. It reacts, for example, with methyl iodide to form quaternary salts which on heating rearrange to 2-methylpyridies and 4-methylpyridies. Molecular orbital calculations indicate that C-3 is the carbon having the highest electron density but, even at this position, the electron density is much lower than that of benzene. Electrophilic substitution, therefore, requires forcing conditions and the reactions which may be useful are summarized in Scheme 6-4. Nitration of pyridine at C-3 may be achieved in good yield (c. 60%) by reaction with N_2O_5 in the presence of SO_2, although this does not follow a normal electrophilic substitution mechanism. Radical phenylation results in the formation of a mixture of all three monophenylpyridines. Nucleophilic substitution results in substitution mainly at the 2-position[11].

Pyridine-*N*-oxide, prepared most readily by treatment of pyridine with a peroxy acid such as peroxyacetic or peroxy benzoic acid, is a much weaker base than pyridine. It is readily nitrated in the 4-position and the directive effect of the *N*-oxide is such that 4-substitution takes place in most cases except those with a hydroxy or dimethyl amino group in the 2-position. If the 4-position is blocked, nitration usually fails. Direct halogenation and sulfonation do not proceed readily and, as with pyridine itself, the Friedel- Crafts reaction fails. Pyridine-*N*-oxide is converted into 2-acetoxypyrdine by reaction with acetic anhydride, and on heating with bromine and acetic anhydride-sodium acetate 3,5-dibromopyridine-*N*-oxide is formed. Chlorination at the 4-position with deoxygenation can be carried out using phosphorus pentachloride. Deoxygenation of *N*-oxides is readily carried out using, for example, phosphorus trichloride[12]. The reactions of pyrrodine are shown in Scheme 6-4.

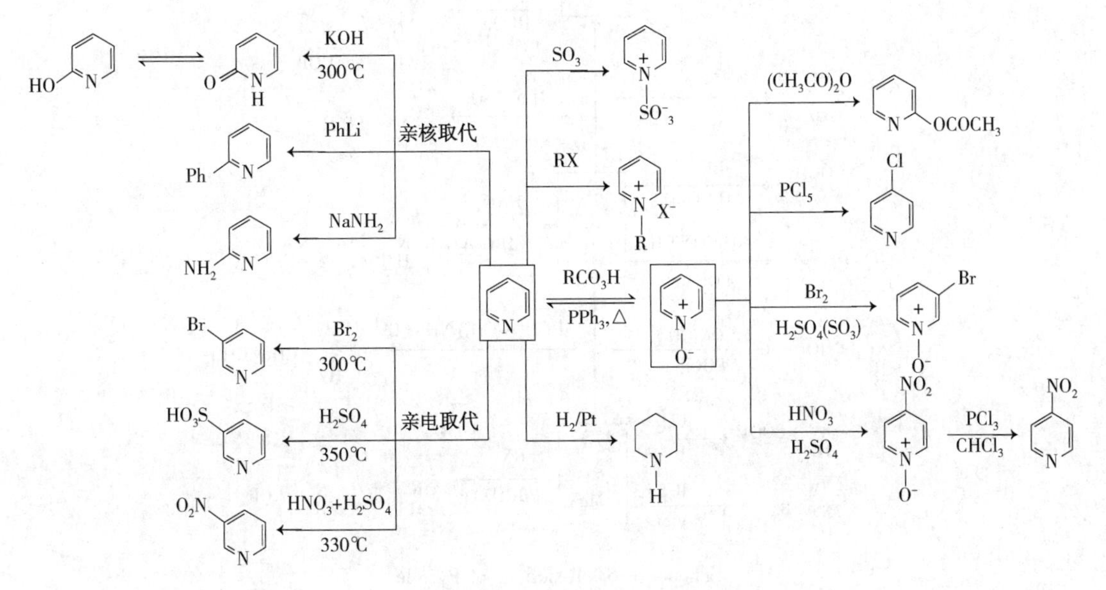

Scheme 6-4 Reactions of Pyrrodine

Furan, pyrrole and thiophene, in contrast to pyridine, are electron-rich molecules which react with electrophiles mainly in the 2-and *S*-positions. However, under acidic conditions furan, and to a lesser extent pyrrole, are polymerized. Direct halogenation of furan, pyrrole and thiophene usually results in the formation of polyhalogenated products. , Scheme 6-5, 6-6 and 6-7 summarize some useful synthetic reactions of these compounds.

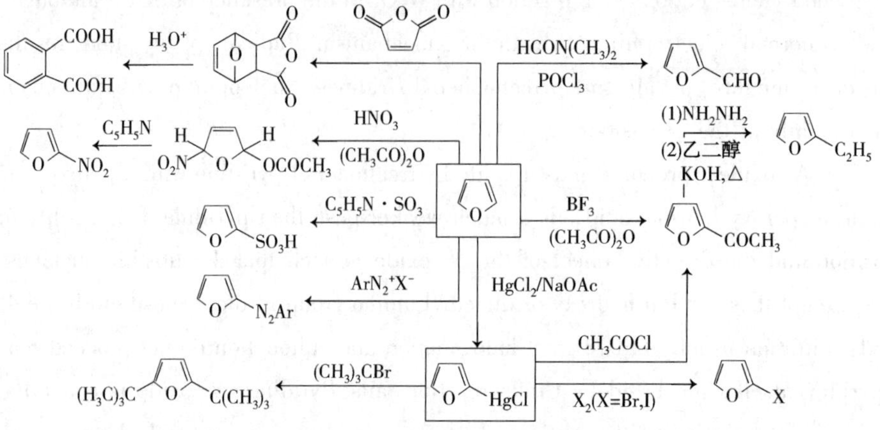

Scheme 6-5 Reactions of Furan

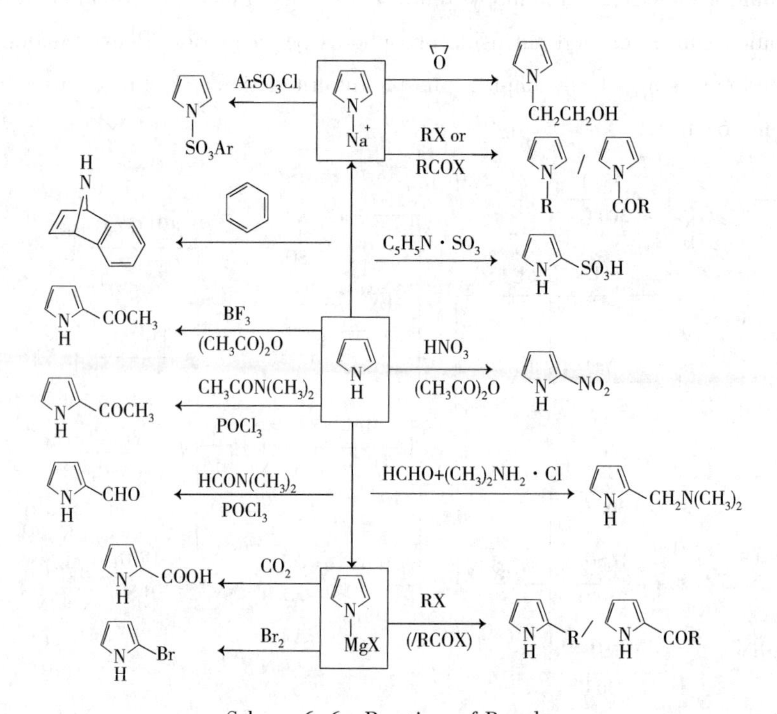

Scheme 6-6 Reactions of Pyrrole

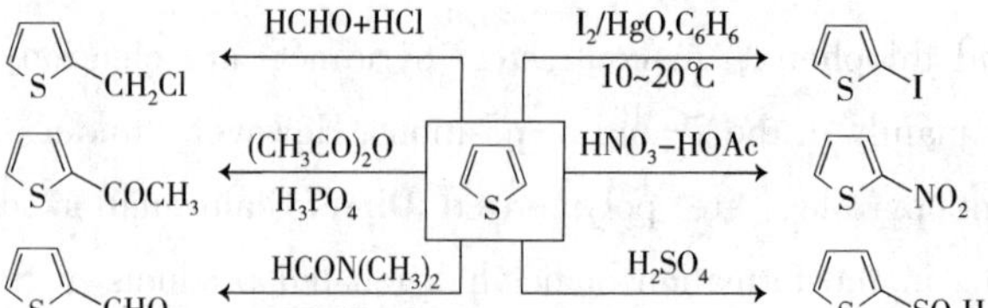

Scheme 6-7 Reactions of Thiophene

6.2 官能团互换 Interconversion of Functional Groups

As we know from organic chemistry, certain functional groups are readily introduced in a specific manner whilst others are not. It is required to interconvert functional groups in such a manner that the remainder of molecule remains unaffected. This section will attempt to show in outline only, how specific functional groups can be interconverted.

6.2.1 羟基的转化 Transformation of Hydroxyl Group

Alcohols are a very important compounds for synthesis. However, because the hydroxide ion is a very poor leaving group, alcohols are not reactive as alkylating agents. They can be activated to substitution by *O*−protonation, but the acidity that is required is incompatible with most nucleophiles except those, such as the halides, that are anions of strong acids. The preparation of sulfonate esters from alcohols is an effective way of installing a reactive leaving group on an alkyl chain. *p*−Toluenesulfonate (*tosylate*) and methanesulfonate (*mesylate*) esters are used most frequently for preparative work, but the very reactive trifluoromethanesulfonates (*triflates*) are useful when an especially good leaving group is required[13].

$$ROLi + ClSO_2-C_6H_4-CH_3 \longrightarrow ROSO_2-C_6H_4-CH_3$$

Alcohols are weak bases which are capable of reacting as nucleophiles. Reaction of alcohols with acid chlorides or anhydride results in the formation of esters. In most cases the reactions is promoted by the addition of tertiary base. The alkoxide ion is a stronger nucleophile, which can react with alkyl halides, sulfonates and sulfates to form ethers. However, elimination competes with substitution in reaction involving secondary and tertiary halides.

The alcohols are the most common precursors for alkyl halides and a variety of procedures have been developed for this transformation. Alkyl halides may be prepared from alcohols using reagents such as thionyl chloride for chlorides, constant boiling hydrobromic acid or phosphorous tribromide for bromides, and iodine with red phosphorous for iodides. Mild conditions must be employed for the preparation of tertiary halides to prevent elimination taking place, e. g. *t*−butanol shaken with concentrated hydrochloric acid gives *t*−butyl chloride. Some additional reagents that are useful when the more common reagents induce rearrangement, racemization or decomposition will be incorporated.

Dehydration of alcohols to alkenes can be carried out using a wide variety of Brönsted acids. With strong acids, acyclic alcohols appear to be dehydrated largely by an E1 mechanism, and the products derived are usually of the Saytzeff type perhaps with skeletal rearrangement of the intermediate carbocation. Some reagents, e. g. phosphorus oxychloride, are regarded as including dehydrations which are highly stereo-specifically trans, consistent with an E2 mechanism. Since E1 elimination may be less stereospecific the E2, choice of reagent may be an important factor in determining product distribution in hydration of alcohols.

Alcohols add to 2,3-dihydropyran under acidic conditions to give mixed acetals which are used to protect hydroxyl groups. The reactions of alcohols with carbonyl and carboxyl groups are discussed later.

Phenols can be alkylated and acylated in ways similar to those used for alcohols. Aryl methyl ethers are often prepared by reaction of phenol with diazomethane. The preparation of aryl halides from phenols is of little preparative significance.

The main transformations using alcohols and phenols are shown in Scheme 6-8.

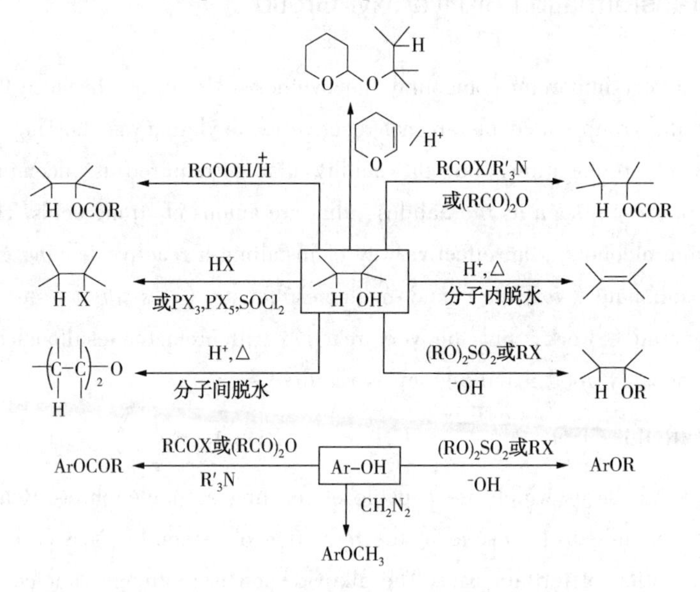

Scheme 6-8 Transformation of Hydroxyl Group

6.2.2 氨基的转化 Transformation of Amino Group

The amine group is basic and reacts as a nucleophile with alkyl halides, giving rise to secondary and tertiary amines and to quaternary ammonium salts. Acid chlorides and anhydrides give amides. The sulfonamide derived from reaction of a primary amine with a sulfonyl chloride has an acidic hydrogen, which may be removed to produce a strongly nucleophilic species, known as Hinsberg reaction:

$$ArNH_2+TsCl \longrightarrow ArNHTs \xrightarrow{NaOH} Ar\bar{N}Ts \xrightarrow{RCl} ArN(R)Ts$$

Hinsberg reaction

For aliphatic amines, reaction of a primary amine with nitrous acid is of little preparative significance due to the formation of a complex mixture of products, except in cases where elimination reactions cannot take place. However, it has been shown that primary aliphatic amines can be transformed into a wide variety of products by converting the amino group into better leaving group such as 2,4,6-triphenylpyridine, which can be displaced by a range of nucleophiles. Examples of these transformations include the following[14]:

2,4,6-triphenylpyrylium Br^- + $CH_3(CH_2)_3NH_2$ ⟶ 1-butyl-2,4,6-triphenylpyridinium Br^- $\xrightarrow{\text{heat at m.p.}}$ 2,4,6-triphenylpyridine + $CH(CH_2)_3Br$ 72%

2,4,6-triphenylpyrylium Br^- + $Ph(CH_2)_2NH_2$ ⟶ 1-(2-phenylethyl)-2,4,6-triphenylpyridinium Br^- $\xrightarrow{\text{AcONa}}$ 2,4,6-triphenylpyridine + $CH_3CO_2(CH_2)_2Ph$ 70%

Treatments of secondary amines with nitrous acid results in the formation of N-nitroso compounds, which can be reduced to N, N-disubstituted hydrazines. The reaction of tertiary aliphatic amines is complex and of no preparative importance.

The reaction of primary aromatic amines with nitrous acid is of considerable significance. The diazonium salts so formed can undergo a wide variety of transformation which are of preparative use. These together with other reactions involving the amino groups are shown in Scheme 6-9. The reactions of amino groups with carbonyl compounds will be considered later.

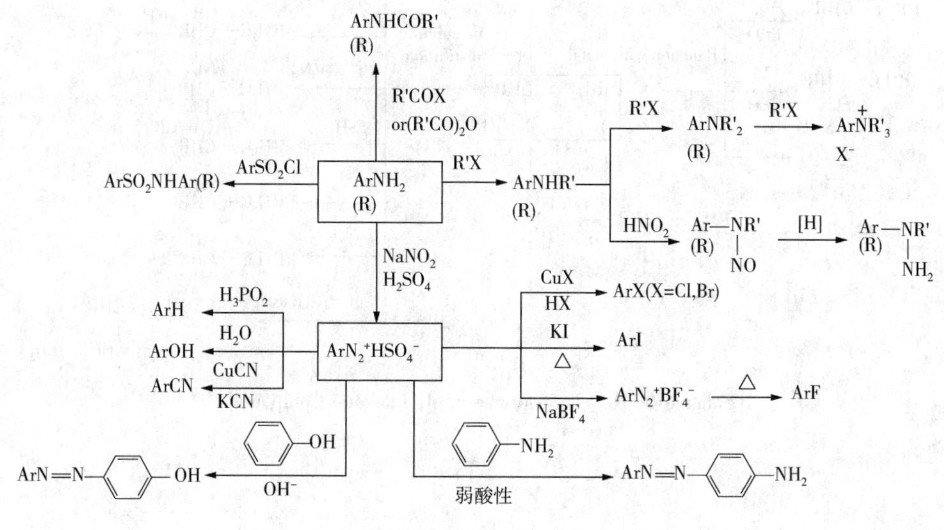

Scheme 6-9 Transformation of Amino Group

6.2.3 卤代物的转化 Transformation of Halogeno Compounds

A halogen, in addition to providing a good leaving group, withdraws electrons from adjacent carbon atom. Hence, alkyl halides participate in a wide variety of nucleophilic substitution reactions. Reactions with alcohols and with amines are most common used, and reactions with thiolate anions, cyanide ions, anions derived from alkynes and other carbanions are all valuable. Elimination reactions may, however, complicate the situation, especially in the case of secondary halides, and for many tertiary halides only elimination products are obtained.

Alkyl halides may be hydrolyzed using sodium hydroxide, but in the case of most secondary and tertiary halides elimination is a competitive reaction. Elimination is favored in the case of strong base reacting in a non-polar solvent at high temperature. Base-catalyzed elimination from secondary and tertiary halides normally obeys the Saytzeff rule. Use a bulky base with a 2° halide to ensure no S_N2, get Saytzeff (most substituted) alkene product. Use a non-bulky base with a 3° halide (no S_N2) to get the Saytzeff alkene product.

Alkyl halides react with certain metals to form metal alkyls. Of particular synthetic importance are alkyl-lithium derivatives and Grignard reagents, RMgX. These reagents are both strong base and good nucleophiles; their synthetic utility are discussed in carbon - carbon bond forming reactions. A summary of the reactions of alkyl halides is given in Scheme 6-10[15].

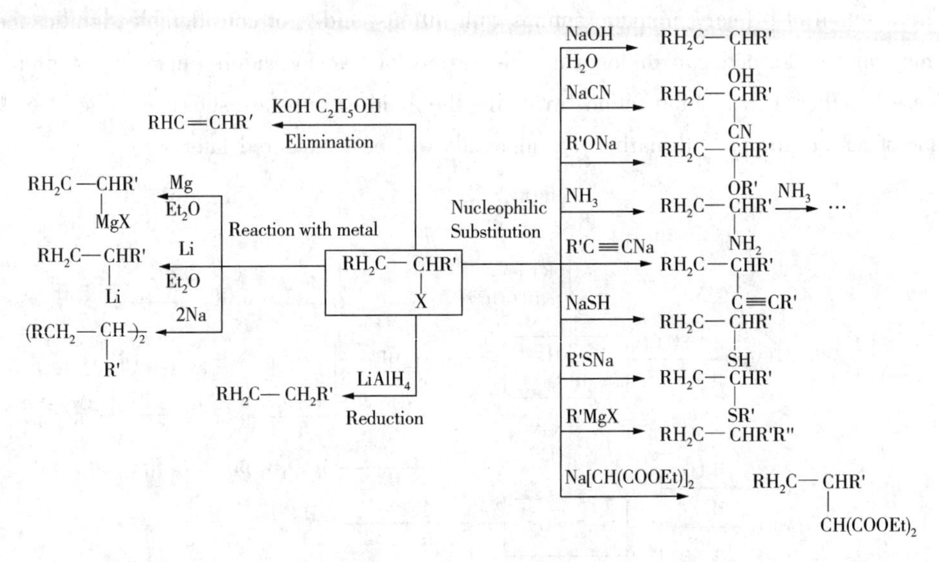

Scheme 6-10 Transformation of Halogeno Compounds

6.2.4 硝基化合物和腈的转化 Transformation of Nitro Compounds and Nitriles

Aliphatic nitro compounds are less of lesser synthetic importance than aromatic nitro compounds. However, a stable carbanion can be used in many of the reactions involving enol-like anions[16].

Due to the case of formation, aryl nitro compounds are of great importance for introducing a nitrogen-containing function to the aromatic ring. Reduction with a wide variety of reagents cause conversion to the amino group whose synthetic versatility has just has been discussed. Reduction to hydroxylamines, azo compounds and *N*, *N*-disubstituted hydrazines is also possible depending on the reagent chosen.

The replacement of a halide or sulfonate by cyanide ion, extending the carbon chain by one atom and providing an entry to carboxylic acid derivatives, has been a reaction of synthetic importance since the early days of organic chemistry. Unsaturated cyanide groups may be reduced to amines and amides. A summary of the reactions of nitro compounds and nitriles is given in Scheme 6-11.

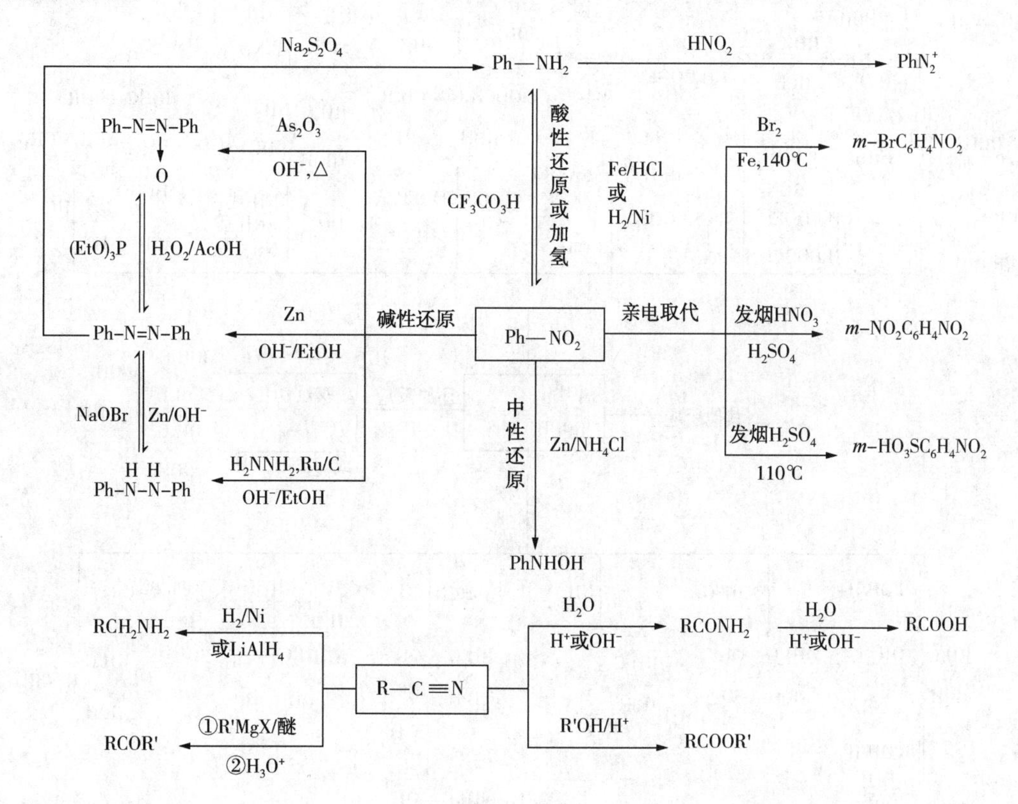

Scheme 6-11 Transformation of Nitro Compounds and Nitriles

6.2.5 醛和酮的转化 Transformation of Aldehyde and Ketone

All the transformations shown there are valuable tools of the synthetic organic chemist. Carbonyl groups provide access to hydrocarbon via Clemmensen or Wolff-Kishner reduction, to alcohols of the same carbon skeleton by a variety of reduction methods, and to alcohols of more complex structure by reaction with organolithium or Grignard reagents. One of the characteristics of carbonyl group is its tendency to undergo nucleophilic addition reactions[17].

A negatively polarized atom or group is transferred to the positively polarized carbon of the carbonyl group in the rate-determining step of these reactions. Grignard reagents, organolithium reagents, lithium aluminum hydride, and sodium borohydride all react with carbonyl compounds by nucleophilic addition. A number of nucleophilic addition reactions and oxidation and reduction reactions of aldehydes and ketones in Scheme 6-12. Some of these are of synthetic interest, others are of mechanistic importance, and a few possess both qualities.

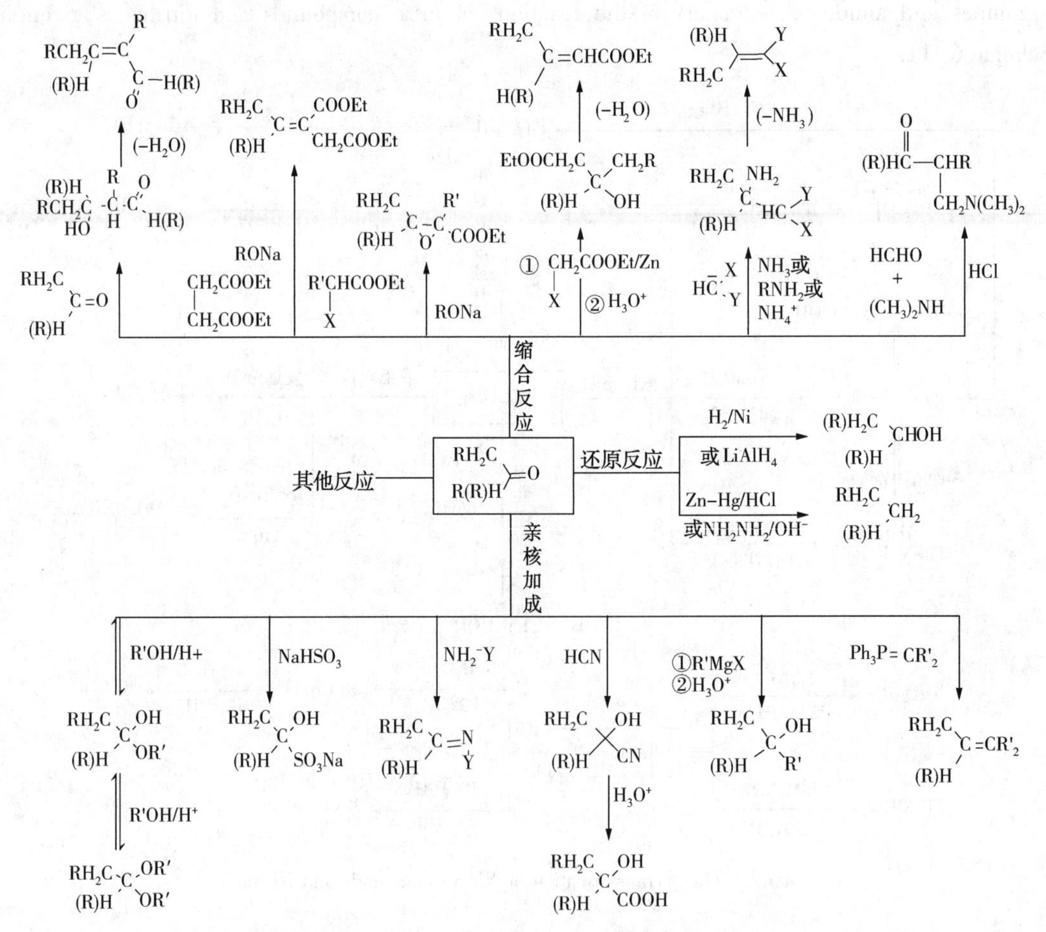

Scheme 6-12 Transformation of Aldehyde and Ketone

Aldehyde and ketone react reversibly under acidic conditions with alcohols and hemi-ketals and then acetals and ketals:

$$\gt C{=}O + ROH \xrightleftharpoons{H^+} \gt C(OR)(OH) \xrightarrow[-H_2O]{ROH,\ H^+} \gt C(OR)(OR)$$

The acetals and ketals derived by reaction of aldehydes and ketones with ethylene glycol are used to protect the carbonyl group during reactions carried out under neutral or alkaline conditions. The analogous dithioketals are used in a conversion of carbonyl groups into methylene groups, the reaction required, however, a large excess of Raney nickel.

$$\gt C{=}O + \begin{array}{c}CH_2SH\\|\\CH_2SH\end{array} \xrightarrow{H^+} \gt C\begin{array}{c}S\\ \\S\end{array}\!\!\rbrack \xrightarrow{\text{Raney nickel}} \gt CH_2$$

6.2.6 羧酸及其衍生物的转化 Transformation of Carboxylic Acid and Its Derivatives

Carboxylic acids are converted by acid-catalyzed reaction with acohols into esters. For methyl ester another convenient method involves the use of diazomethane. For more complex esters, reaction of the alcohol with the acid chloride or with anhydride may be more satisfactory; another method involves reaction of an acyl halide with silver salt of the carboxylic acid. Many of the procedures used for amide formation will also serve in esterification.

Acid chlorides are usually prepared by reaction of the acid with thionyl chloride. They are converted into anhydrides by reaction with the sodium salt of the acid. Reaction of acid chlorides with diazomethane results in the formation of dizoketones, which are converted by treatment with moist silver oxide into the carboxylic acids containing an additional methylene group. Reduction of carboxylic acid and its derivatives is incorporated.

Amides can be prepared by reaction of ammonia or the appropriate amine with anhydrides, esters or acid chlorides. The methods of amide formation are used widely in peptide syntheses. Primary amides can be dehydrated to nitriles, which can also be prepared by reaction of alkyl halides with potassium cyanide. A useful synthetic reaction of amides is their conversion into amines on treatment with bromide and alkali. Alternative procedures for converting acids and their derivatives into amines are thermal degradation of acid azides in alcoholic solvents and the treatment of carboxylic acids with hydrazoic acid.

The interconversions are summarized in Scheme 6-13.

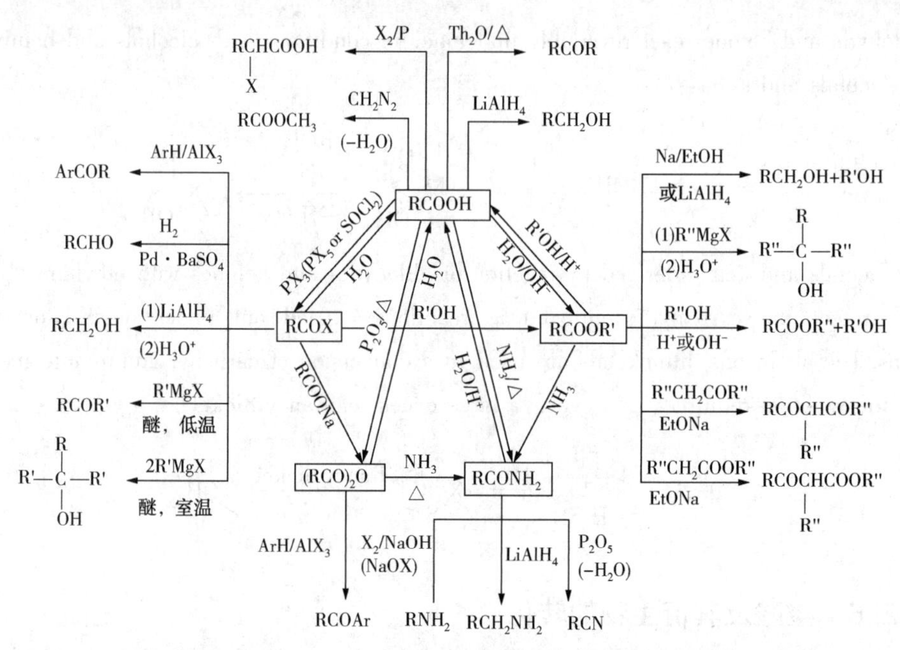

Scheme 6-13 Transformation of Carboxylic Acid and Its Derivatives

参考文献
References

1. Ouellette R J, Rawn J D. Principles of Organic Chemistry. Boston: Elsevier, 2015. 65-94.
2. Imamoto T. Comprehensive Organic Synthesis. Oxford: Pergamon, 1991. 793-809.
3. Hill R A. Comprehensive Organic Functional Group Transformations. Oxford: Elsevier Science, 1995. 1-40.
4. Altenbach H J. Comprehensive Organic Synthesis. Oxford: Pergamon, 1991. 829-871.
5. Kelly S E. Comprehensive Organic Synthesis. Oxford: Pergamon, 1991. 729-817.
6. Hegedus L S. Comprehensive Organic Synthesis. Oxford: Pergamon, 1991. 571-583.
7. Grimshaw J. Electrochemical Reactions and Mechanisms in Organic Chemistry. Amsterdam: Elsevier Science B. V., 2000. 27-53.
8. Verbruggen W P N, Blanco-Ania D, et, al. Comprehensive Heterocyclic Chemistry IV. Oxford: Elsevier, 2022. 485-515.
9. Bruckner R. Advanced Organic Chemistry. San Diego: Academic Press, 2002. 169-219.
10. Ikawa T. Comprehensive Aryne Synthetic Chemistry. Amsterdam: Elsevier, 2022. 15-56.
11. Zaikin P A, Borodkin G I. Late-Stage Fluorination of Bioactive Molecules and Biologically-Relevant Substrates. Amsterdam: Elsevier, 2019. 105-135.
12. Asif M, Imran M. Handbook of Greener Synthesis of Nanomaterials and Compounds. Amsterdam: Elsevier, 2021. 69-108.
13. Prinn R G. Treatise on Geochemistry (Second Edition). Oxford: Elsevier, 2014. 1-18.
14. Breitwieser K, Munz D. Advances in Organometallic Chemistry. New York: Academic Press, 2022. 79-132.
15. Spargo P L. Comprehensive Organic Functional Group Transformations. Oxford: Elsevier Science, 1995. 1-36.
16. Nishiwaki N. Comprehensive Organic Synthesis (Second Edition). Amsterdam: Elsevier, 2014. 100-130.
17. Vargel C. Corrosion of Aluminium (Second Edition). Amsterdam: Elsevier, 2020. 733-736.

7 官能团保护与脱保护

Chapter 7 Protection and Deprotection of Functional Groups

In some cases, many molecules contain more than one functional group. It is often necessary to manipulate only one (or some) of these groups at a time. This is only possible by choosing a variety of protecting groups, which can be manipulated using different reaction conditions. When does one use a protecting group and how is that group chosen? When to use a protecting group is not an easy question to answer. Such groups can be used to provide directing effects, influence solubility and other physical properties, make it easier to isolate certain products, and they can increase the crystallinity of a compound. In addition, installation and removal of protective groups (保护基的添加与移除) are common in the pharmaceutical industry as prodrugs. The classical uses for protecting groups are described in this chapter.

Protecting groups play a key role in the synthesis of complex naturally occurring substances—peptides, carbohydrates, and oligonucleotides. When the synthetic target is a relatively complex molecule, a sequence of reactions that would be expected to lead to the desired product must be devised. At the present time, syntheses requiring steps are common and many that are even longer have been completed. In the planning and execution of such multistep syntheses, an important consideration is the compatibility of the functional groups that are already present with the reaction conditions required for subsequent steps[1,2]. It is frequently necessary to modify a functional group in order to prevent interference with some reaction in the synthetic sequence. A protective group can be put in place and then subsequently removed in order to prevent an undesired reaction or other adverse influence. For example, alcohols are often protected as trisubstituted silyl ethers and carbonyl groups as acetals. The silyl group masks both the acidity and nucleophilicity of the hydroxy group. An acetal group can prevent both unwanted nucleophilic additions or enolate formation at a carbonyl group.

Three considerations are important in choosing an appropriate protective group[3,4]:

(ⅰ) the nature of the group requiring protection.

(ⅱ) the reaction conditions under which the protective group must be stable.

(ⅲ) the conditions that can be tolerated for removal of the protecting group.

No universal protective groups exist.

When a chemical reaction is to be carried out selectively at one reactive site in a multifunctional compound, other reactive sites must be temporarily blocked. Many protective groups have been, and are being, developed for this purpose. A protective group must fulfill a number of requirements:

(ⅰ) It must react selectively in good yield to give a protected substrate that is stable to the projected reactions.

(ⅱ) The protective group must be selectively removed in good yield by readily available, preferably nontoxic reagents that do not attack the regenerated functional group.

(ⅲ) The protective group should form a derivative (without the generation of new stereogenic centers) that can easily be separated from side products associated with its formation or cleavage.

(ⅳ) The protective group should have a minimum of additional functionality to avoid further sites of reaction. All things considered, no protective group is the best protective group.

7.1 羟基保护基 Hydroxy-Protecting Groups

Many functional groups will interfere in reactions being carried out elsewhere in the molecule. This affects (usually detrimentally) yields and levels of regio-, chemo- and stereoselectivity. Alcohols provide a good example of such functionality (other functional groups which can be a problem include amines and carbonyl groups). For example, the acidic proton of a hydroxy group will destroy one equivalent of a strongly basic organometallic reagent and possibly adversely affect the reaction in other ways. Therefore, we can minimise problems by *protecting* the alcohol. This protection strategy removes the acidic proton on the alcohol, and also reduces the nucleophilicity and basicity of the oxygen atom by steric hindrance and /or electronic effects[3]. In some cases, protection of the hydroxy group also improves the solubility of alcohols in nonpolar solvents.

Conversion of alcohols to ethers usually proceeds in high yield to give a relatively inert product. Due to this lack of reactivity, the reaction conditions required for cleavage of the ether may be very harsh. Specialized ether protecting groups have therefore been developed that can be removed under mild and selective conditions. Another solution is to convert alcohols to acyclic acetals or ketals. These groups possess much of the inert character of ethers, but are easily converted back to the alcohol. It is also possible to convert alcohols to esters under mild conditions, and hydrolysis easily converts them back to the alcohol.

7.1.1 醚 Ether

ROMe — Methyl Ether (Me)

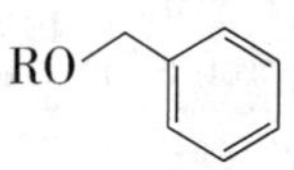

RO–CH₂Ph — Benzyl Ether (Bn)

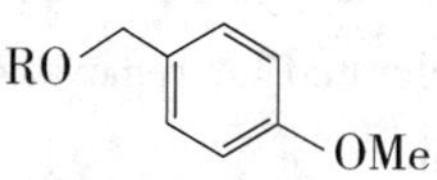

RO–CH₂C₆H₄–OMe — p-Methoxybenzyl Ether (PMB)

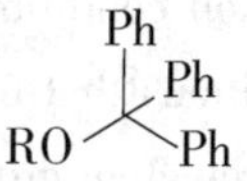

RO–CPh₃ — Trityl Ether (Tr)

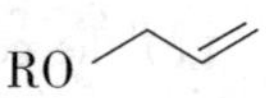

RO–CH₂CH=CH₂ — Allyl Ether (All)

（ⅰ）Methyl ether

$$ROH \xrightarrow[NaOH]{Me_2SO_4} ROCH_3 \xrightarrow[\text{several days}]{BF_3/RSH} ROH$$

$$ROH \xrightarrow[THF,0℃]{NaH} \xrightarrow{CH_3I} ROCH_3 \xrightarrow[\substack{CH_2Cl_2 \\ \text{aq.acid workup}}]{TMSI} ROH$$

Formation:

A simple way to protect the OH of an alcohol is by conversion to its methyl ether (-OMe). Reaction of an alcohol with a base [often sodium hydride (NaH) in THF or DMF] gives an alkoxide. Subsequent reaction with iodomethane gives the methyl ether via an S_N2 reaction called the Williamson ether synthesis (Williamson 醚合成反应). Methyl ethers can also be prepared by treatment of an alcohol with trimethyloxonium tetrafluoroborate (Meerwein's reagent), or with dimethyl sulfate (Me_2SO_4), a base and a phase-transfer (相转移) agent (such as tetrabutylammonium iodide).

Cleavage:

Methyl ethers are stable to strong bases, nucleophiles, organometallics, ylides, hydrogenation (except for benzylic ethers), oxidizing agents, and hydride reducing agents. They are stable in the pH range 1~14. Methyl ethers can be cleaved with concentrated HI, but trimethylsilyl iodide unmasks the alcohol in chloroform at ambient temperatures. Lewis acids such as boron tribromide (BBr_3) also cleave methyl ethers to alcohols.

The *t*-butyl group is an exception and has found some use as a hydroxy-protecting group. Owing to the stability of the *t*-butyl cation, *t*-butyl ethers can be cleaved under moderately acidic conditions. *t*-Butyl ethers can also be cleaved by acetic anhydride-$FeCl_3$ in ether.

（ⅱ）Benzyl ether

$$ROH \xrightarrow[\text{(b)}PhCH_2Br]{\substack{\text{Formation} \\ \text{(a)NaH,THF}}} ROCH_2Ph \xrightarrow[\substack{\text{or} \\ Ra\text{-}Ni,EtOH \\ \text{or} \\ Na,NH_3(l), EtOH}]{\substack{\text{Cleavage} \\ Pd/C,\ H_2,EtOH}} ROH$$

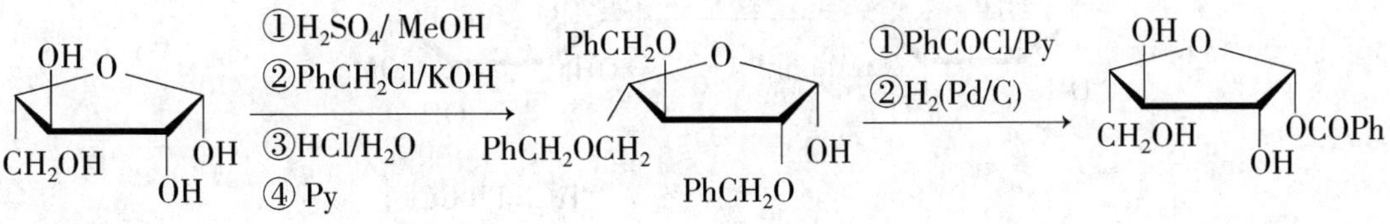
OH
O
CH2OH
OH
OH
①H2SO4/ MeOH
②PhCH2Cl/KOH
③HCl/H2O
④ Py
PhCH2O
O
PhCH2OCH2
OH
PhCH2O
①PhCOCl/Py
②H2(Pd/C)
OH
O
CH2OH
OCOPh
OH

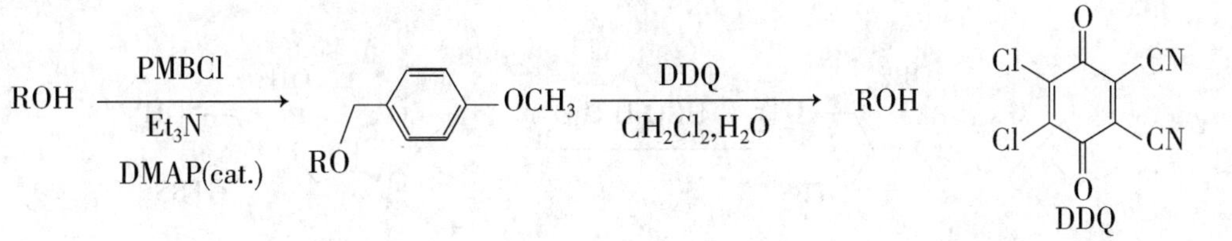
ROH
PMBCl
Et3N
DMAP(cat.)
RO
OCH3
DDQ
CH2Cl2,H2O
ROH
O
Cl
CN
Cl
CN
O
DDQ

Formation:

Benzyl ether (—OCH_2Ph, OBn), is prepared by reaction of the nucleophilic alkoxide of an alcohol with benzyl bromide or chloride. Phenyldiazomethane can also be used to introduce benzyl groups. Benzyl ethers are stable to a wide range of reagents including pH 14, carbanions, and organometallics, nucleophiles, hydrides and some oxidizing agents.

Cleavage(裂解):

Benzyl ether may be cleaved by n-BuLi in the presence of TMEDA or HMPA. Another common method is catalytic hydrogenation with a palladium catalyst, but hydrogenolysis with sodium or potassium in ammonia is also used. Treatment of benzyl ether with sodium and ammonia led to deprotection, and formation of alcohol in 91% yield.

4-Methoxyphenyl (PMP) ethers find occasional use as hydroxy protecting groups. Unlike benzylic groups, 4-Methoxyphenyl (PMP) ethers cannot be made directly from the alcohol. Benzyl groups having 4-methoxy (PMB) or 3,5-dimethoxy (DMB) substituents can be removed oxidatively by dichlorodicyanoquinone (DDQ) and CAN.

Allyl ethers can be removed by conversion to propenyl ethers, followed by acidic hydrolysis of the resulting enol ether.

(ⅲ)三苯基甲基乙醚

Trityl Ether

In general, trityl ethers are used selectively to protect primary alcohols in carbohydrate chemistry. Trityl ethers are stable to bases, nucleophiles. They are cleaved by acids, H_2/Pd. Treatment of trityl ethers with CF_3CO_2H in *t*-BuOH or HCO_2H in water also lead to deprotection, and formation of alcohol. Notably, they may be selectively cleaved in the presence of other protecting groups.

OH, OH → (Ph_3CCl; Et_3N,DMAP(cat.), CH_2Cl_2) → OH, OTr 100%

CH_2OH, O, OH, OMe, OH, OH → (TrCl/Py) → CH_2OTr, O, HO, OMe, OH, OH

TrCl=Ph_3CCl

OBz, O, O, OTr, OTBS → (HCO_2H,H_2O) → OBz, O, O, OH, OTBS

7.1.2 硅醚

Silyl Ether

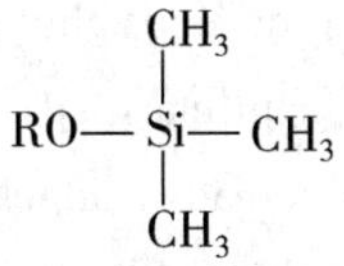

$$RO-Si(CH_3)_3$$
Trimethylsilyl(TMS)

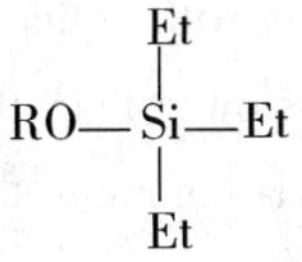

$$RO-Si(Et)_3$$
Triethylsilyl(TES)

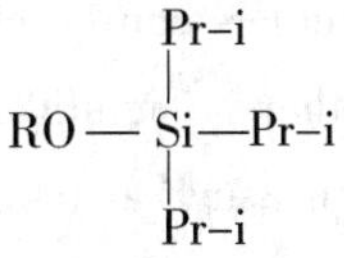

$$RO-Si(Pr\text{-}i)_3$$
Triisopropylsilyl(TIPS)

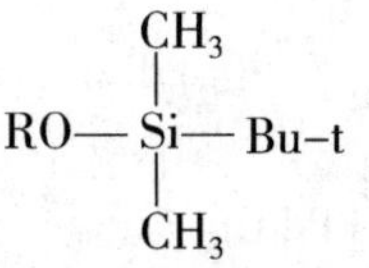

$$RO-Si(CH_3)_2-Bu\text{-}t$$
t-Butyldimethylsilyl (TBS or TBDMS)

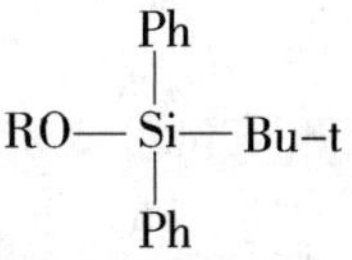

$$RO-Si(Ph)_2-Bu\text{-}t$$
t-Butyldiphenylsilyl (TPS or TBDPS)

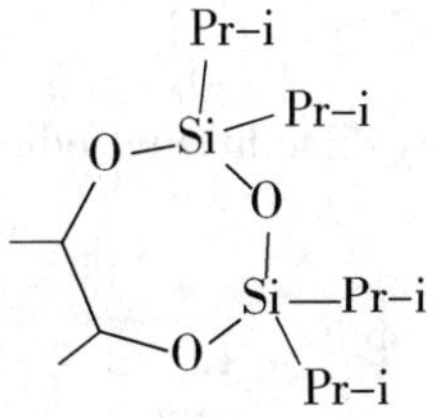

Tetraisopropyldisilylene (TIPDS)

(ⅰ) 甲基丙烯酰氧乙基醚

Methyl/Ethyl Silyl Ether

$$ROH \xrightarrow[Et_3N,THF]{TMSCl} RO-SiMe_3$$

$$ROH \xrightarrow[2,6\text{-Lutidine},\ CH_2Cl_2]{Et_3SiOTf} RO-SiEt_3$$

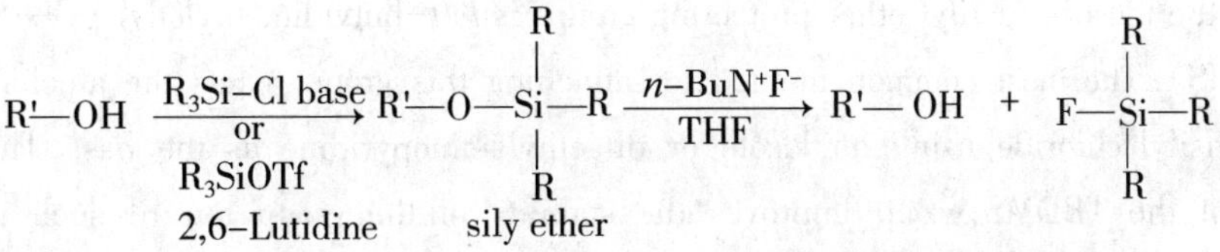

$$R'-OH \xrightarrow[R_3SiOTf,\ 2,6\text{-Lutidine}]{R_3Si\text{-}Cl\ base\ or} R'-O-SiR_3 \xrightarrow[THF]{n\text{-}BuN^+F^-} R'-OH + F-SiR_3$$

sily ether

R_3SiOTf= trialkylsilyl trifluoromethanesulfonate

Formation:

Silyl ethers have become extremely important for the protection of alcohols with the generic structure $OSiR_3$. Variation in the R group leads to significant differences in the stability of the protecting group. The prototype of this class of protecting groups is the trimethylsilyl ether (O-$SiMe_3$, OTMS). The group is usually attached to the alcohol by reaction with chlorotrimethylsilane (Me_3SiCl) in the presence of an amine base such as triethylamine or pyridine. Trimethylsilyl (三甲基硅醚) ethers are very susceptible to acids, or bases in protic media. The group is not very stable to organometallics such as Grignard reagents, or to nucleophiles, hydrogenation or hydrides, but it can be used with some oxidants. A more stable silyl derivative is the triethylsilyl group (O-

$SiEt_3$, OTES). This group is attached by reaction of the alcohol with chlorotriethylsilane (Et_3SiCl) and pyridine. The OTES group is generally more stable than OTMS, and in particular it is less sensitive to water.

Cleavage:

In general, silyl ethers can be cleaved with aqueous base or acid, but the rate of hydrolysis for a secondary silyl ether is significantly slower than that of a primary silyl ether. Under aprotic conditions, fluoride ion (such as tetrabutylammonium fluoride) readily attacks silicon and cleaves the O—Si bond. In addition, alkyl alcohols are more reactive than phenolic derivatives, and the order of reactivity for other silane-protected functional groups is $COOSiR_3 > NHSiR_3 > CONHSiR_3 > SSiR_3$.

(ii) tert-butyldimethylsilyl ether

① $^tBu(Me)_2SiCl$, Imidazole, DMF, 10h, 35℃, 96%
② DIBAL(tolune), 10min, −78℃, N_2, 94%

$H_2(Pd/C)$, 1h, r.t. — 96%

$Bu_4N^+F^-/THF$ — 92%

Formation:

One of the most used silyl ether protecting groups is *tert*-butyldimethylsilyl [O—Si(Me)$_2$*t*—Bu (OTBDMS). The most common method for attaching this group reacts the alcohol with *tert*-butyldimethylsilyl chloride using imidazole or dimethylaminopyridine as the base. The increased steric bulk of the TBDMS group improves the stability of the group toward such reactions as hydride reduction and Cr(VI) oxidation. The group is stable to base and reasonably stable to acid (pH 4 ~ 12) than the other silyl protecting groups, and also to nucleophiles, organometallics, hydrogenation (except in acidic media), hydrides or oxidizing agents.

Cleavage:

It can be removed with aqueous acid, but fluoride ion is the most common cleavage method (fluoride tetrabutylammonium fluoride, tetrabutylammonium chloride, KF, and aqueous HF).

In general, the stability of silyl ethers towards acidic media increases as indicated: TMS(1) <TES(64) <TBS(20,000) <TIPS(700,000) <TBDPS(5,000,000) (relative reaction rate).

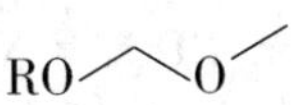

Methoxymethyl Ether (MOM)　　Benzyloxymethyl Ether (BOM)　　*p*–Methoxybenzyl Ether (PMBM)　　2,2,2–Trichloroethoxymethyl Ether

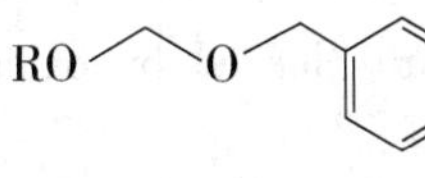

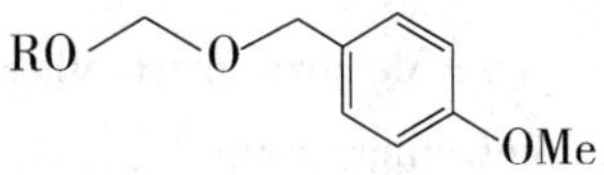

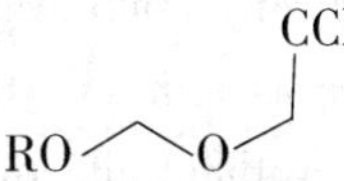

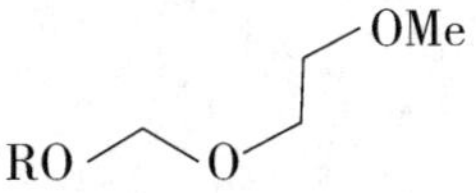

2–Methoxyethoxymethyl Ether (MEM)　　2–(Trimethylsilyl)ethoxymethyl Ether (SEM) 　　Methylthiomethyl Ether (MTM)　　Tetrahydropyranyl Ether (THP)

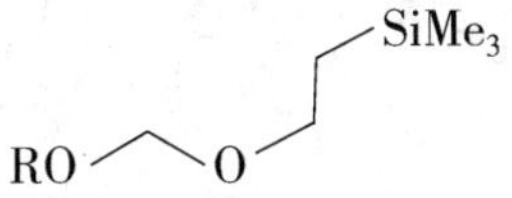

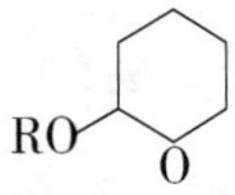

7.1.3 缩醛 Acetal

(ⅰ) MOM

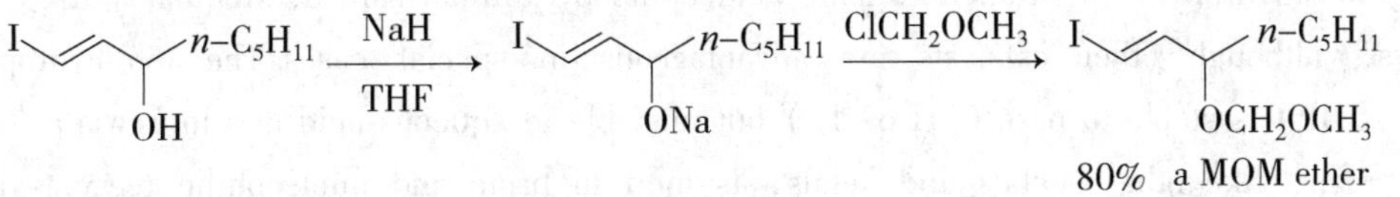

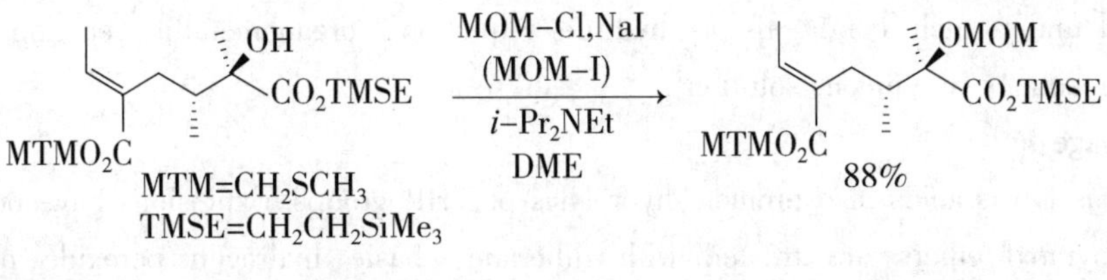

NaH, THF; $ClCH_2OCH_3$ → OCH_2OCH_3, 80% a MOM ether

MOM–Cl, NaI (MOM–I), *i*–Pr_2NEt, DME → 88%

MTM=CH_2SCH_3
TMSE=$CH_2CH_2SiMe_3$

Formation:

Several ether-like protecting groups have been developed that show greater selectivity in reactions with acids or Lewis acids, allowing more latitude in the choice of a method to remove them. The methoxymethyl (MOM) and β–methoxyethoxymethyl (MEM) groups are used to protect alcohols and phenols as formaldehyde acetals. Methoxymethyl ether (O–CH_2OMe OMOM) can be formed by reaction of the alcohol with a base, and then with $ClCH_2OCH_3$ (chloromethyl methyl ether). Sometimes a source of iodide ion ($Bu_4N^+I^-$, LiI, or NaI) is added to enhance the reactivity of the alkylating reagent.

Cleavage:

The MOM and MEM groups can be cleaved by pyridinium tosylate (吡啶对甲苯磺酸盐) in moist organic solvents. The MOM group is somewhat sensitive to acid, but is generally stable in the pH 4 ~ 12 range. It is important to note that HCl in an aqueous THF solution (pH≈1) does not cleave the BnO group, whereas it does cleave the O–MOM group. Reagents such as zinc bromide, magnesium bromide, titanium tetrachloride, dimethylboron bromide, or trimethylsilyl iodide permit MOM group removal. Benzyloxymethyl ethers (BOM) are cleaved more easily by hydrogenolysis with $Pd(OH)_2$. *p*-Methoxybenzyl ethers are cleaved by oxidation DDQ, $(NH_4)_2Ce(NO_3)_6$ without affecting other ether groups.

(ii) Tetrahydropyranyl Ether (THP)

ROH + (dihydropyran, δ⁻/δ⁺) $\xrightarrow{H^+}$ (tetrahydropyranyl)–OR

$HC\equiv CCH_2OH$ $\xrightarrow[H^+]{\text{dihydropyran}}$ THP–$OCH_2C\equiv CH$ $\xrightarrow[THF]{C_2H_5MgBr}$ THP–$OCH_2C\equiv CMgBr$

$\xrightarrow[②H_3O^+]{①CO_2}$ THP–$OCH_2C\equiv CCOOH$

Formation:

The THP group is introduced by an acid-catalyzed addition of the alcohol to the vinyl ether moiety in dihydropyran. *p*-Toluenesulfonic acid or its pyridinium salt are frequently used as the catalyst, although other catalysts are advantageous in special cases. The tetrahydropyranyl ether (THP) is stable to base (pH 6~12) but unstable to aqueous acid and to Lewis acids. The THP group, like other acetals and ketals, is inert to basic and nucleophilic reagents and is unchanged under such conditions as hydride reduction, organometallic reactions, or base-catalyzed reactions in aqueous solution[5].

Cleavage:

Various Lewis acids also promote hydrolysis of THP groups. Explosions have occurred when tetrahydropyranyl ethers are treated with diborane: basic hydrogen peroxide or with 40% peroxyacetic acid.

7.1.4 酯和碳酸盐 Ester and Carbonate

Another class of alcohol protecting groups are the *O*-esters (*O*-COR), where COR is an acyl group. Ester protecting groups are limited in scope due to their susceptibility to nucleophilic acyl substitution, hydrolysis or reduction.

Acetate (Ac) | Chloroacetate | Dichloroacetate | Trichloroacetate | Trifluoroacetate (TFA)

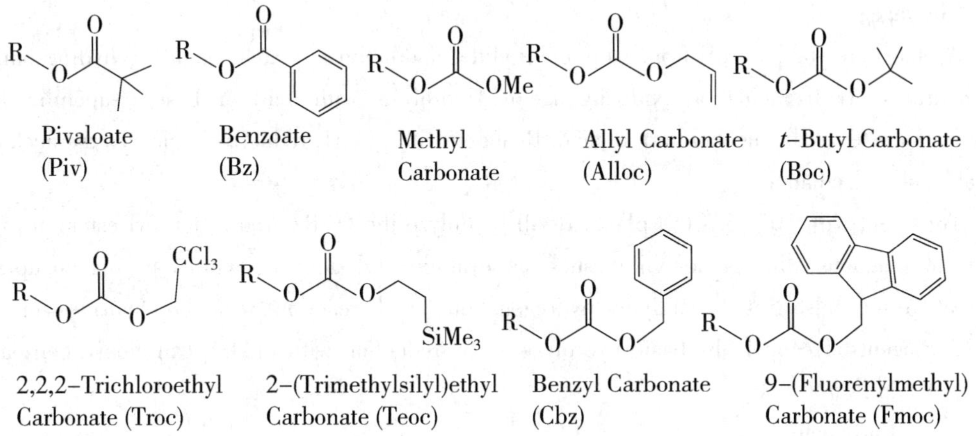
Pivaloate
(Piv)
Benzoate
(Bz)
Methyl
Carbonate
Allyl Carbonate
(Alloc)
t-Butyl Carbonate
(Boc)
2,2,2-Trichloroethyl
Carbonate (Troc)
2-(Trimethylsilyl)ethyl
Carbonate (Teoc)
Benzyl Carbonate
(Cbz)
9-(Fluorenylmethyl)
Carbonate (Fmoc)

(ⅰ) 酯

Ester

Acetates, benzoates, and pivalates, which are the most commonly used derivatives, can be conveniently prepared by reaction of unhindered alcohols with acetic anhydride, benzoyl chloride, or pivaloyl chloride, respectively, in the presence of pyridine or other tertiary amines.

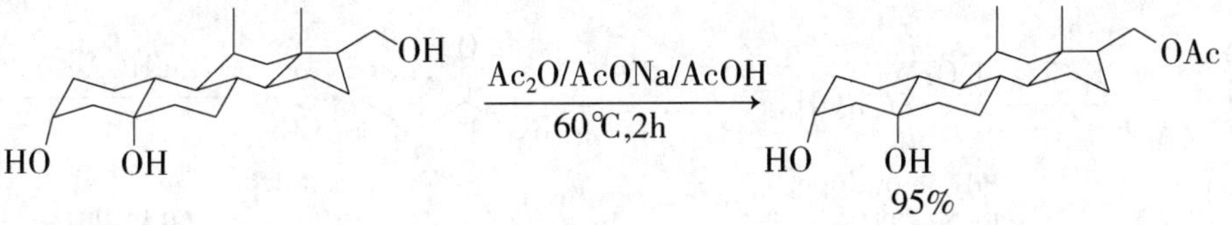
OH
HO
OH
Ac2O/AcONa/AcOH
60℃,2h
OAc
HO
OH
95%

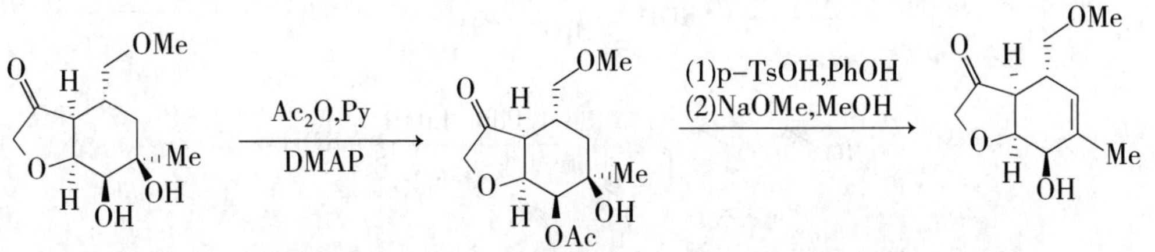
OMe
O
H
Me
O
H
OH
OH
Ac2O,Py
DMAP
OMe
O
H
Me
O
H
OH
OAc
(1)p-TsOH,PhOH
(2)NaOMe,MeOH
OMe
O
H
O
H
OH
Me

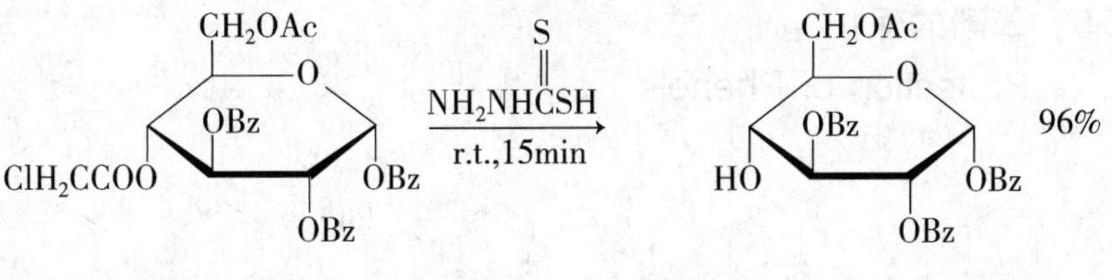
CH2OAc
O
OBz
ClH2CCOO
OBz
OBz
S
NH2NHCSH
r.t.,15min
CH2OAc
O
OBz
HO
OBz
OBz
96%

Formation:

Acetates, benzoates, and pivalates, which are the most commonly used derivatives, can be conveniently prepared by reaction of unhindered alcohols with acetic anhydride, benzoyl chloride, or pivaloyl chloride, respectively, in the presence of pyridine or other tertiary amines. Acetates (OAc) are formed by reaction of an alcohol with acetic anhydride or acetyl chloride and pyridine or triethylamine. Alcohols react with benzoyl chloride ($PhCOCl$) or benzoic anhydride [$(PhCO)_2O$], with pyridine or NEt_3, to give the benzoyl derivative (benzoate ester, OBz). Benzoates are more stable to hydrolysis than acetates, with a pH stability from pH 1 up to 10~11.

Cleavage:

Acetates are stable to pH from 1 up to 8, ylides and organocuprates, catalytic hydrogenation, borohydrides, Lewis acids or oxidizing agents. Hydrolysis with acid or base (saponification) cleaves the ester to the alcohol and an acid. Reaction with $LiAlH_4$ reductively cleaves the acetate to the alcohol and ethanol.

Treatment with 10% K_2CO_3 (pH 11) will hydrolyze the O–Bz group. Benzoyl esters are more stable to reactions with nucleophiles such as cyanide and acetate, ylides or organocuprates. Benzoate esters also resist catalytic hydrogenation, and reaction with borohydrides or with oxidizing agents. Treatment by basic hydrolysis or by reduction with $LiAlH_4$ can cleave benzoates.

(ii) 碳酸盐

Carbonate

Alkoxycarbonyl groups are stable protectors of the amino function in amino acid chemistry but they are seldom used to protect hydroxy groups in carbohydrates. An example shows that alkoxycarbonyl groups can be easily installed on carbohydrates by reaction with the appropriate chloroformate in the presence of TMEDA at low temperature. The method has been used to append Cbz, Troc, Aloc and Fmoc groups.

RO–C(=O)–OMe —(K_2CO_3, MeOH)→ ROH

RO–C(=O)–O–CH_2CCl_3 —(Zn, HOAc)→ ROH

RO–C(=O)–O–allyl —($Pd_2(dba)_3$, dppe, Et_2NH, THF)→ ROH

RO–C(=O)–O–$CH_2CH_2SiMe_3$ —(TBAF, THF)→ ROH

RO–C(=O)–O–CH_2Ph —(H_2, Pd/C, EtOH)→ ROH

7.1.5 苯酚的保护

Protection of Phenols

Methyl Ether (PhOMe) t–Butyl Ether Benzyl Ether Allyl Ether

Silyl Ether ($PhOSiR_3$) Phenyl Ester Phenyl Carbonate Acetal

Phenols can be transformed into ether and carbonate in ways similar to those used for alcohols. The following examples show the representative protecting and deprotecting procedures[6].

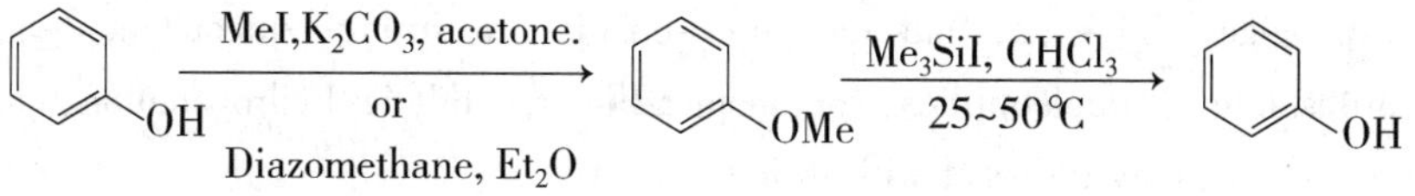

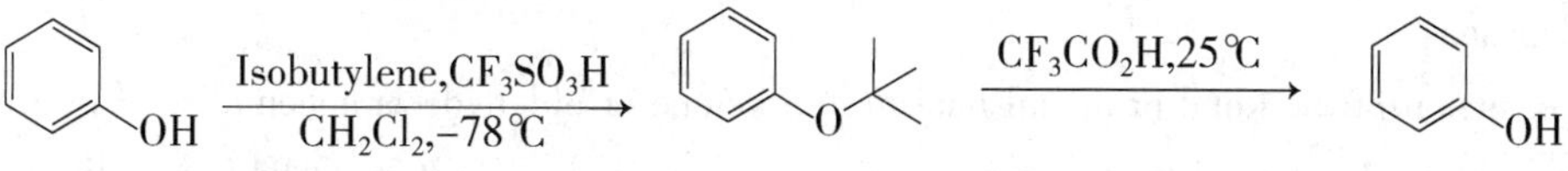

MeI, K_2CO_3, acetone. or Diazomethane, Et_2O ; Me_3SiI, $CHCl_3$, 25~50℃

Isobutylene, CF_3SO_3H, CH_2Cl_2, −78℃ ; CF_3CO_2H, 25℃

7.2 羰基保护基 Carbonyl-Protecting Groups

The major method for protecting ketones and aldehydes is conversion to a ketal or acetal by nucleophilic acyl addition, using alcohol or diol reactants. A variation of this reaction uses thiol or dithiol reactants to generate dithioketals or dithioacetals[7].

7.2.1 非环状的缩醛和酮 Acyclic Acetals and Ketals

HCl/MeOH, r.t. ; HC≡CH, KNH_2, NH_3(liq.) ; $CH_2(COOH)_2$, r.t., 8h ; $DDQ/C_6H_6/Et_2O$

$Bi(NO_3)_3 \cdot 5H_2O$(0.25%), CH_2Cl_2, r.t., 45min, 76%

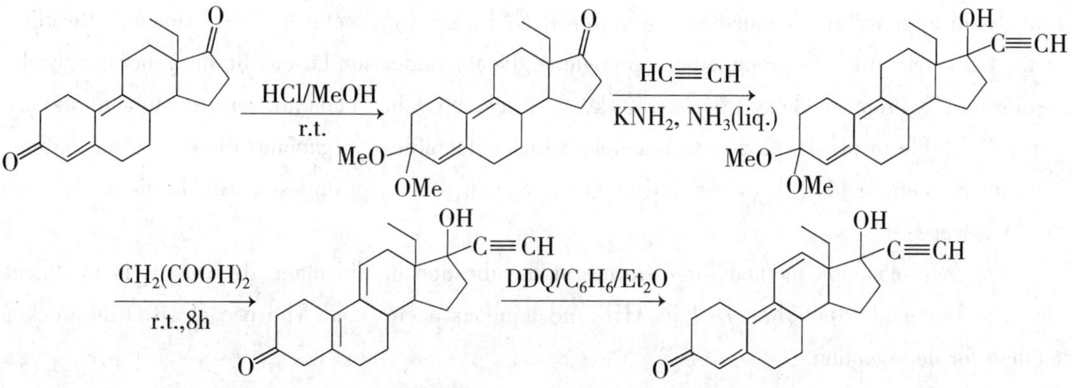

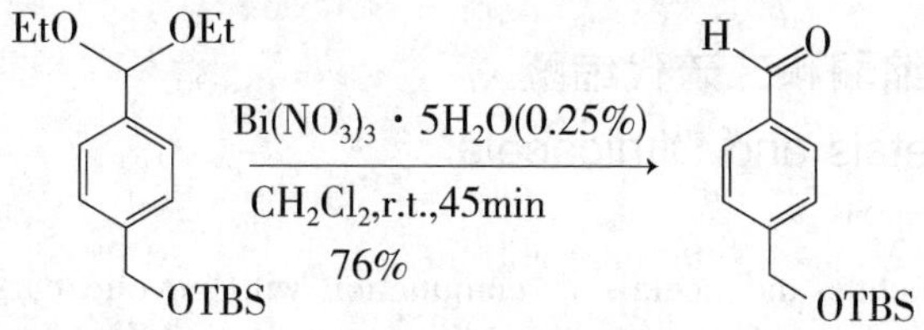

Formation:

Acyclic acetals and ketals are formed by reaction of the carbonyl with alcohols such as methanol or ethanol under anhydrous conditions, in the presence of an acid catalyst. Methanol and ethanol are probably the most common ones used because the yields of product with these reagents are high, and the lower molecular weight alcohol by products are easily removed after deprotection. In general, dry HCl (gas), sulfuric acid, $BF_3 \cdot OEt_2$ or *p*-toluenesulfonic acid (PPTS) are the catalysts of choice. Diethyl and dimethyl ketals and acetals are generally stable in a pH range of 4 ~ 12, but they are sensitive to strong aqueous acid and to Lewis acids. They are stable to nucleophiles, organometallics, catalytic hydrogenation, hydrides and most oxidizing agents (they do react with ozone).

Cleavage:

Conversion of the ketal or acetal back to the ketone or aldehyde is accomplished by treatment with aqueous acids such as trifluoroacetic acid or oxalic acid (HOOC−COOH). Acyclic acetals of ketones and aldehydes can be deprotected with a catalytic amount of bismuth nitrate pentahydrate.

7.2.2 环状缩醛和缩酮 Cyclic Acetals and Ketals

CO_2Me ... $\xrightarrow[\text{TsOH}]{HOCH_2CH_2OH}$ ... CH_2OH ... $\xrightarrow[Et_2O]{LiAlH_4}$ ... CH_2OH ... $\xrightarrow{HClO, Me_2CO/H_2O}$ ... CH_2OH

Formation:

A carbonyl group reacts with 1, 2-ethanediol (ethylene glycol) and PPTS, $BF_3 \cdot OEt_2$, or oxalic acid to give the ethylenedioxy (乙二醇半缩醛) ketal or acetal (1,3-doxolane). Reaction of the carbonyl with 1,3-propanediol (propylene glycol) under similar conditions generates the 1, 3-dioxane, a propylenedioxy (缩丙二醇) ketal or acetal. Cyclic ketals are sensitive to acid and are generally stable to pH 4~12 and to reactions with nucleophiles, organometallics, hydrogenation (except in acetic acid), hydrides and oxidizing agents, but they do react with Lewis acids[8].

Cleavage:

The most common method for cleavage of the dioxane or dioxolane derivatives is treatment with aqueous acid, including HCl in THF and aqueous acetic acid. The use of DDQ as another method for deprotection.

7.2.3 二硫代缩酮和二硫代缩醛 Dithioketals and Dithioacetal

The use of cyclic ketals and acetals in conjunction with acyclic carbonyl derivatives offers some selectivity for protection of more than one carbonyl. Expanding the methodology to include the sulfur analogs [dithioketals and dithioacetals, $R_2C(SR')_2$] gives even more flexibility.

$$R(C{=}O)R' \xrightarrow[\text{or } RSSi(CH_3)_3, ZnI_2, Et_2O. \text{ or } RSH, BF_3.Et_2O, CH_2Cl_2.]{RSH, HCl, 20℃} RC(SR)_2R'$$

Formation:

A ketone or aldehyde reacts with two equivalents of a thiol (mercaptan) and an acid (typically HCl and BF_3). Boron trifluoride etherate ($BF_3 \cdot OEt_2$) is commonly used as an acid catalyst to form dithianes and dithiolanes from ketones and aldehydes. Cyclic dithioketal and dithioacetal derivatives are relatively insensitive to acid, showing stability from pH 1 ~ 12. They are stable to nucleophiles, organometallics, hydrides and many oxidizing agents, although CrO_3 can oxidize the sulfur.

Cleavage:

Dithioacetals and ketals are particularly reactive with mercuric and silver salts, and mercuric chloride under aqueous conditions is the most commonly used method for conversion back to the carbonyl.

7.2.4 烯醇醚和烯胺 Enol Ether and Enamine

CO$_2$Me; $HC(OEt)_3/H_2SO_4$ /EtOH; EtO; CO$_2$Me; 90%

①$LiAlH_4$ ② $HCl/MeOH/H_2O$; CHOH; 65%

α, β-Unsaturated ketone react with trialkyl orthofomates in ethanol or dioxane to give enol ether. Saturated ketones are inert in this reaction conditions. Enol ether may selectively protect α, β-unsaturated ketone[9,10].

HO; H; H; O; NH benzene; reflux; N; 100%

Enamine usually is involved in the protection of α, β-unsaturated ketone for the synthesis of steroid compounds. Refluxing secondary amines and carbonyl compounds in benzene, followed by removing solvents, gives enamines. They are stable to $LiAlH_4$, Grignard reagents and other organometallic compounds. The enamine is easily cleaved by dilute acid.

Malononitrile occasionally is used to protect carbonyl compounds.

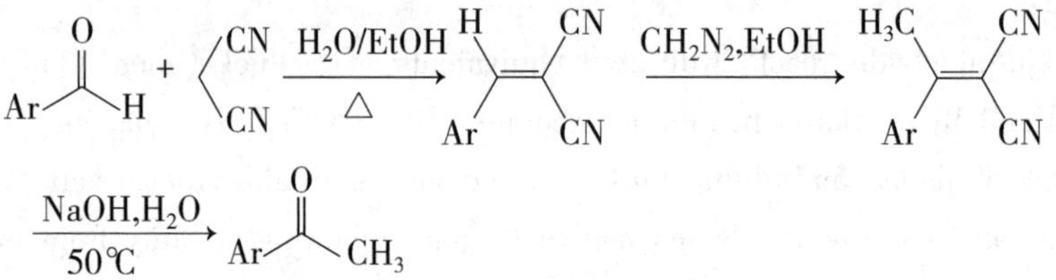

7.3 二醇保护基

Diol-Protecting Groups

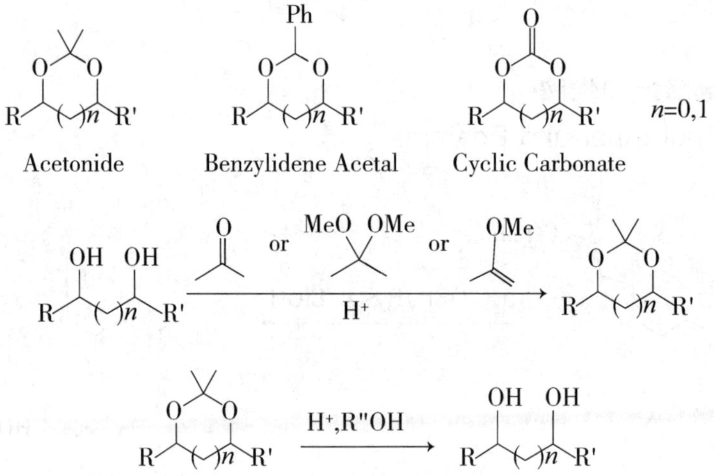

Acetonide　　Benzylidene Acetal　　Cyclic Carbonate　　n=0,1

Formation:

The vicinal nature of the hydroxyls of 1, 2-diols allows them to be protected as cyclic ketals. When a 1, 2-diol such as 2, 3-butanediol reacts with a ketone such as acetone in the presence of an acid catalyst, a 1, 3-dioxolane is formed. A 1, 3-diol will generate a six-membered ring acetonide, which is a 1, 3-dioxane derivative. This acetal protective group is resistant to basic and nucleophilic reagents, but is readily removed by aqueous acid.

Formaldehyde, acetaldehyde, and benzaldehyde are also used as the carbonyl component in the formation of cyclic acetals, and they function in the same manner as acetone. A disadvantage in the case of acetaldehyde and benzaldehyde is the possibility of forming a mixture of diastereomers, because of the new stereogenic center at the acetal carbon. Benzylidene acetals display reversed selectivity and discrimination between 1,2 and 1,3-diols of a triol group occurs.

Cleavage:

The group is stable to base, but not to acid (pH 4 ~ 12). It is stable to nucleophiles, organometallics, catalytic hydrogenation, hydrides, and oxidizing agents. It can be cleaved with aqueous HCl, with acetic acid or with p-toluenesulfonic acid in methanol. This group has been used extensively in the manipulation of carbohydrates.

Selective protection（选择性保护）：

Benzylidene acetals formation with 1,3-diols occurs in preference protection to 1,2-diols.

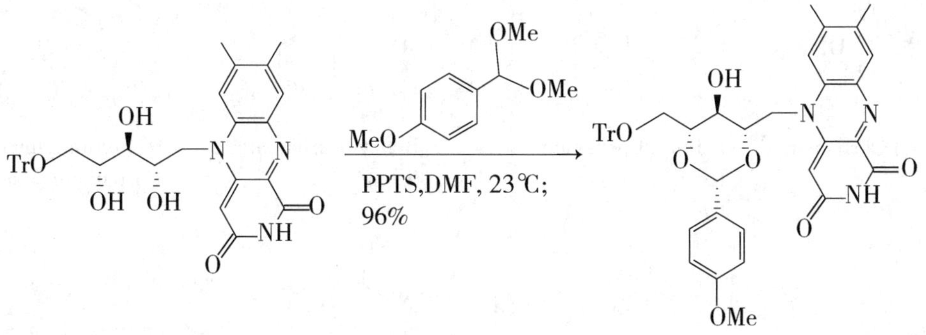

In general, acetonide formation with 1,2-diols occurs in preference protection to 1,3-diols; Benzylidene acetals display reversed selectivity.

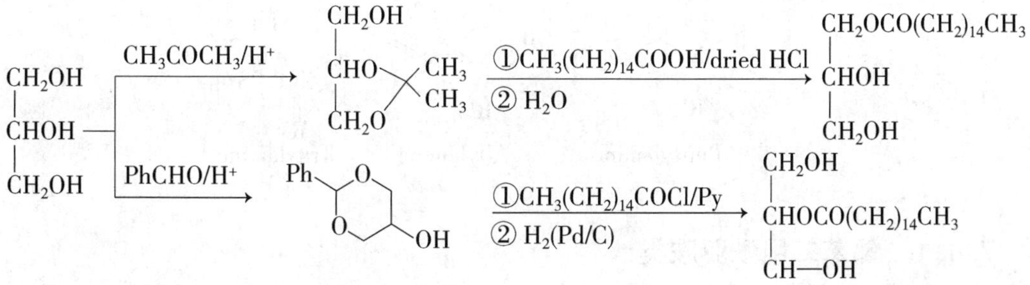

Selective deprotection（选择性脱保护）：

Hydrolysis of the less substituted dioxane and dioxolane（二氧戊环）occurs preferentially in substrates bearing two such groups.

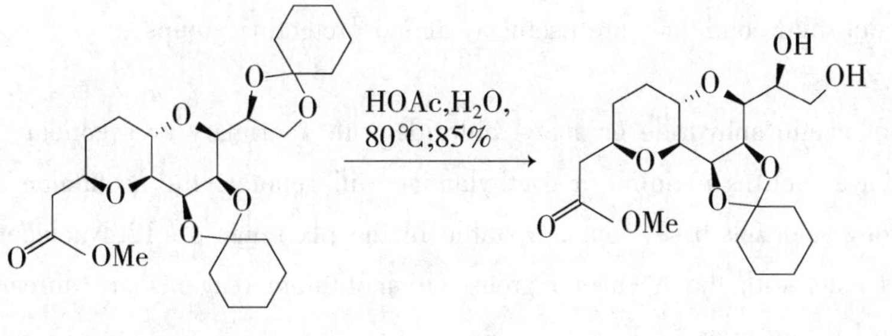

7.4 氨基保护基 Amino-Protecting Groups

Amines are nucleophilic and easily oxidized. Primary and secondary amino groups are also sufficiently acidic that they are deprotonated by many organometallic reagents. If these types of reactivity are problematic, the amino group must be protected. The most general way of masking nucleophilicity is by acylations, and carbamates are particularly useful[11].

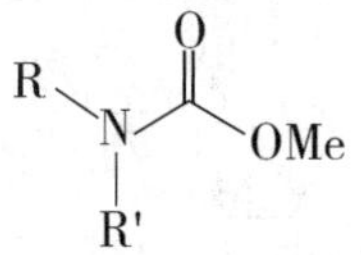

Methyl Carbamate 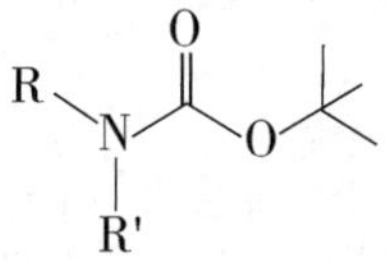　　t-Butyl Carbamate (Boc) 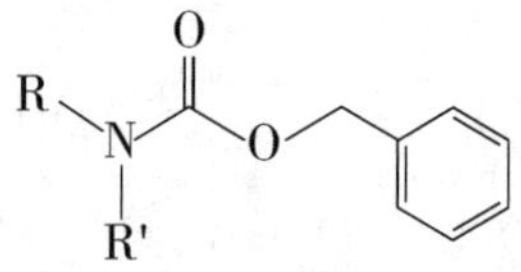　　Benzyl Carbamate (Cbz) 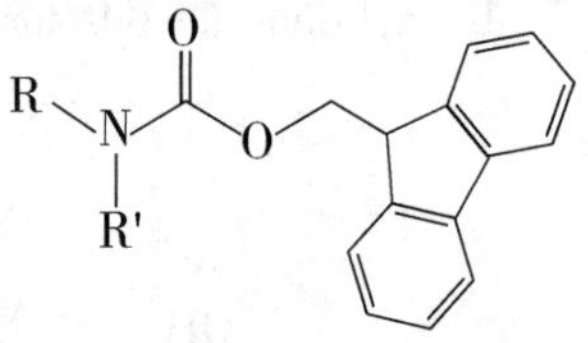　　9-Fluorenylmethyl Carbamate (Fmoc)

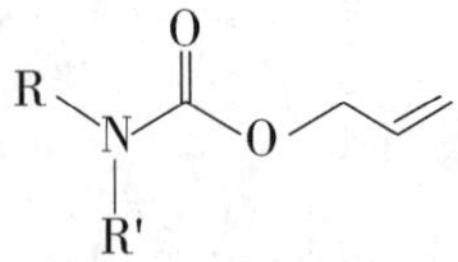

Allyl Carbamate (Alloc) 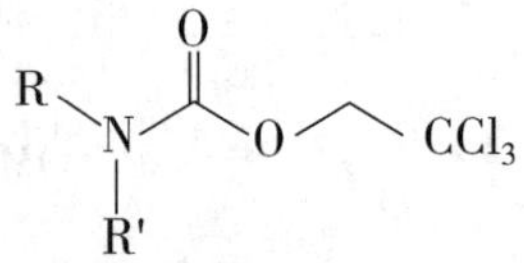　　2,2,2-Trichloroethyl Carbamate (Troc) 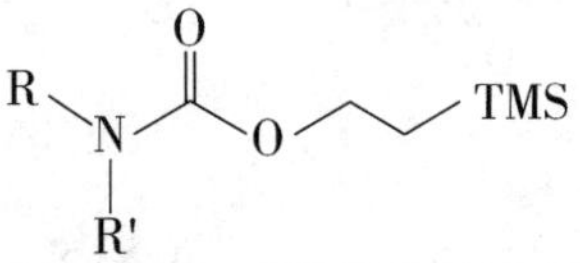　　2-(Trimethylsilyl)ethyl Carbamate (Teoc) 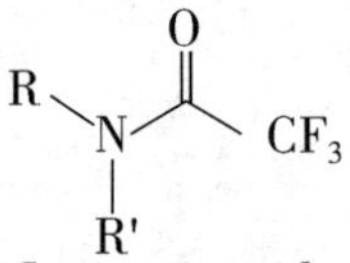　　Trifluoroacetamide

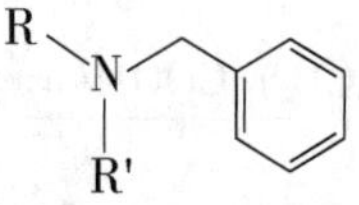

Benzylamine 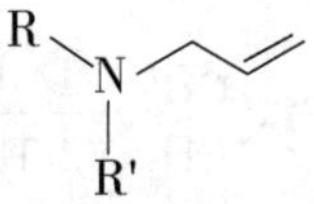　　Allylamine 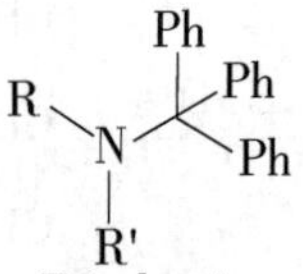　　Tritylamine

7.4.1 邻苯二甲酰亚胺类 Amides

Simple amides are satisfactory protecting groups only if the rest of the molecule can resist the vigorous acidic or alkaline hydrolysis necessary for removal. For this reason, only amides that can be removed under mild conditions are useful as amino protecting groups.

Formation:

Reaction of acetic anhydride or acetyl chloride with a primary or secondary amine, in the presence of a base such as pyridine or triethylamine will generate the acetamide. Acetamides are sensitive to strong acid and base, but are stable in the pH range 1 ~ 12. Nucleophiles and many organometallics react with the N-acetyl group. Organolithium reagents are unreactive, although reactive Grignard reagents are.

Cleavage:

The two major methods for conversion of N-acetamides to the amine are treatment with aqueous acid and treatment with triethyloxonium tetrafluoroborate. This group can also be reduced by catalytic hydrogenation, borane or borohydride reducing agents (not $LiAlH_4$), and can be oxidized by many oxidizing agents.

Benzamide is stable to pH 1 ~ 4 and is cleaved by strong acid (6N HCl, HBr) or diisobutylaluminum hydride

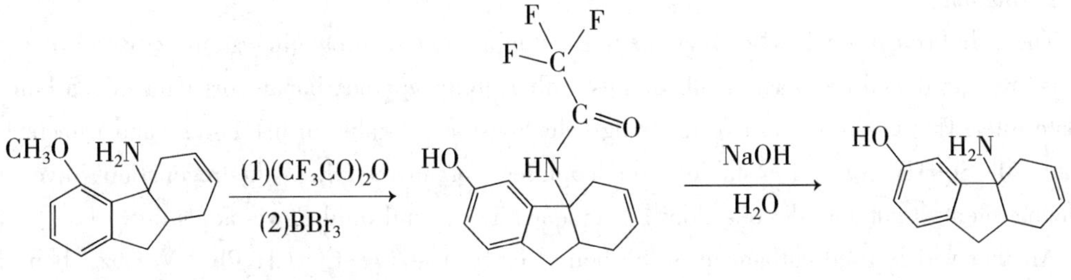

F F F—C C=O HN

CH_3O H_2N —(1)$(CF_3CO)_2O$, (2)BBr_3→ HO HN —NaOH, H_2O→ HO H_2N

Trifluoroacetyl groups are good amino-protecting groups. Owing to the strong EWG effect of the trifluoromethyl group, trifluoroacetamides are subject to hydrolysis under mild conditions (温和条件下水解).

7.4.2 氨基甲酸酯类 Carbamates

The carbamates (N-COOR) are a related class of protecting groups for nitrogen. Several different carbamates have been used for the protection of amino acids in peptide synthesis. One of the most popular is the *tert*-butyl carbamate (*tert*-butoxycarbonyl, [$NCOCMe_3$], N-Boc).

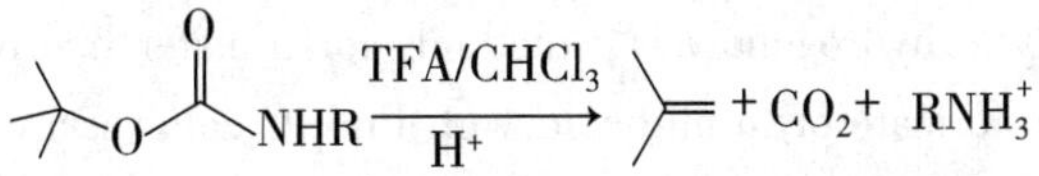

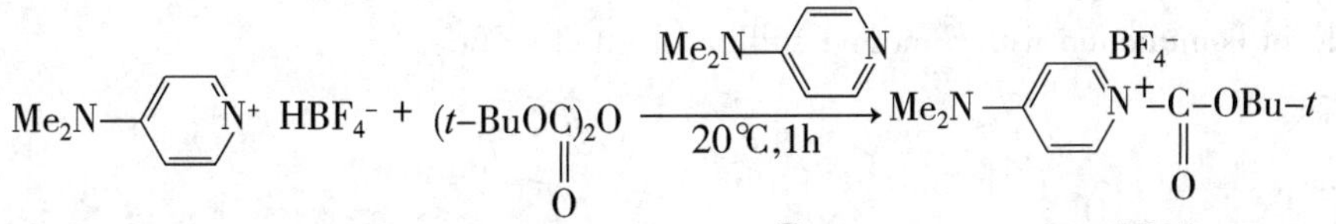

O, NHR —TFA/$CHCl_3$, H^+→ + CO_2 + RNH_3^+

Me_2N—(pyridinium)N^+ HBF_4^- + $(t\text{-}BuOC)_2O$ (O) —Me_2N-pyridine, 20℃,1h→ Me_2N—(pyridinium)N^+—C(=O)—OBu-t BF_4^-

(3) (95%)

H_2NCH—C(=O)ONa (R) —(1)(3)/H_2O, (2)HCl/H_2O→ t-BuO—C(=O)—NHCH(R)COOH

Cl_3C O C(=O) NHR —Zn-AcOH, $+2e^-$,$+H^+$→ Cl^- + Cl C(=O) Cl + CO_2 + RNH_2

AcO, AcO, AcO, O, OAc, $NH_3^+Cl^-$ —(1)$(Tfac)_2O$/Py, (2)HBr/AcOH→ AcO, AcO, AcO, O, TfacNH, Br —(1) Me_3Si, N, O=, N, O, Me_3Si; (2)NH_3/MeOH→ OH, OH, OH, O, H_2N, N, O=, N, H, O

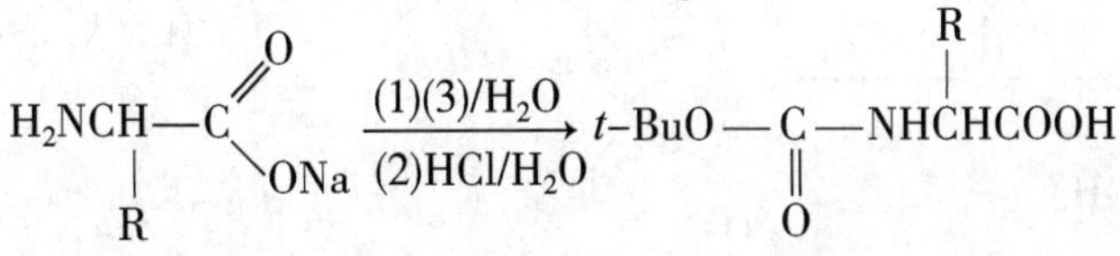

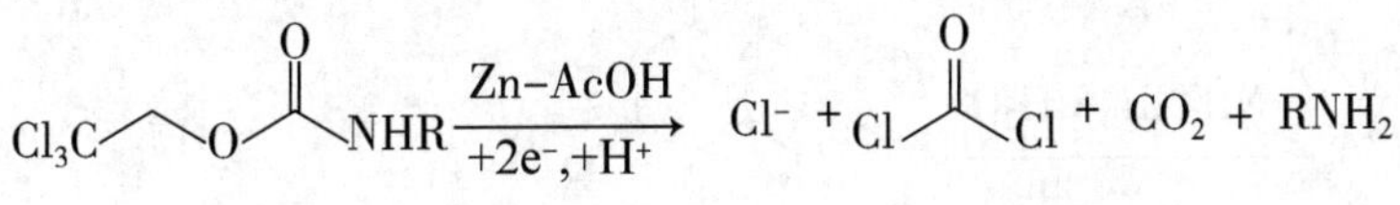

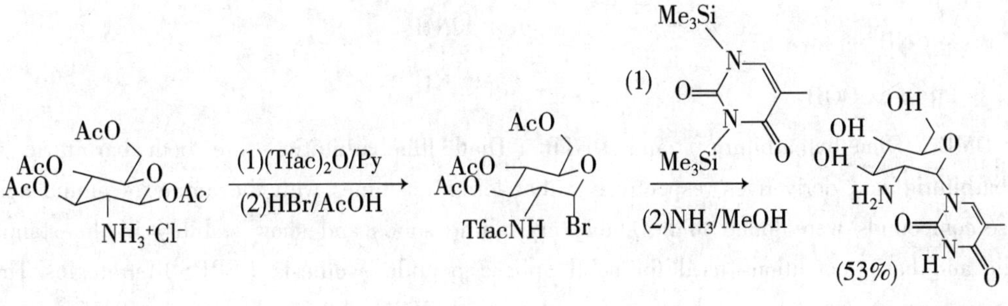

(53%)

Formation:

The *t*-butoxycarbonyl (*t*Boc) group is a valuable amino-protecting group. *t*-Butoxycarbonyl groups are introduced by reaction of amines with *t*-butoxypyrocarbonate or a mixed carbonate-imidate ester. This group is sensitive to strong anhydrous acid (stable to pH 1~12) and trimethylsilyl triflate ($Me_3Si-OTf$), but it is stable to nucleophiles, organometallics, hydrogenation, hydrides, oxidizing agents (not Jones' conditions), aqueous acid, and mild Lewis acids.

Another widely used carbamate is the benzyl carbamate ($N-CO_2CH_2Ph$, *N*-Cbz, benzyloxy carbonyl). The group is attached by reaction of an amine with benzyl chloroformate($PhCH_2O_2CCl$) in the presence of base (aqueous carbonate or triethylamine). The group (*N*-Cbz) is very stable to acid and base (pH 1 ~ 12), to nucleophiles, to milder organometallics (it reacts with organolithium reagents and Grignard reagents), to milder Lewis acids, and to most hydrides (it reacts with $LiAlH_4$). The group is sensitive to hydrogenolysis by catalytic hydrogenation, which is the primary means of cleaving this group (Pd on carbon is the usual catalyst).

Cleavage:

The Boc group is usually removed by treatment with aqueous HCl or with anhydrous trifluoroacetic acid.

The amine can be regenerated from a Cbz derivative by hydrogenolysis of the benzyl C—O bond, which is accompanied by spontaneous decarboxylation of the resulting carbamic acid. In addition to standard catalytic hydrogenolysis, methods for transfer hydrogenolysis using hydrogen donors such as ammonium formate or formic acid with Pd—C catalyst are available. The Cbz group also can be removed by a combination of a Lewis acid and a nucleophile: for example, boron trifluoride in conjunction with dimethyl sulfide or ethyl sulfide.

O R HN COOH N H + R H_2N COOH

O N N O O NH R COOH $NH_2NH_2 \cdot H_2O$ O N N O O NH NH_2 + R H_2N COOH

DMB (Dimethylbarbituric) and DETB (Diethylthiobarbituric) are both barbituric and thiobarbituric acid derivatives respectively, that forms enamines with the amine of amino acids. These compounds were found to be stable crystalline solids and show stability in the standard acidic and basic conditions used for solid-phase peptide synthesis (SPPS) strategies. These protecting groups are cleaved by a mild solution of 2% hydrazine hydrate in DMF and 2% hydroxylamine in DMF, both at short reaction times. Their use in SPPS showed that DMB-protected amino acids allow the preparation of peptides and therefore could be an alternative to the Fmoc strategy currently used. A further advantage of these protecting groups is that their preparation does not involve the concourse of phosgene derivatives and therefore they could be considered greener protecting groups than the carbamate-based one.

[1,3-Dimethyl-2,4,6-(1*H*, 3*H*,5*H*)-trioxypyrimidin-5-ylidene] methyl (DTPM) group was developed as an amine participating protecting group for aminosugars. The advantage of the DTPM group is the relatively mild protection/deprotection conditions.

AcO, AcO, AcO, O, Br, NHDTPM + HO (2-naphthylmethanol) —AgOTf / $(CH_2Cl)_2$→ AcO, AcO, AcO, O, O, NHDTPM

7.4.3 邻苯二甲酰亚胺类 Phthalimides

Phthalimides, which are used to protect primary amino groups, can be cleaved by treatment with hydrazine, as in the Gabriel synthesis of amines. This reaction proceeds by initial nucleophilic addition at an imide carbonyl, followed by an intramolecular acyl transfer.

N—COOEt + RNH_2 —Na_2CO_3/H_2O, 15~30min→ [NHR, NHCOOEt]

—$-H_2NCOOEt$→ N—R —NH_2NH_2→ RNH_2 + NH, NH

7.5 羧基保护基 Carboxy-Protecting Groups

Methyl Ester　*t*-Butyl Ester　Allyl Ester　Benzyl Ester

Phenyl Ester　Silyl Ester　Ortho Ester　Oxazoline

If only the O—H, as opposed to the carbonyl, of a carboxyl group has to be masked, it can be readily accomplished by esterification. Alkaline hydrolysis is the usual way for regenerating the acid. *t*-Butyl esters, which are readily cleaved by acid, can be used if alkaline conditions must be avoided. 2,2,2-trichloroethyl esters, which can be reductively cleaved with zinc, are another possibility. Some esters can be cleaved by treatment with anhydrous TBAF. These reactions proceed best for esters of relatively acidic alcohols, such as 4-nitrobenzyl, 2,2,2-trichloroethyl, and cyanoethyl[4].

$$RCOOH + R'OH \xrightarrow[\text{or DCC,DMAP}]{\text{EDC·HCl}} RCOOR'$$

HCl·N(CH₃)₂–CH₂CH₂CH₂–N=C=N–Et

1-[3-(dimethylamino)propyl]-3-ethyl carbodiimide hydrochloride (EDC)

dicyclohexyldiimide(DCC)

EDC · HCl is more expensive, but the urea by product is water soluble and simplifies the purification of products.

$$RCOOH \xrightarrow{\text{TMSCHN}_2\text{, MeOH,benzene.}} RCOOMe$$

This is considered a safe alternative to using diazomethane.

$$RCOOH \xrightarrow[\text{or Allyl alcohol,TsOH, benzene.}]{\text{Allyl bromide,Cs}_2\text{CO}_3\text{,DMF}} RCOOCH_2CH{=}CH_2 \xrightarrow{\text{Pd(Ph}_3\text{P)}_4\text{,RSO}_2\text{Na,CH}_2\text{Cl}_2} RCOOH$$

$$RCOOH \longrightarrow RCOOC(CH_3)_3$$

Formation:

(1) Isobutylene, H_2SO_4, Et_2O, 25℃.

(2) 2,4,6-trichlorobenzoyl chloride, Et_3N, THF; *t*-BuOH, DMAP, benzene, 20℃.

(3) *t*-BuOH, EDC · HCl, DMAP, CH_2Cl_2.

(4) *i*-PrN=C(O—Bu-*t*)NH—Pr-*i*, toluene, 60℃.

Cleveage:

(1) CF_3CO_2H, CH_2Cl_2.

(2) Bromocatechol borane.

Carboxylic acids can also be protected as ortho esters. Ortho esters derived from simple alcohols are very easily hydrolyzed, and the 4-methyl-2,6,7-trioxabicyclo[2.2.2]octane structure is a more useful ortho ester protecting group.

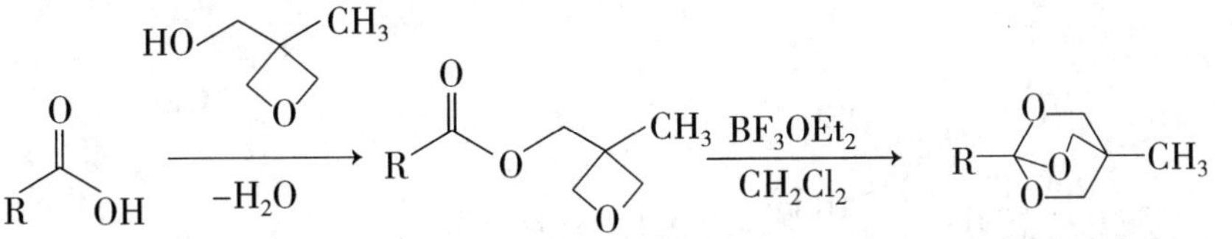

an OBO ortho ester

The more difficult problem of protecting the carbonyl group can be accomplished by conversion to an oxazoline derivative. One example is the 4,4-dimethyl derivative, which can be prepared from the acid by reaction with 2-amino-2-methylpropanol or with 2,2-dimethylaziridine. The heterocyclic derivative successfully protects the acid from attack by Grignard or hydride-transfer reagents. The carboxylic acid group can be regenerated by acidic hydrolysis or converted to an ester by acid-catalyzed reaction with the appropriate alcohol.

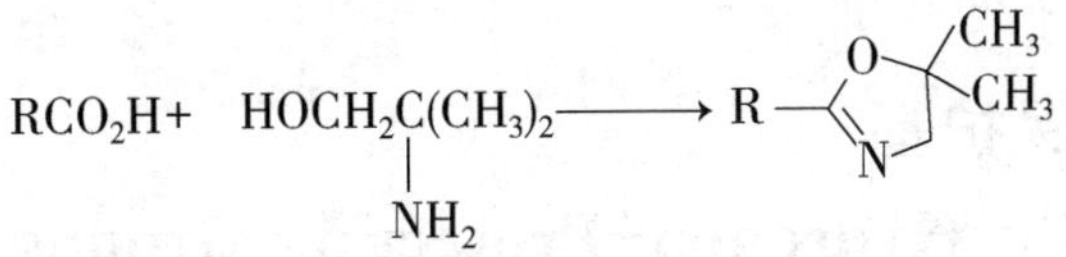

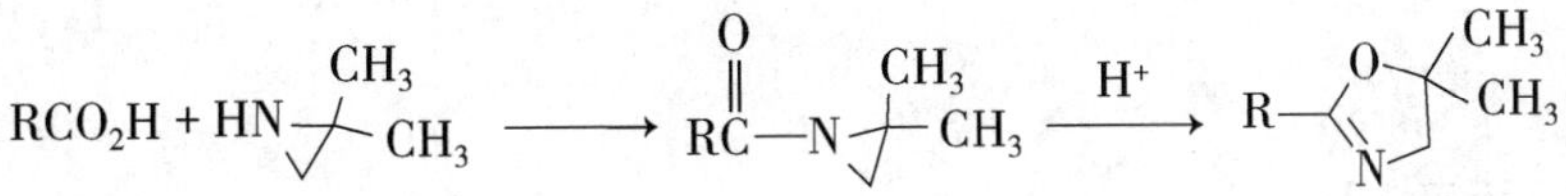

7.6 磷酸盐保护基 Phosphate-Protecting Groups

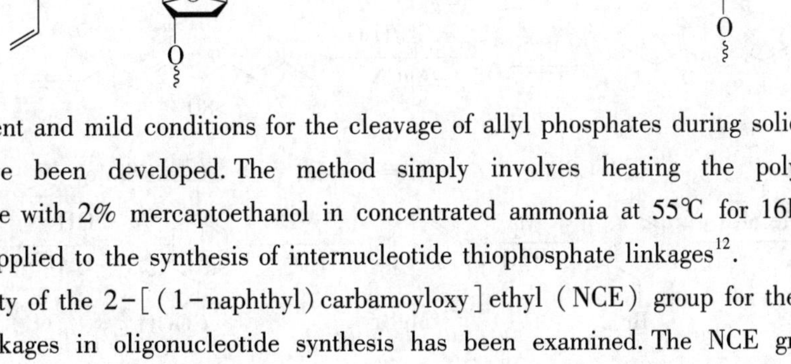

Convenient and mild conditions for the cleavage of allyl phosphates during solid-phase DNA synthesis have been developed. The method simply involves heating the polymer-bound oligonucleotide with 2% mercaptoethanol in concentrated ammonia at 55℃ for 16h. The method can also be applied to the synthesis of internucleotide thiophosphate linkages[12].

The utility of the 2-[(1-naphthyl)carbamoyloxy]ethyl (NCE) group for the protection of phosphate linkages in oligonucleotide synthesis has been examined. The NCE group is stable towards standard reagents used in DNA synthesis and it is quantitatively removed under basic conditions to release the phosphate (as its ammonium salt) and the inert oxazolidinone.

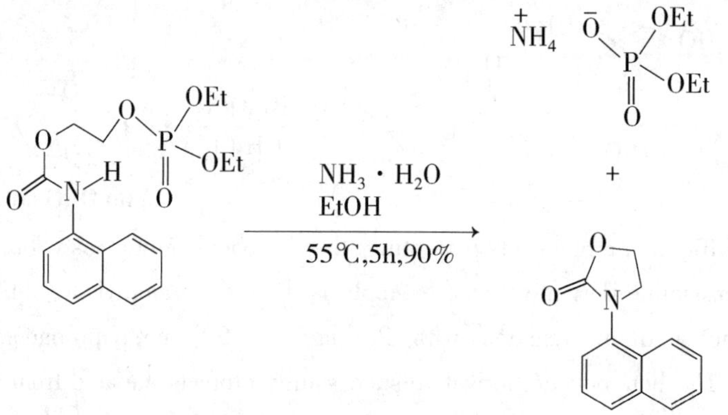

$NH_3 \cdot H_2O$
EtOH
55℃,5h,90%

The phosphoramidite strategy has been and is being most widely employed in both liquid phase and solid-phase syntheses of various kinds of oligos.

7.7 碳氢保护基 Carbon Hydrogen-Protecting Groups

7.7.1 末端炔 Alk-1-yne

Silyl alkynes have become most common for the protection of alk-1-yne with the generic structure C—SiR_3. They are prepared by the reaction of unhindered alk-1-yne with Grignard reagents, following by the exchange with TMSCl. The group is very stable to organometallics such as Grignard reagents, or to oxidants[13].

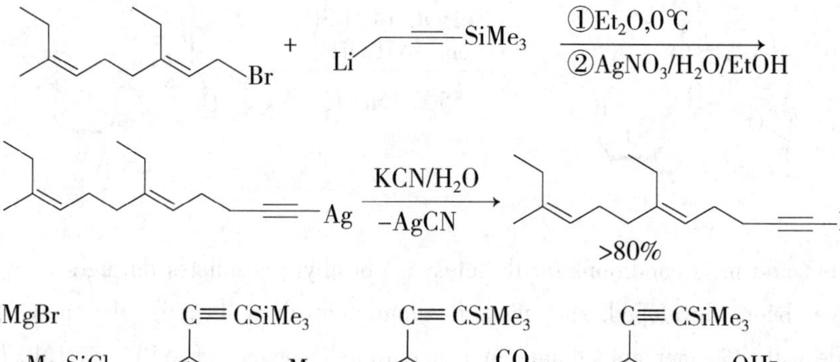

①Et_2O,0℃
②$AgNO_3/H_2O/EtOH$

KCN/H_2O
-AgCN
>80%

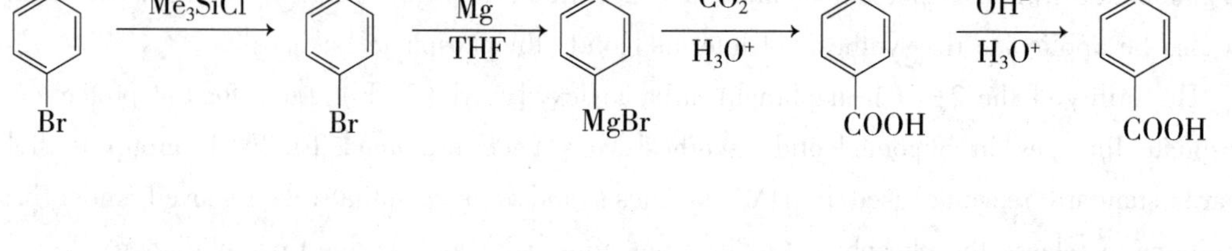

Me_3SiCl; Mg, THF; CO_2, H_3O^+; OH^-, H_3O^+

7.7.2 芳香族碳氢键 Aromatic C—H Bond

The decarboxylation of the aromatic acids is easily carried out than that of aliphatic acids. A base - promoted carboxylation introduce a carboxy group via Kolbe - Schmitt reaction. After completing substitution reaction, carboxy group is released at refluxing temperature.

OH, OH $\xrightarrow[\text{Kolbe reaction}]{CO_2/NaHCO_3}$ OH, OH, COOH $\xrightarrow[30\sim35℃]{Br_2/AcOH}$ Br, OH, OH, COOH $\xrightarrow[100℃,\text{reflux}]{H_2O}$ Br, OH, OH

Sulfonation is an easily reversible reaction and this makes the sulfonic acid group a useful blocking group in synthesis.

CH_3 $\xrightarrow[100℃]{\text{concentrated } H_2SO_4}$ CH_3, SO_3H $\xrightarrow[Fe]{Cl_2}$ CH_3, Cl, SO_3H $\xrightarrow[150℃]{H_3O^+}$ CH_3, Cl

Transformation of the nitro group into an amine and furtherdiazonium salts is a common reaction process. After bromination of the benzene ring, diazonium salts are removed and give back the unlocked products.

$C(CH_3)_3$ $\xrightarrow[H_2SO_4]{HNO_3}$ $C(CH_3)_3$, NO_2 $\xrightarrow[Fe]{Br_2}$ Br, $C(CH_3)_3$, NO_2 $\xrightarrow{Fe+HCl}$ Br, $C(CH_3)_3$, NH_2 $\xrightarrow[②H_3PO_2/H_2O]{①NaNO_2+H_2SO_4}$ Br, $C(CH_3)_3$

Friedel-Crafts alkylation leads to an easy introduction of *t*-butyl at benzene ring. Steric effects facilitate the regioselectivity. Since alkyl substituents activate the arene substrate, dichloride substitution is ready to occur. *t*-Butylbenzene is more reactive at high temperature in the presence of acidic catalysts and *t*-butyl group is eliminated at reaction conditions.

OH $\xrightarrow{,H_2SO_4}$ OH, $C(CH_3)_3$ $\xrightarrow{Cl_2}$ OH, Cl, Cl, $C(CH_3)_3$ $\xrightarrow[\text{activated clay}]{275\sim350℃}$ OH, Cl, Cl

7.7.3 脂肪族碳氢键 Aliphatic C—H Bond

α-H of carbonyl compounds is reactive in the presence of acidic or basic catalysts. Formation of aldehyde group at α-H of ketones can selectively protect CH_2 of ketones. Methylation at α-H of CH position is ready to occur when aldehyde group is eliminated at reaction conditions.

HCOOEt/EtONa, CHO, $C_6H_5NHCH_3$, H, O, $CHNC_6H_5$, CH_3

①$NaNH_2$ ②CH_3I, H_3C, $CHNC_6H_5$, CH_3, H_3O^+, △, CH_3

trans and cis isomers

参考文献 References

1. Weintraub P M, Turnbull K, et al. Annual Reports in Organic Synthesis. Boston: Academic Press, 1995. 315-330.
2. Stick R V, Williams S J. Carbohydrates: The Essential Molecules of Life (Second Edition). Oxford: Elsevier, 2009. 35-74.
3. Smith M B. Organic Synthesis (Third Edition). Oxford: Academic Press, 2010. 587-622.
4. The Role of Protective Groups in Organic Synthesis [M]. Protective Groups in Organic Synthesis. 1999: 1-16.
5. Aghapour G. Tandem and Selective Conversion of Tetrahydropyranyl and Silyl Ethers to Oximes Catalyzed with Trichloroisocyanuric Acid. *J. Phosphorus, Sulfur, and Silicon and the Related Elements*. 2015, 190 (9): 1464-1470.
6. Ganaie B A, et al. $BF_3 \cdot OEt_2$ Mediated Ethylation of Phenols: A New Protection Group Strategy. *J. Polycyclic Aromatic Compounds*. 2022.
7. Smith M B. Organic Synthesis (Fourth Edition). Boston: Academic Press, 2017. 185-213.
8. Stick R V. Carbohydrates. London: Academic Press, 2001. 37-65.
9. Robertson J, Stafford P M. Carbohydrates. Oxford: Academic Press, 2003. 9-68.
10. The Role of Protective Groups in Organic Synthesis [M]. Greene's Protective Groups in Organic Synthesis. 2006: 1-15.
11. Protection for the Amino Group [M]. Protective Groups in Organic Synthesis. 1999: 494-653.
12. Sontakke V A, et al. Synthesis and Stability of Nucleoside 3', 5'-Cyclic Phosphate Triesters Masked with Enzymatically and Thermally Labile Phosphate Protecting Groups. *J. European Journal of Organic Chemistry*. 2015, 2015 (2): 389-394.
13. Li W, Yu B. Advances in Carbohydrate Chemistry and Biochemistry. Oxford: Academic Press, 2020. 1-69.

第三部分 设计有机合成

Part Ⅲ Designing Organic Synthesis

8 逆合成分析战略

Chapter 8 Retrosynthetic Analysis Strategy

Retrosynthetic analysis as an imaginative process is introduced. Concepts of retrosynthetic analysis are presented and disconnection and functional group interconversion of target compounds are discussed. Some examples on synthons and synthetic equivalents are presented. Generic strategies in retrosynthesis are overviewed. The strategy to design and devise a synthesis of target molecule is indicated.

8.1 逆合成分析导论 Introduction to Retrosynthetic Analysis

8.1.1 合成策略 Synthetic Strategy

The construction of a complex organic structure, defined *target molecule* (目标分子), requires first of all the identification of the smaller fragments that can be used to build up the final target. In the case of oligomers such as peptides, oligosaccharides or oligonucleotides, the choice is obvious: the constituent monomers are the building blocks, and the synthesis requires the junction of those monomers by condensation. Two functional groups, one for each monomer, are involved in the reaction, whereas the other functional groups of the molecule, which can interfere in the reaction, must be protected. In order to perform the condensation, it is then required to activate one of the two functional groups that must react together. Nowadays peptide and oligonucleotide synthesis can be performed in an automated manner due to the repeatable procedure[1]. In the case of oligosaccharides, one additional problem must be solved, when a sugar links another sugar or an aglycon, a stereogenic center (the anomeric center) is involved in the reaction, and therefore two possible stereoisomers (defined α - and β - anomers) can be formed. Therefore, the stereochemical outcome of the reaction must be controlled in order to obtain the desired stereoisomer. This problem, together with the fact that carbohydrates present more than one hydroxyl group, and only one must react with the second sugar, makes oligosaccharide synthesis not trivial[2].

When the target molecule presents a complex skeleton mainly made by carbon atoms chains, no constituent monomers or building blocks can be immediately envisaged. In this case the identification of the starting materials for the synthesis requires much more fantasy and some rules: in other word, a retrosynthetic analysis[3].

8.1.2 逆合成分析基本概念 Basic Concepts in Retrosynthetic Analysis

The concept of retrosynthetic analysis has been developed by E. J. Corey who received for this reason the Nobel Prize in chemistry in 1990. In Corey's words "Retrosynthetic (or antithetic) analysis is a problem solving technique for transforming the structure of a synthetic target (TGT) molecule to a sequence of progressively simpler structures along a pathway which ultimately leads to simple or commercially available starting materials for a chemical synthesis".

The retrosynthetic analysis is based on a sequence of ***disconnections* (切断)**. Each disconnection is a mental process in which a molecule is fragmented into two pieces that in the mind of the synthetic organic chemist can generate the molecule under examination by known reactions. To make a simple example, Scheme 8-1 shows the possible disconnections of a simple target molecule. The cleavage of the linkage between the carbon atom bearing the oxygen and the ethyl (a) or the methyl (b) group makes two possible disconnections. The carbon atom bearing a hydroxyl group can be generated from a carbonyl function (electrophile), by reaction with a carbanion (nucleophile). In Scheme 8-1 the electrophile is the carbonyl group of acetone (disconnection a) or propanone (disconnection b) and the nucleophile an organometallic reagent such as ethyl magnesium bromide (disconnection a) or methyl lithium (disconnection b). The carbanion is defined ***synthon* (合成子)** whereas the organometallic reagent from which it is generated is defined ***synthetic equivalent*[4](等价试剂)**.

H_3C, OH, H_3CH_2C, CH_3 —a⇒ H_3C, H_3C C=O + $\ominus CH_2CH_3$ (synthon) - - -→ [C_2H_5MgBr synthetic equivalent]

—b⇒ H_3C, H_3CH_2C C=O + $\ominus CH_3$ (synthon) - - -→ [CH_3Li synthetic equivalent]

Scheme 8-1 Represented scheme of retrosynthetic analysis

The retrosynthetic analysis: a simple disconnection that presents two possibilities. Often, the target molecule is formed by a skeleton made not only by carbon atoms, but also involving some heteroatoms. The heteroatom represents and/or is part of functional groups. The carbon-heteroatom bonds of the functional group are ideal position for a disconnection. To make an example, esters and lactones, or amides and lactams, can be easily disconnected into the fragments that can generate those bonds, one containing the carboxylic group and the other containing the hydroxyl or the amino group. The presence of functional group in the skeleton of a target molecule will direct the choice of the disconnection[5].

As mentioned above, retrosynthetic analysis is an imaginative process in which the target molecule (TM) is disconnected into less complex structures, the next generation of target molecules. This procedure is repeated down to simple, easily available starting compounds. Methodical breaking apart of the target molecule leading to more simple structures that can be prepared by known or conceivable reactions is the basis of retrosynthetic analysis. Application of this procedure requires a basic knowledge of organic chemistry and rather strict adherence to certain rules. First we accentuate that the most important retrosynthetic rule is related to the basic property of the C—C bond, electronic structure and electronic charges of the fragments that emerge on disconnection of this bond. The rule states that disconnection should follow the correct mechanism. Products of disconnection are synthons - anionic or cationic fragments or radicals. Behind synthons, however, real molecules should exist, denoted as reagents or synthetic equivalents. Before consideration of the electronic structure of synthons and properties of their acceptable synthetic equivalents, let us see the general scheme that illustrates retrosynthetic analysis (Scheme 8-1). In Scheme 8-1 the waved line and bent arrow over the line representing the critical C—C bond indicate the site of disconnection. The broad arrow indicates the disconnection process from target molecule TM to the charged species, anionic synthon and cationic synthon. The dashed arrows then indicate the conceptual connection of synthons with real compounds, reagents or synthetic equivalents. This retro-analysis process continues until simple, commercially available compounds are reached. Consequently, any new target molecule along the retrosynthetic scheme should have more easily available synthetic equivalents than the previous one.

In the example reported in Scheme 8-2 the amide is generated from benzoic acid and a secondary amine, which in turn can be obtained by Michael addition to an α,β-unsaturated ester. In both cases, a carbon-heteroatom linkage is involved. Another important instrument in the retrosynthetic strategy is the ***functional group interconversion*** (**官能团互换**)[6]. Interconversion of a functional group (FGI) is one of the possible transformations of the functional group in the target molecule and includes change of the oxidation state or exchange of a heteroatom in this group. There are three main transformations of the functional group: interconversion (FGI), removal(FGR) and addition (FGA), assigned by double arrows (⇒).

target molecule　　synthons　　synthetic equivalent

$(H_3C)_2C(OH)CO_2H \Longrightarrow (H_3C)_2C^{\oplus}{-}OH \equiv (H_3C)_2C{=}O$

$\ominus CO_2H \equiv \ominus CN$

A simple disconnection：the cyanide ion is the synthetic equivalent of the−COOH synthon.

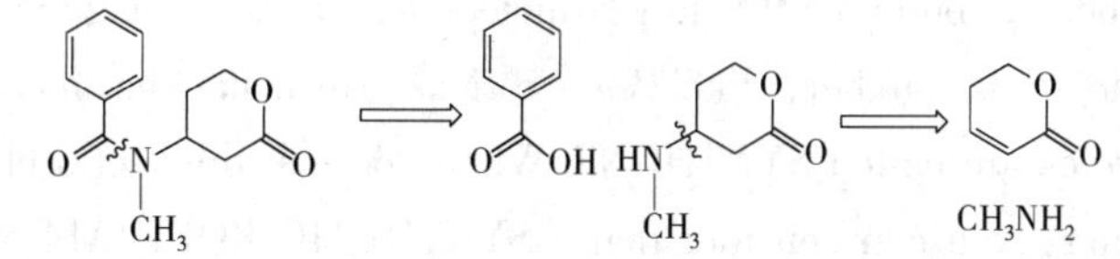

O N CH$_3$ O O ⟹ O OH HN CH$_3$ O O ⟹ O O CH_3NH_2

Scheme 8−2　Disconnection of a molecule containing an amidic bond

Definitions（定义）

Target Molecule（TM）（定义）	what you need to make
Retrosynthetic Analysis（逆合成分析）	the process of deconstructing the TM by breaking it into simpler molecules until you get to a recognisable SM
Starting Material（SM）（起始原料）	an available chemical that you can arrive at by retrosynthetic analysis and thus probably convert into the target molecule
Disconnection（切断）	taking apart a bond in the TM to see if it gives a pair of reagents
Functional Group Interconversion（FGI）（官能团变换）	changing a group in the TM into a different one to see if it gives an accessible intermediate
Synthon（合成子）	conceptual fragments that arise from disconnection
Synthetic Equivalent（等价试剂）	chemical that reacts as if it was a synthon

8.1.2.1　Examples

Target molecule：

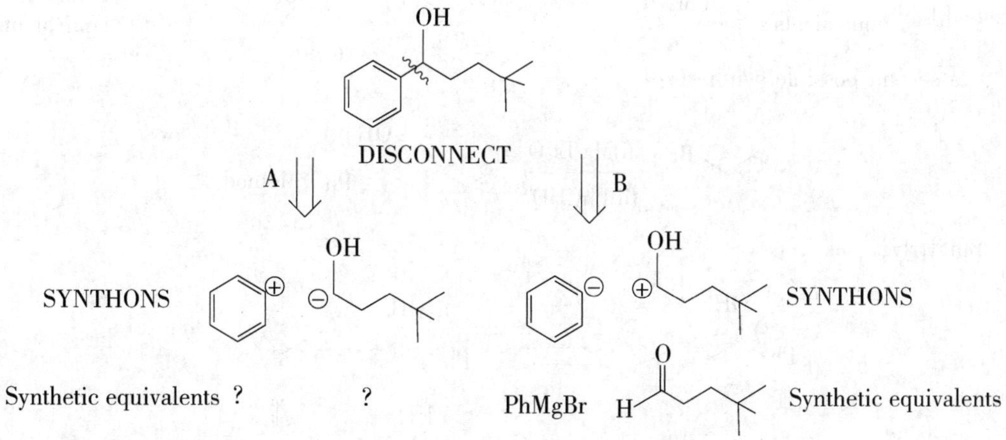

OH

DISCONNECT

A　　B

OH　　OH

SYNTHONS ⊕ ⊖ 　　⊖ ⊕ SYNTHONS

Synthetic equivalents ?　　?　　PhMgBr　H O　Synthetic equivalents

Therefore，the target molecule could be synthesised as follows：

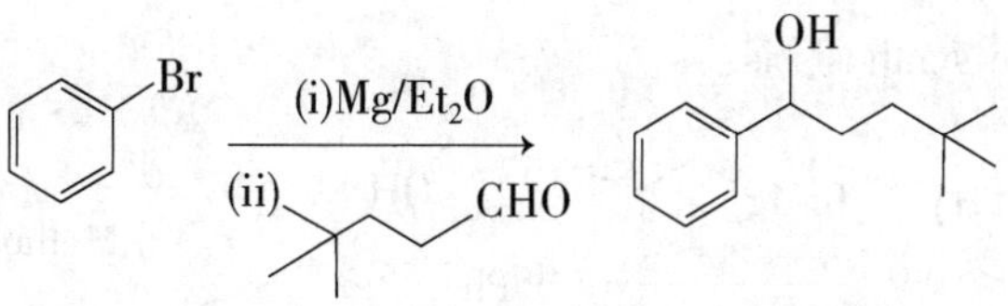

Br $\xrightarrow[\text{(ii)} \quad \text{CHO}]{\text{(i)Mg/Et}_2\text{O}}$ OH

What is a synthon?

When we disconnect a bond in the target molecule, we are imagining a pair of charged fragments that we could stick together, like Lego® bricks, to make the molecule we want. These imaginary charged species are called SYNTHONS. When you can think of a chemical with polarity that matches the synthon, you can consider that a SYNTHETIC EQUIVALENT of the synthon.

Thus, R–C⊕(OH)–H ≡ R–C(=O$^{\delta-}$)$^{\delta+}$–H an aldehyde is a synthetic equivalent for the above synthon[2].

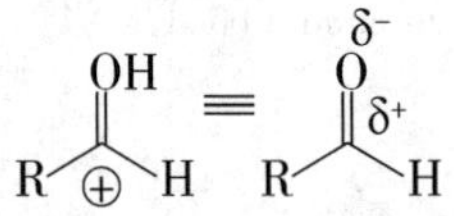

There can be more than one synthetic equivalent for a given synthon, but if you can't think of one... try a different disconnection.

Always consider alternative strategies:

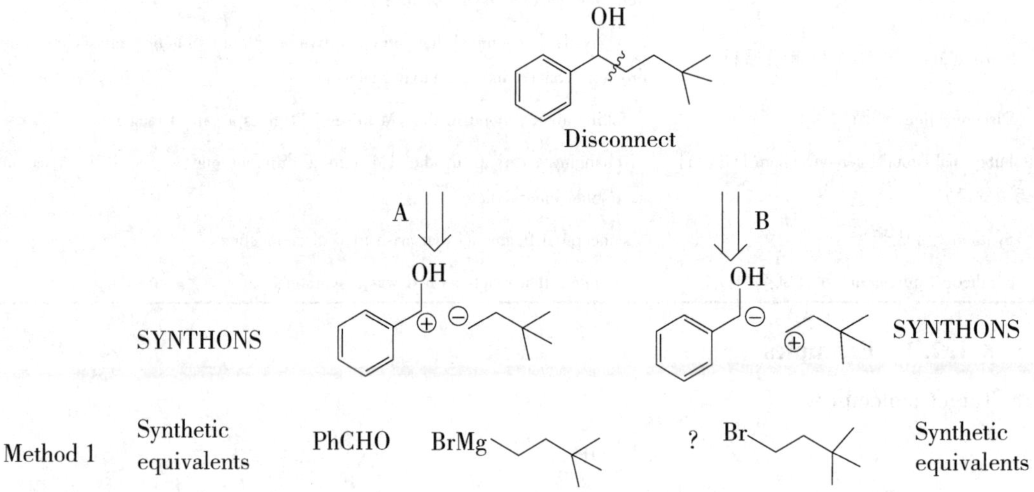

OH

Disconnect

A B

OH OH

SYNTHONS SYNTHONS

Method 1 Synthetic equivalents PhCHO BrMg ? Br Synthetic equivalents

∴a second possible synthesis:

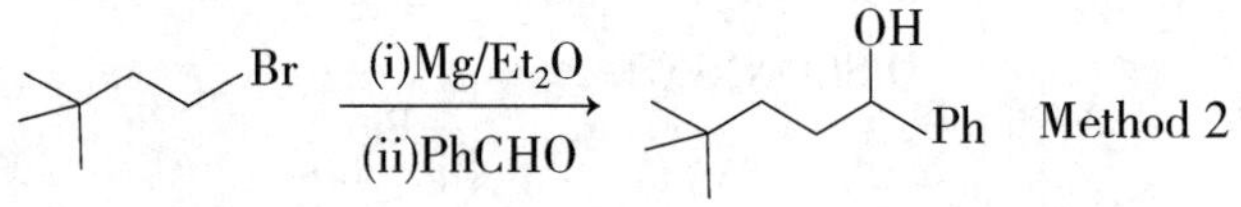

Br (i)Mg/Et_2O (ii)PhCHO OH Ph Method 2

Similarly,

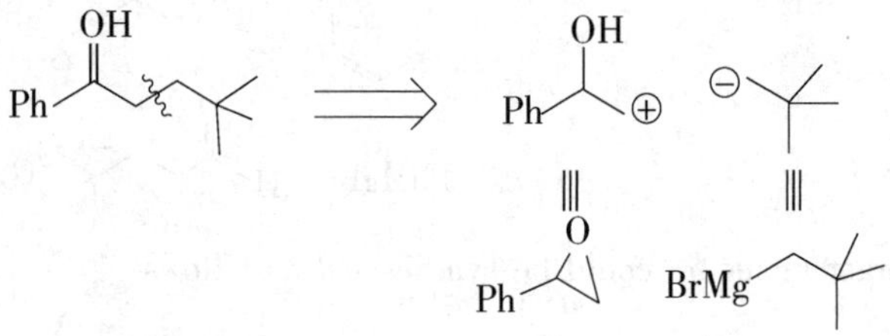

OH Ph ⟹ OH Ph ⊕ ⊖

O Ph BrMg

thus a third possible synthesis is:

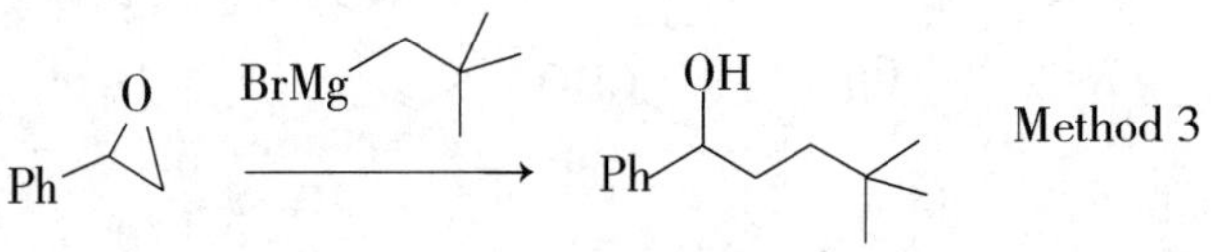

O Ph BrMg ⟶ OH Ph Method 3

Besides disconnections, we can also consider functional group interconversion. Our target molecule is a secondary alcohol, which could be prepared by reduction of a ketone.

This is represented as follows:

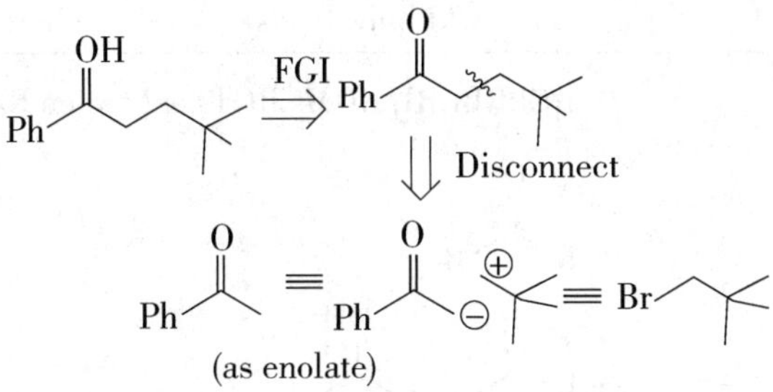

OH Ph ⟹ FGI Ph O

⇓ Disconnect

Ph O ≡ Ph O ⊖ (as enolate) ⊕ ≡ Br

∴synthesis number four:

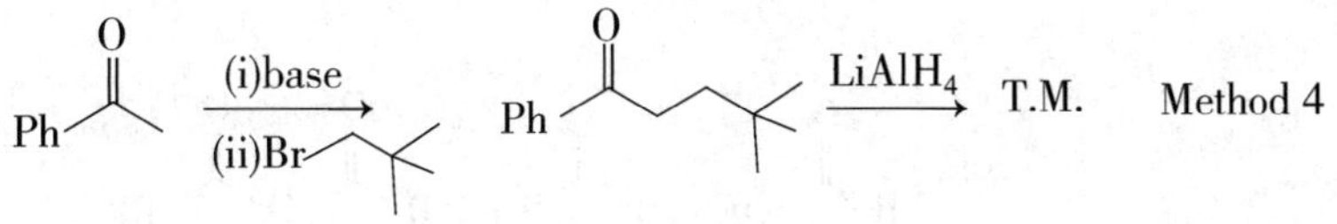

Ph O —(i)base, (ii)Br→ Ph O —$LiAlH_4$→ T.M. Method 4

Analysis number five:

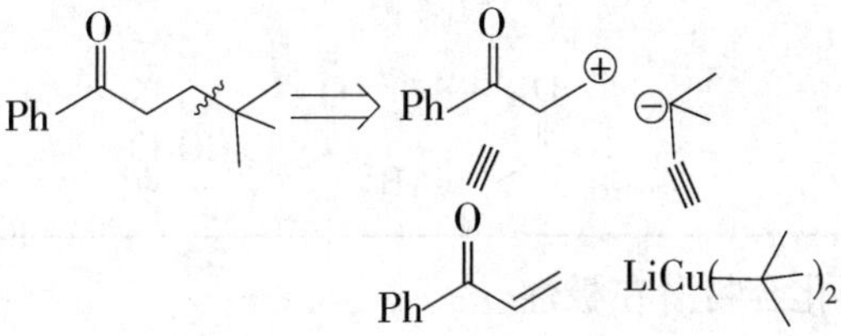

Ph O ⟹ Ph O ⊕ ⊖

≡ Ph O ≡ $LiCu(\text{t-Bu})_2$

Synthesis number five:

Ph O —t-Bu_2CuLi→ Ph O —$NaBH_4$→ T.M. Method 5

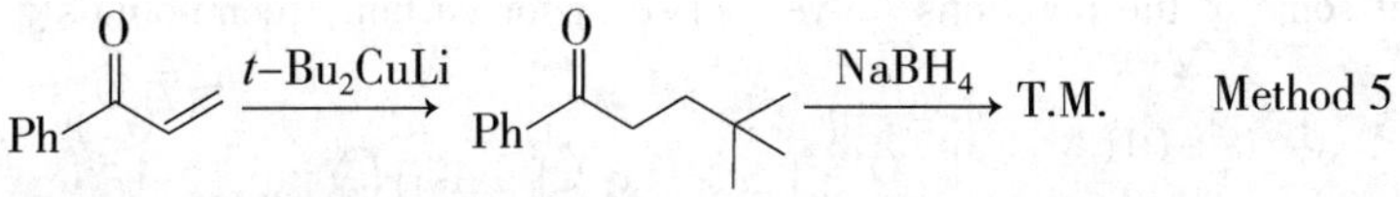

Disconnecting heteroatoms can also be a good idea:

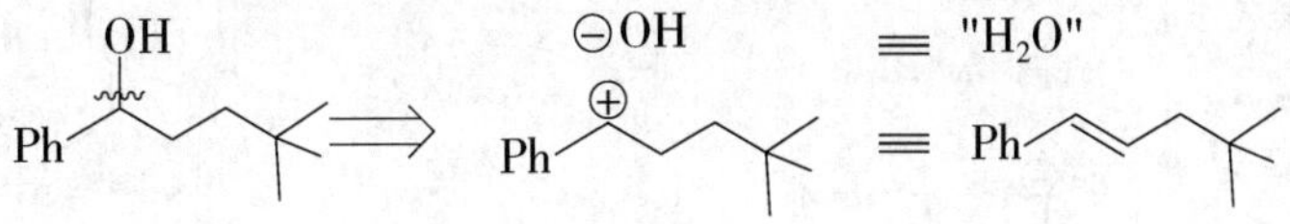

OH Ph ⟹ ⊖OH ≡ "H_2O" Ph ⊕ ≡ Ph

6th approach:

Ph —(i) $Hg(OAc)_2$, (ii) $NaBH_4$→ Ph OH Method 6

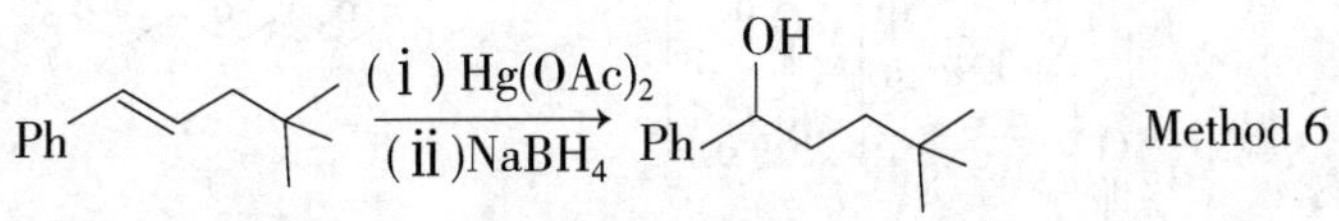

There are other possibilities, but let's not bother with any more.

How do you choose which method?

Personal choice. If you have a favourite reagent, or if you are familiar with a particular reaction (or if you have a strong aversion to a reaction/reagent) then this will affect your choice.

Also you need to bear in mind the efficiency of the reactions involved, and any potential side reactions (for example, self-condensation of PhCOMe in Method 4).

8.1.2.2 合成子和等价试剂
Some synthons and synthetic equivalents

Synthon	Equivalent(s)
$R^{\oplus}$	RCl,RBr,RI, ROMs,ROTs only when R=Alkyl
R–C⊕(OH)–R	R–CO–R
R–CH(OH)–CH2⊕	epoxide of R–CH=CH2; R–CH(OH)–CH2Br
R–CO–CH2–CH2⊕	R–CO–CH=CH2
R–C⊕=O	R–CO–OEt; R–CO–Cl; R–CO–O–CO–R
$R^{\ominus}$ (alkyl;NOT"RH+base")	RMgBr,RLi,R_2CuLi,other organometallic reagents
R–CO–CH2⊖	R–CO–CH3; R–CO–CH2–CO_2Et

8.1.2.3 单官能团化合物潜在极性
Latent polarity in monofunctional compounds

Think about some of the reactions we've looked at for carbonyl compounds:

A: ketone + $Nu^{\ominus}$ → R2C(OH)Nu; [C=O: O δ^-, C δ^+]

B: ketone α-H + :base → enolate ($O^{\ominus}$) + $E^{\oplus}$ → α-E ketone; [O δ^-, C δ^+, α-C δ^-]

C: enone + $Nu^{\ominus}$ → enolate with β-Nu; + $E^{\oplus}$ → ketone with α-E and β-Nu; [O δ^-, C δ^+, α-C δ^-, β-C δ^+]

i.e. δ^+ δ^- δ^+(C=O, O δ^-) δ^- δ^+ δ^- δ^+ etc.

The partial positive and negative charges indicate the latent polarity of the bonds in a molecule. They help us choose the synthons for key disconnections in a retrosynthetic analysis. *viz.*

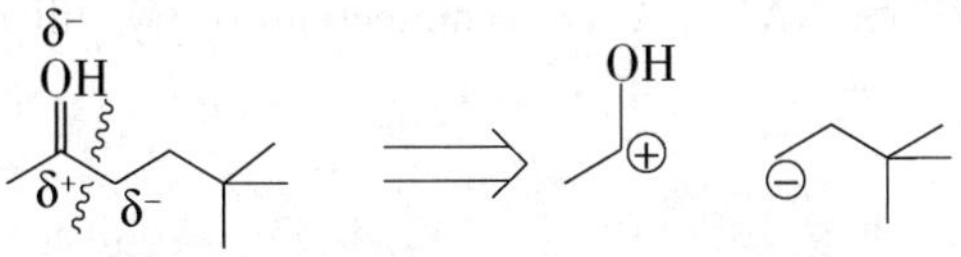

one of the disconnections we saw earlier.

8.1.2.4 双官能团化合物潜在极性 Latent polarity in bifunctional compounds

Consider a 1,3-disubstituted molecule, e. g.

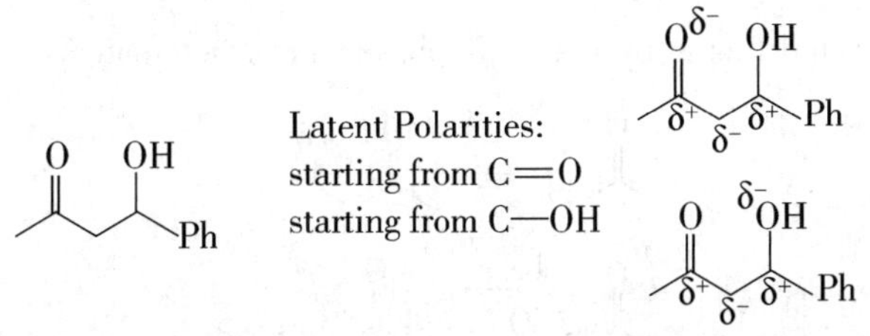

Latent Polarities:
starting from C=O
starting from C—OH

When the latent polarities in a bifunctional molecule overlap they reinforce each other, this is termed Consonant Polarity (辅助极性). In these circumstances the analysis is straightforward.

thus:

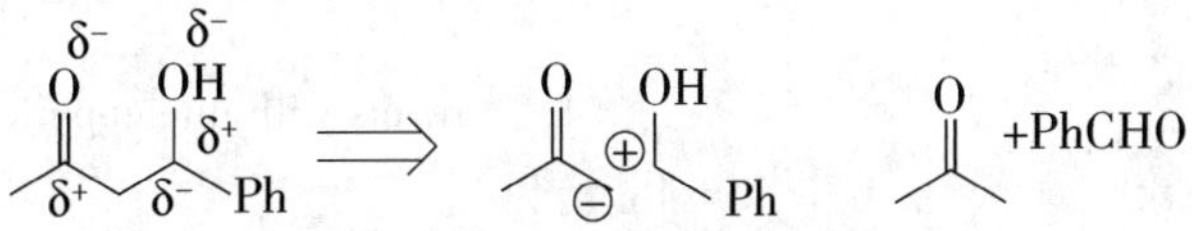

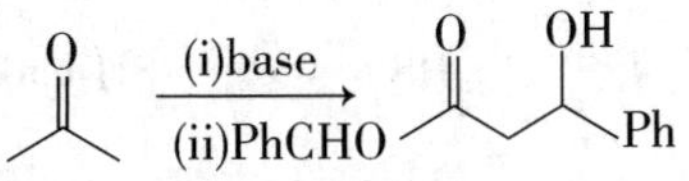

+PhCHO

(i)base
(ii)PhCHO

The same applies to 1,5-disubstitution:

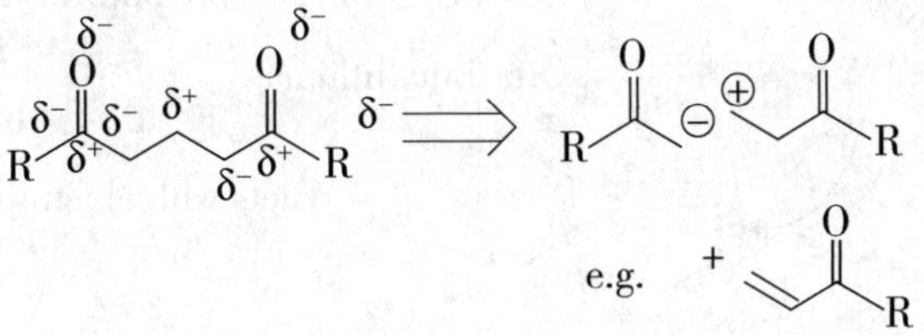

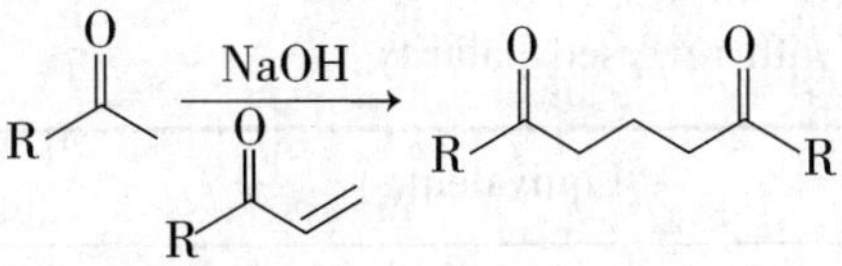

e.g. +

NaOH

But what about 1,4-disubstitution?

The polarities don't overlap and are termed DISSONANT. Any disconnection we try will result in a synthon that has the "wrong" polarity.

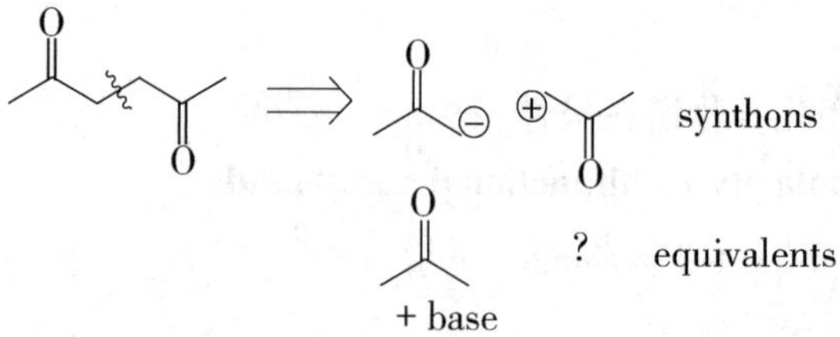

synthons

? equivalents

+ base

One way to get around this is by judicious placement of heteroatoms:

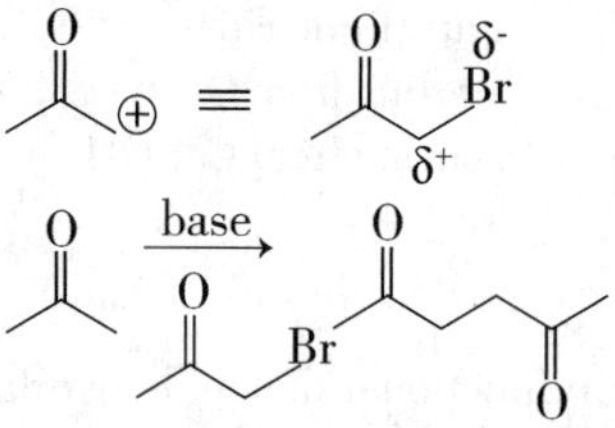

δ- Br δ+

base

Br

The German word UMPOLUNG (极性反转), meaning *polarity reversal* is used to describe the situation where the polarity in a compound is deliberately changed to facilitate a particular reaction. For example,

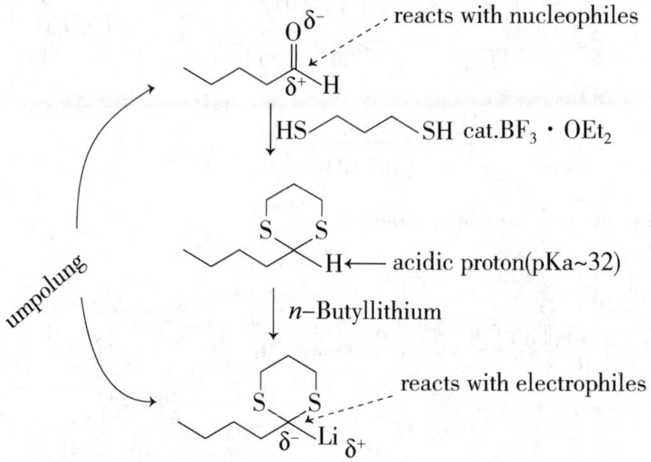

reacts with nucleophiles

$O^{\delta-}$ δ^{+} H

HS SH cat.$BF_3 \cdot OEt_2$

S S H← acidic proton(pKa~32)

umpolung

n–Butyllithium

reacts with electrophiles

S S δ^{-} Li δ^{+}

Equivalents for synthons with reversed polarity:

Synthon	Equivalent(s)
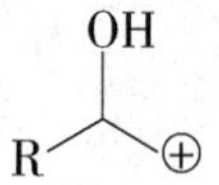 R–CH(OH)–⊕	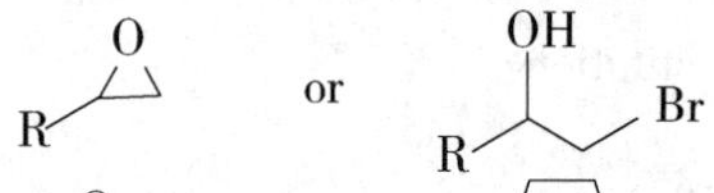 R-epoxide or R–CH(OH)–CH₂Br
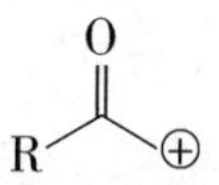 R–C(=O)–⊕	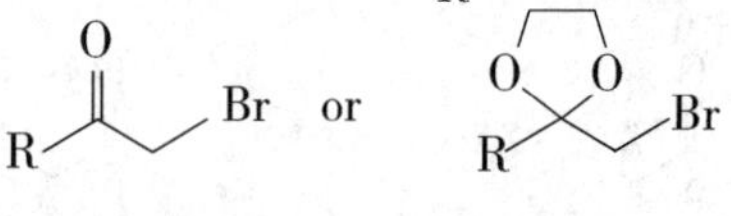 R–C(=O)–CH₂Br or R-dioxolane–CH₂Br

续表

Synthon	Equivalent(s)
MeCO⁻ (acyl anion)	ethoxyethylene (OEt) + *sec*–BuLi
RCO⁻ (acyl anion)	2-R-1,3-dithiane + *n*–BuLi "Corey–Seebach reaction"
HCO⁻ (formyl anion)	1,3,5-trithiane + *n*–BuLi or $MeNO_2$ +base ("Nef reaction")
HOCO⁻	NaCN

ethoxyvinyllithium EVL:

OEt-ethylene —s–BuLi (VERY strong base)→ [α-lithio ethoxyethylene] —$E^{\oplus}$→ CH₂=C(OEt)E —H_3O^+→ MeCOE

similarly from acetylene:

acetylene —(i)base, (ii)$E^{\oplus}$→ HC≡CE —H_3O^+, HgO→ [CH₂=C(OH)E] —tautom.→ MeCOE

Latent polarity and FGI (a quick consideration):

PhCOCH=CHPh —FGI→ Mismatched(dissonant) / Matched(consonant)

Leads to obvious disconnection ⟹ PhCOCH₂⁻ + ⁺CH(O)Ph {PhCOMe&PhCHO}

8.2 逆合成分析战略 Strategy in Retrosynthesis

8.2.1 通用的逆合成分析战略 Generic Strategy in Retrosynthesis

(ⅰ) Consider different possibilities. Try a number of disconnections and FGI's. Try to keep the number of steps down, and stick to known & reliable reactions. In real life, a synthesis has to be economically viable[7].

(ⅱ) Whenever possible, go for a convergent route rather than a linear one, as this will lead to a higher overall yield. e. g.

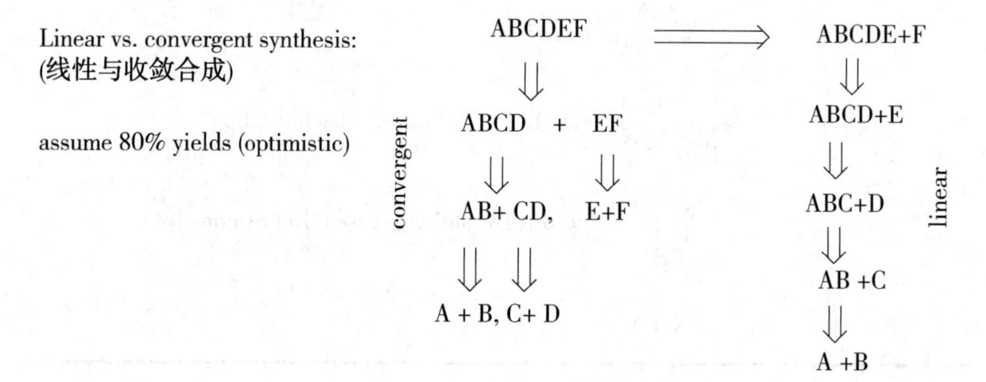

Linear:

	step1		2		3		4		5		...10		...15	
A	→	AB	→	ABC	→	ABCD	→	ABCDE	→	ABCDEF	→	A...K	→	A......P
approx overall yield:		80%		64%		51%		40%		32%		...10%		...3.5%

Convergent:

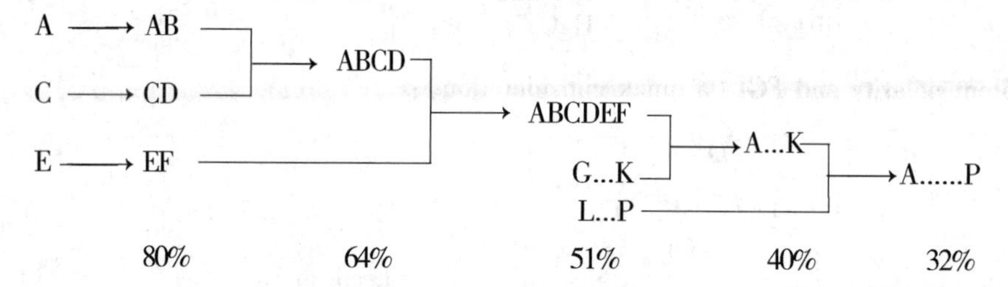

The purely convergent synthesis is ideal; virtually all real syntheses are linear to some degree.

(ⅲ) Aim for the greatest simplification

Make disconnections towards the middle of the molecule (this is more convergent anyway).

Disconnect at branch points.

Use symmetry where possible.

e. g. (towards the middle)

Ph ⟹ Ph ⊖ ⊕

Ph —base / MVK→ Ph

MVK methyl vinyl ketone

e. g. (at branches)

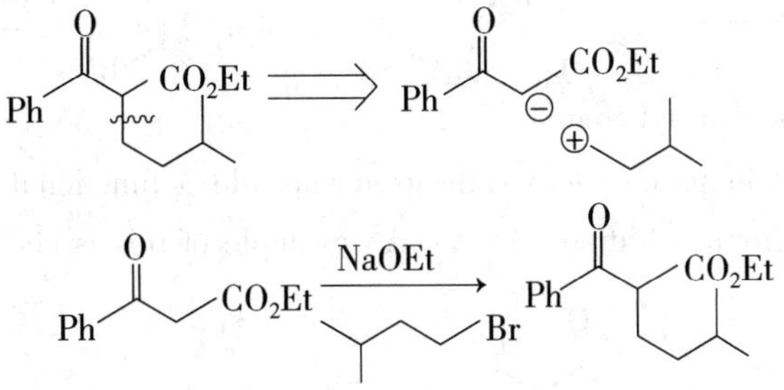

O Ph CO_2Et ⟹ O Ph ⊖ CO_2Et ⊕

O Ph CO_2Et —NaOEt, Br→ O Ph CO_2Et

e. g. (look for symmetry)

O ⟹ HO O ⟹ OH ⊕ O ⊖

O —NaOEt, self-condensation→ [HO O] —$-H_2O$→ O

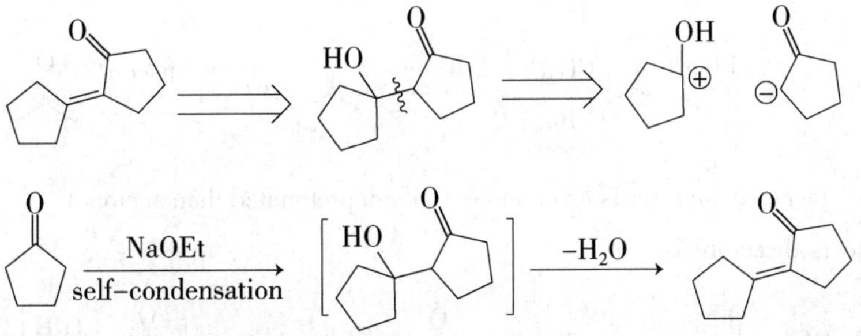

(ⅳ) Add reactive functional groups at a late stage in the synthesis, so they aren't carried through steps where they could react to give side products.

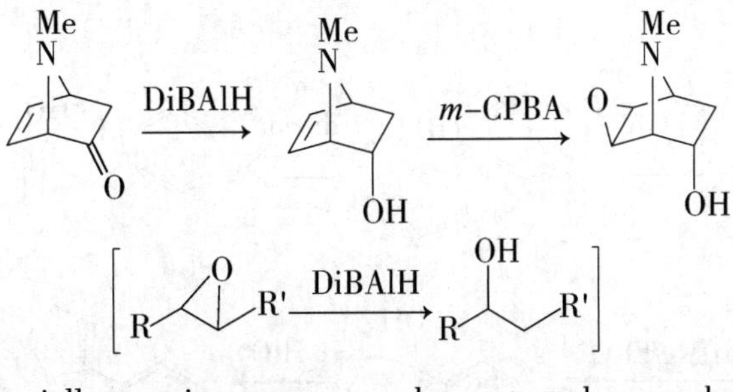

Me N O —DiBAlH→ Me N OH —*m*-CPBA→ Me N O OH

[R O R' —DiBAlH→ R OH R']

Alternatively, potentially reactive groups can be protected or masked so they don't react, e. g. reduction of an ester in the presence of a ketone.

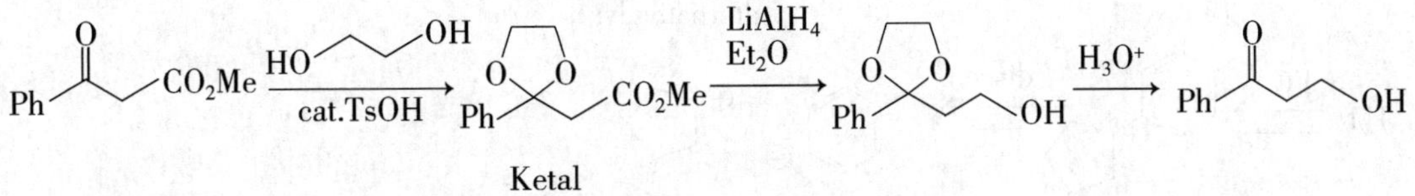

O Ph CO_2Me —HO OH, cat.TsOH→ O O Ph CO_2Me —$LiAlH_4$, Et_2O→ O O Ph OH —H_3O^+→ O Ph OH

Ketal
(stable to bases and nucleophiles)

Note that protection strategy requires two extra steps (must be efficient); better syntheses minimise the use of protecting groups.

A masked group is a functional group that is introduced and can be converted into a different one at a later stage.

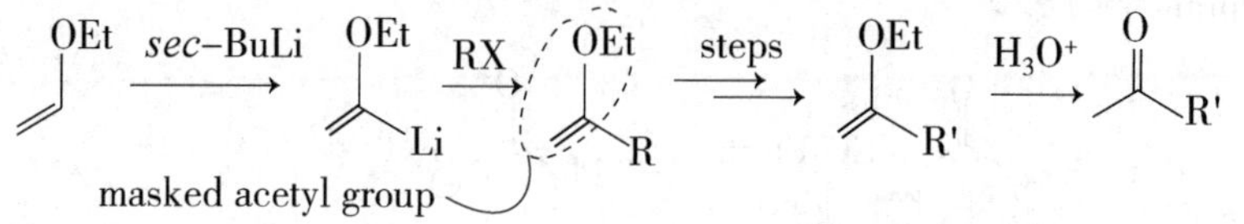

OEt —sec–BuLi→ OEt, Li —RX→ OEt, R —steps→ OEt, R' —H_3O^+→ O, R'

masked acetyl group

(V) Sometimes it helps the retrosynthesis if you add a functional group to facilitate bond formation (Functional Group Addition, FGA). An example of this is acetoacetic ester synthesis:

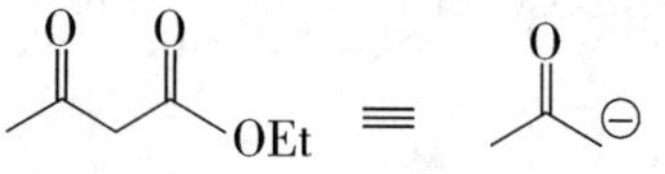

OEt ≡ ⊖

Thus:

FGA ⟹ CO_2Et, Bu; discon. ⟹ CO_2Et; discon. ⟹ CO_2Et

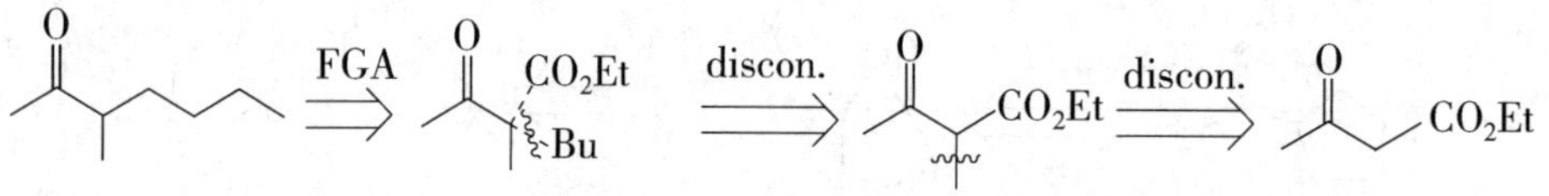

(acetoacetic ester is much more easily deprotonated than acetone)

The synthesis therefore is:

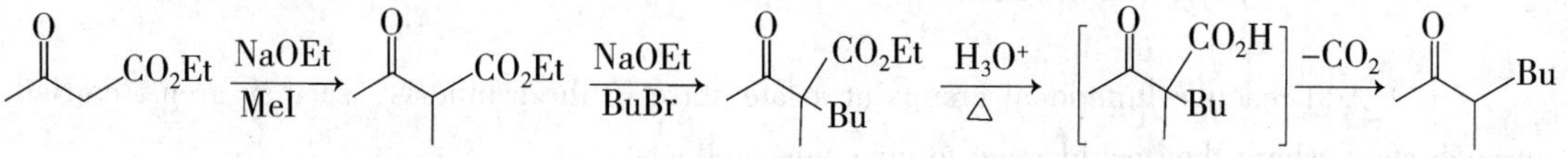

CO_2Et —NaOEt / MeI→ CO_2Et —NaOEt / BuBr→ CO_2Et, Bu —H_3O^+ / △→ [CO_2H, Bu] —$-CO_2$→ Bu

The strategy of FGA applies especially in the case of molecules containing no reactive functional groups:

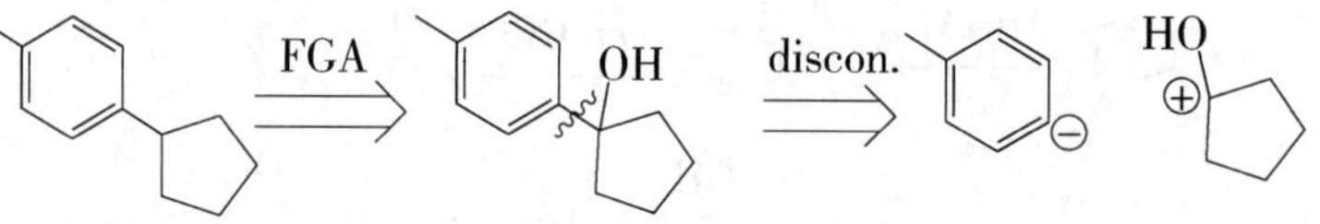

FGA ⟹ OH; discon. ⟹ ⊖ + HO ⊕

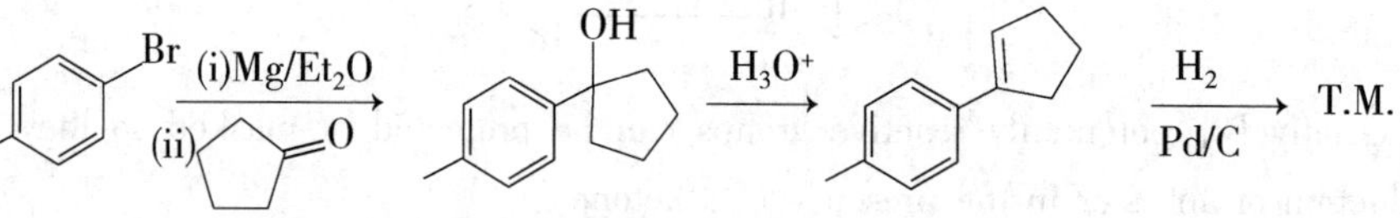

Br —(i)Mg/Et_2O (ii) cyclopentanone→ OH —H_3O^+→ —H_2 / Pd/C→ T.M.

alternatively:

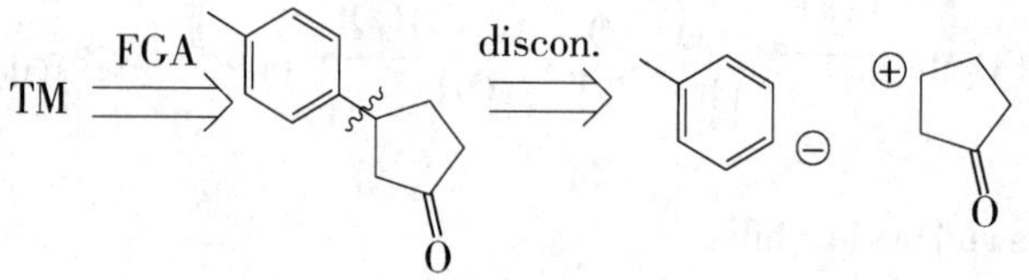

TM —FGA⟹ O; discon. ⟹ ⊖ + ⊕, O

O —$(p\text{-Tol})_2CuLi$→ O —Zn—Hg / HCl→ T.M.

$2\ ArLi + CuCl \longrightarrow Ar_2CuLi + LiCl$

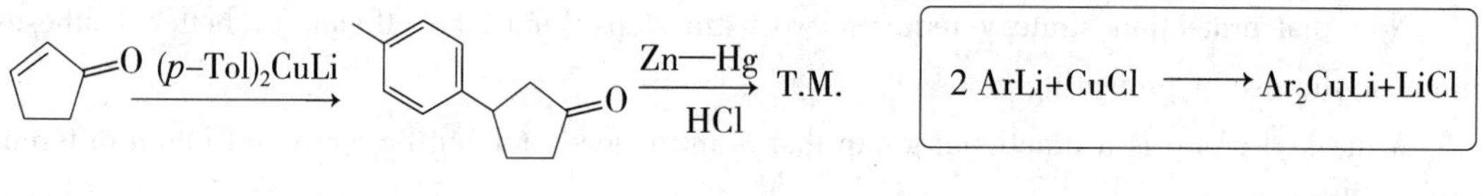

8.2.2 环闭合反应 Ring Closing Reactions

Synthesis of carbocyclic molecules is same with the approaches as to acyclic systems. The probability of reaction between two functional groups is higher if:

(a) reaction is intramolecular (faster reaction).

(b) the distance between the two groups is shorter.

e. g. Intramolecular alkylation[8]:

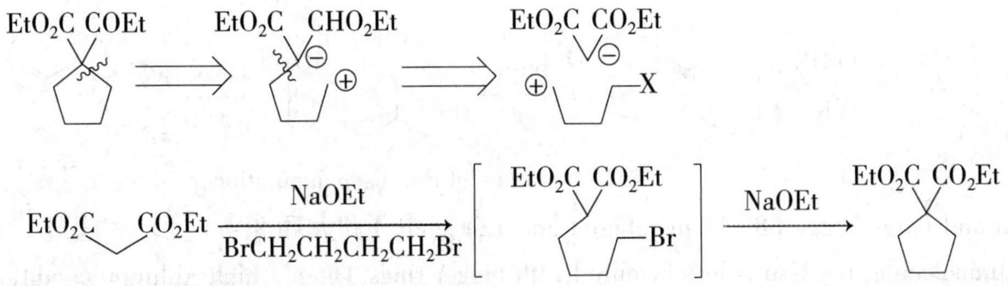

EtO_2C COEt ⟹ EtO_2C CHO_2Et ⟹ EtO_2C CO_2Et X

EtO_2C CO_2Et $\xrightarrow[BrCH_2CH_2CH_2CH_2Br]{NaOEt}$ [EtO_2C CO_2Et Br] $\xrightarrow{NaOEt}$ EtO_2C CO_2Et

Intramolecular acylation e. g. the Dieckmann cyclisation; especially good for 5-membered rings:

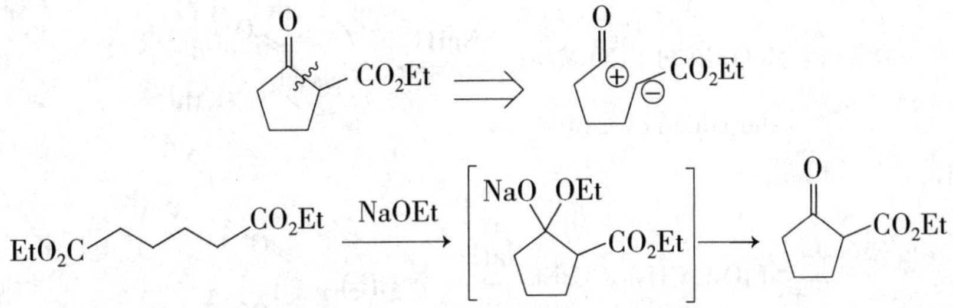

O CO_2Et ⟹ O CO_2Et

EtO_2C CO_2Et $\xrightarrow{NaOEt}$ [NaO OEt CO_2Et] ⟶ O CO_2Et

condensation:

O ⟹ O OH ⟹ O OH

O O $\xrightarrow{t\text{-BuOK}}$ [O OH] ⟶ O

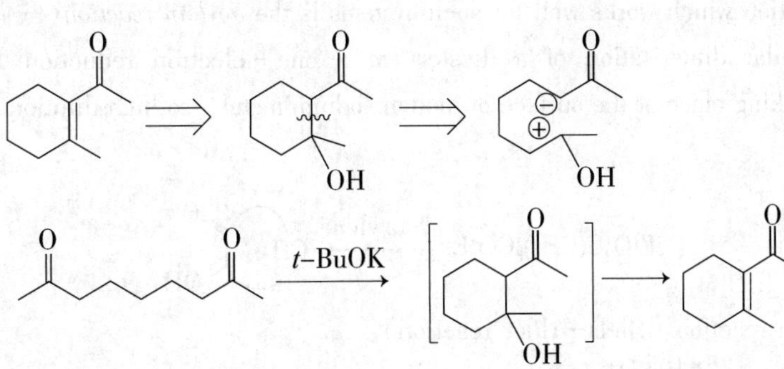

Bicyclic molecules are prepared from cyclic precursors following similar principles.

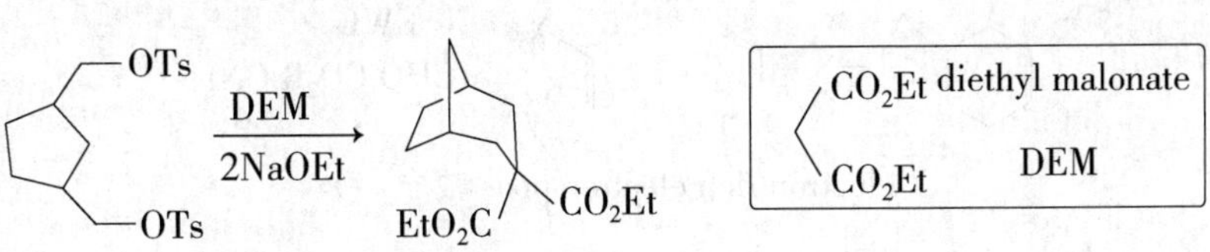

OTs OTs $\xrightarrow[2NaOEt]{DEM}$ EtO_2C CO_2Et

CO_2Et diethyl malonate
CO_2Et DEM

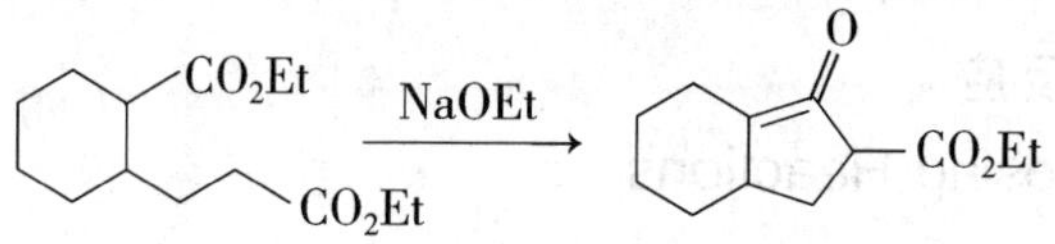
CO2Et
NaOEt
CO2Et
O
CO2Et

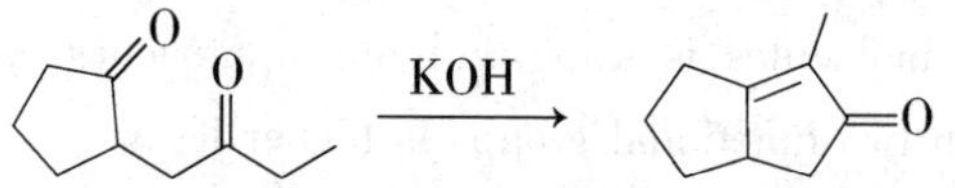
O
O
KOH
O

A special example of condensation is the *Robinson annulation* (opinions vary as to the spelling) . It has been widely used in classical steroid synthesis. It involves Michael addition followed by intramolecular cyclisation[9]:

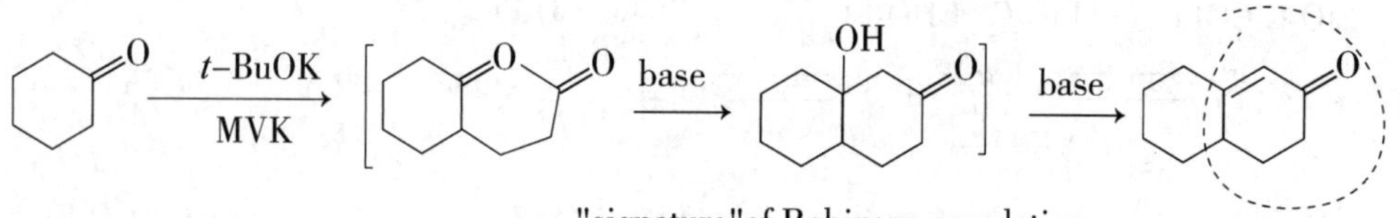
O
t-BuOK
MVK
O
O
base
OH
O
base
O

"signature"of Robinson annulation

Medium and Large Rings (8~11 membered and 12+)(中环和大环 8~11 元环和超过 12 元环)

Intramolecular reaction is less favoured with bigger rings. Often, high-dilution conditions and slow addition can be used to suppress intermolecular reaction and hence promote ring closure.

e. g.

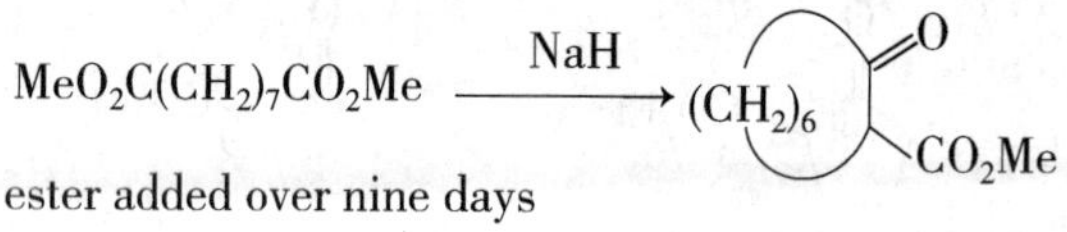
MeO2C(CH2)7CO2Me
NaH
(CH2)6
O
CO2Me
ester added over nine days

Similarly,

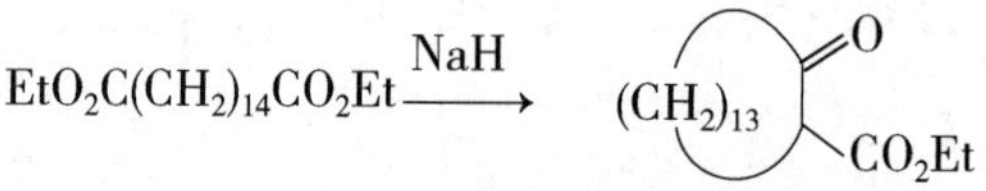
EtO2C(CH2)14CO2Et
NaH
(CH2)13
O
CO2Et

Another reaction which works well for such systems is the *acyloin reaction*(酰偶姻反应). This is the intramolecular dimerisation of a diester *via* a one - electron reduction. The reaction is heterogeneous, taking place at the surface of molten sodium metal, so high dilution is not required.

e. g.

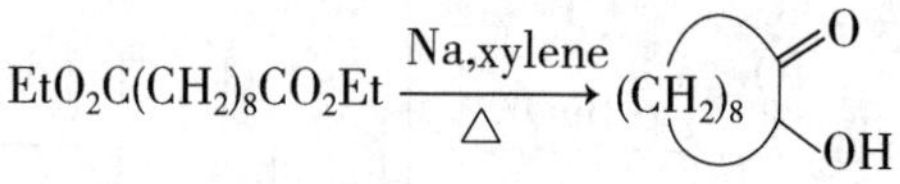
EtO2C(CH2)8CO2Et
Na,xylene
△
(CH2)8
O
OH

Cycloaddition reaction (Diels-Alder reaction):

Generic reaction (通用反应) (in retrosyntheic terms):

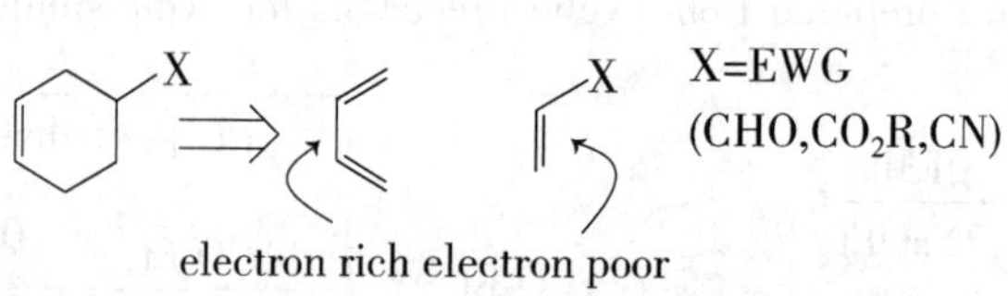
X
X
X=EWG
(CHO,CO2R,CN)
electron rich electron poor

e. g.

concerted reaction

&

These reactions are concerted reactions, usually they are highly stereospecific. This is because the reactions are governed by Frontier Orbital Theory. The actual rules of frontier orbital theory don't interest us at the moment, all we need is a simple guideline we can remember:

Unsymmetrical Diels-Alder reactions:

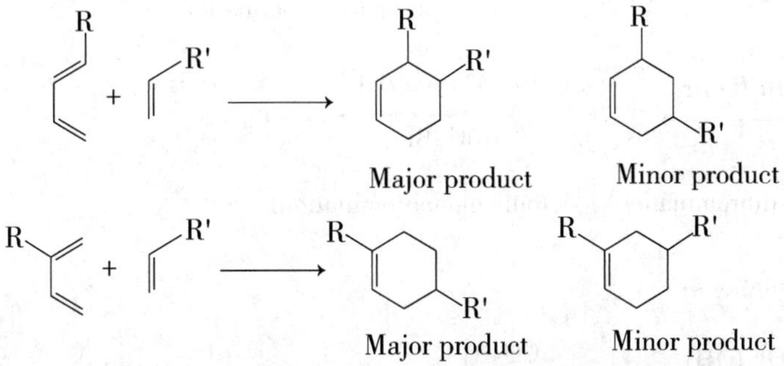

Major product   Minor product

Major product   Minor product

Note that the 1,3-disubstituted product is the **minor product** in both cases.

specific example:

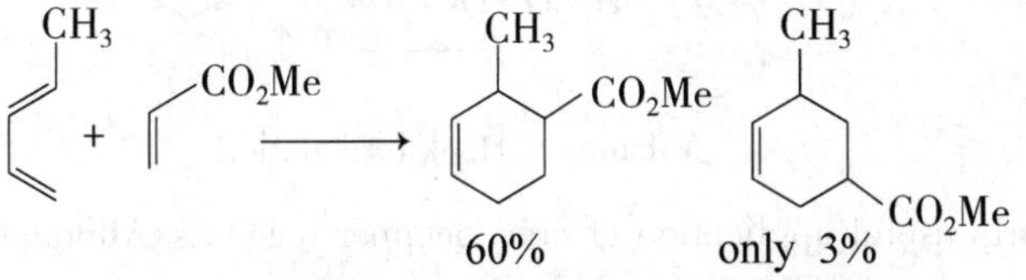

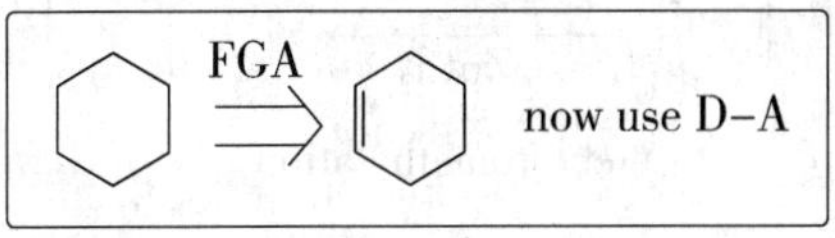

60%   only 3%

FGA   now use D-A

8.2.3 芳香体系中切断和官能团互换 Disconnections & Functional Group Interconversion in Aromatic Systems

Some reactions used in aliphatic systems don't apply for aromatic systems (S_N1 and S_N2 reactions, for example, are extremely unfavourable for ArX. There is a whole bunch of other reactions that apply for aromatic systems[10].

e. g.

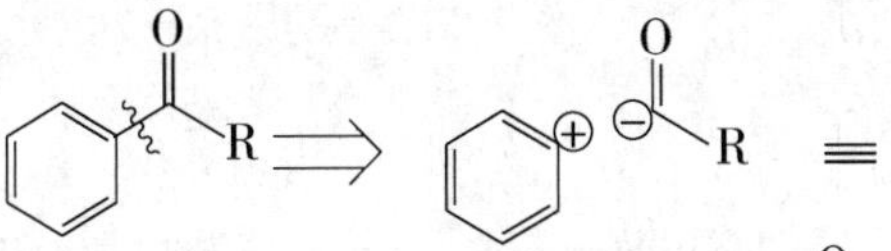

$\equiv$ $PhH+RCOCl+AlCl_3$

Friedel-Crafts acylation

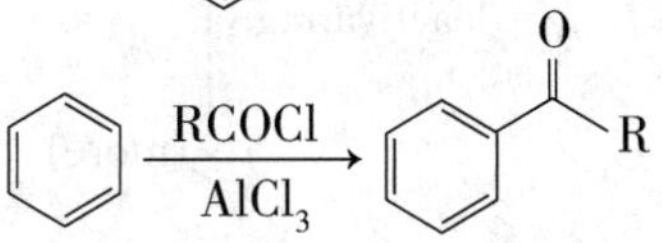

$\xrightarrow[AlCl_3]{RCOCl}$

NH_2 ⇒ NO_2 ⇒ ⊖ ⊕NO_2 ≡ $PhH+HNO_3+H_2SO_4$
aromatic nitration

fuming HNO_3 / H_2SO_4 → NO_2 → Sn/HCl or H_2,Pd/C → NH_2

Br ⇒ ⊕ $Br^{\ominus}$ ≡ $PhN_2X+CuBr$
Sandmeyer reaction

⇒ ⊖ ⊕Br ≡ $PhH+Br_2+ FeBr_3$
aromatic bromination

$Br_2/FeBr_3$ → Br ← (i)$NaNO_2$/HCl (ii)CuBr — NH_2

(can get dibromination) (only monobromination)

Some other reactions:

$KMnO_4$, $(OH)^-$ / H_2O-*t*-BuOH → CO_2H

I → R_2CuLi → R

H_2CO/HCl / $ZnCl_2$ → Cl
chloromethylation

$H_2CO/POCl_3$/DMF → CHO
Vilsmeier-Haak formylation

The last reaction above is a particularly useful application of organocopper reagents. Although the mechanism is quite complicated, it's the result we're interested in at the moment. It's a transformation that is not always easy to achieve by more conventional means.

In planning synthesis of polysubstituted aromatics, the order of reactions is important to ensure that the reagents are compatible and to take advantage of the directing effect of existing substituents:

Group	Directs	Activation	
NH_2,NR_2 OH,O^- NHAc, OR alkyl/aryl/vinyl	*ortho/para**	activating	↑ (more)
CO_2^- X(halogen)	*ortho/para**	neutral	
CO_2H CN COR,CHO SO_3H CX_3 NO_2	*meta*	deactivating	↓ (more)

note that *ortho para*-mixtures can be formed and may have to be separated.

Examples:

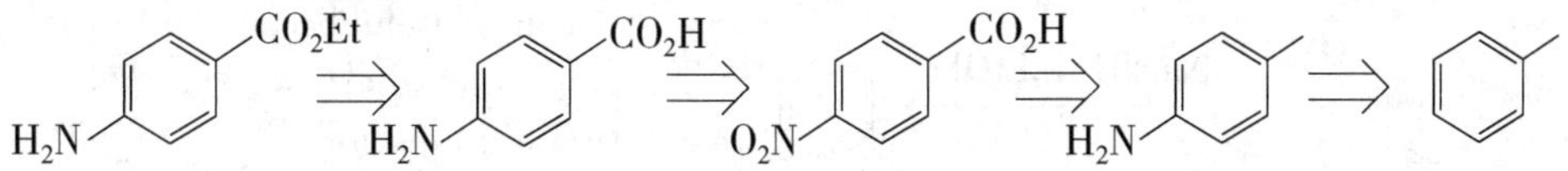

H_2N–C6H4–CO_2Et ⇒ H_2N–C6H4–CO_2H ⇒ O_2N–C6H4–CO_2H ⇒ H_2N–C6H4–Me ⇒ C6H5–Me

benzocaine (painkiller) from toluene

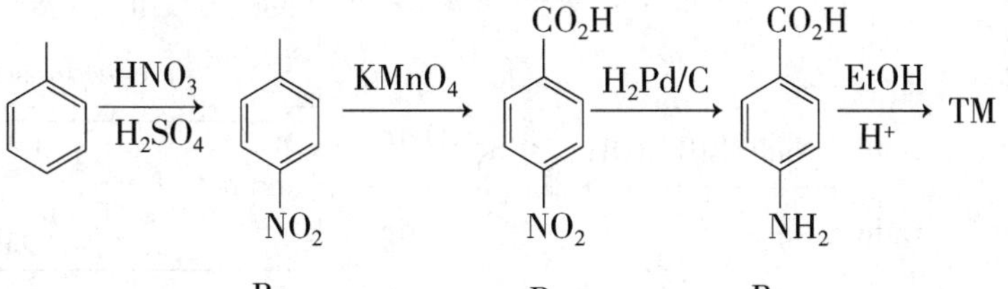

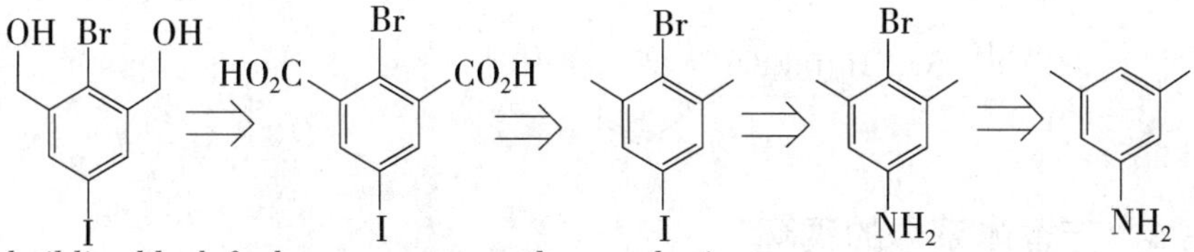

toluene $\xrightarrow[H_2SO_4]{HNO_3}$ p-nitrotoluene (NO_2) $\xrightarrow{KMnO_4}$ (CO_2H, NO_2) $\xrightarrow{H_2Pd/C}$ (CO_2H, NH_2) $\xrightarrow[H^+]{EtOH}$ TM

(OH, Br, OH, I) ⇒ (HO_2C, Br, CO_2H, I) ⇒ (Br, I) ⇒ (Br, NH_2) ⇒ (NH_2)

building block for homogeneous catalyst synthesis

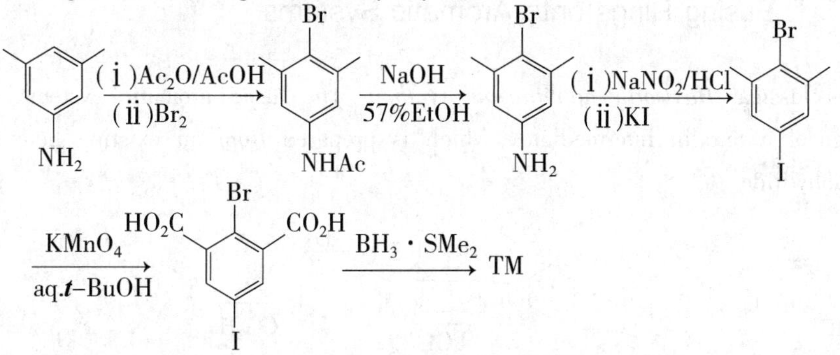

(NH_2) $\xrightarrow[(ii)Br_2]{(i)Ac_2O/AcOH}$ (Br, NHAc) $\xrightarrow[57\%EtOH]{NaOH}$ (Br, NH_2) $\xrightarrow[(ii)KI]{(i)NaNO_2/HCl}$ (Br, I)

$\xrightarrow[aq.t\text{-}BuOH]{KMnO_4}$ (HO_2C, Br, CO_2H, I) $\xrightarrow{BH_3 \cdot SMe_2}$ TM

Birch Reduction:

Partial reduction of aromatic systems by (usually) sodium in liquid ammonia. It's an example of dissolving metal reduction. Such methods used to be quite popular but most applications have been replaced by modern hydride reagents. Dissolving metal reduction does still have it's uses though, and the Birch reduction is one of them[11] (also recall the specific reduction of alkynes to *trans*-alkenes).

The typical conditions involve liquid ammonia (bp. −33℃) and sodium metal, in the presence of a proton source (usually an alcohol, EtOH).

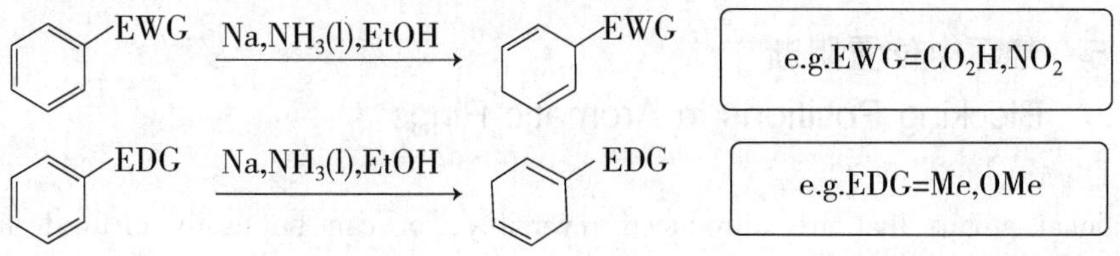

C6H5–EWG $\xrightarrow{Na,NH_3(l),EtOH}$ 1,4-cyclohexadiene–EWG | e.g.EWG=CO_2H,NO_2

C6H5–EDG $\xrightarrow{Na,NH_3(l),EtOH}$ 1,4-cyclohexadiene–EDG | e.g.EDG=Me,OMe

Examples:

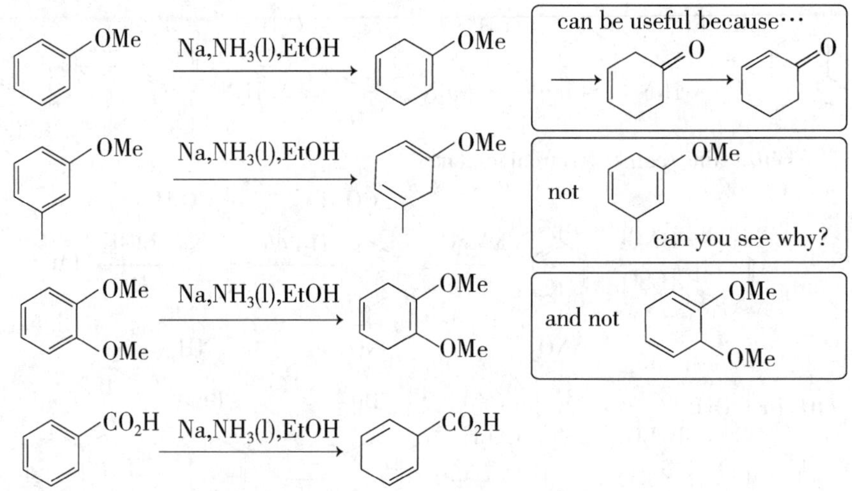

OMe Na,NH_3(l),EtOH OMe

can be useful because…

O O

OMe Na,NH_3(l),EtOH OMe

OMe not

can you see why?

OMe Na,NH_3(l),EtOH OMe

OMe OMe

and not OMe OMe

CO_2H Na,NH_3(l),EtOH CO_2H

8.2.4 芳香体系中的稠环 Fusing Rings onto Aromatic Systems

The classical *Hayworth naphthalene synthesis*. The fused aromatic system is formed by dehydration of a tetralin intermediate, which is prepared from an existing benzene ring and succinic anhydride[12,13].

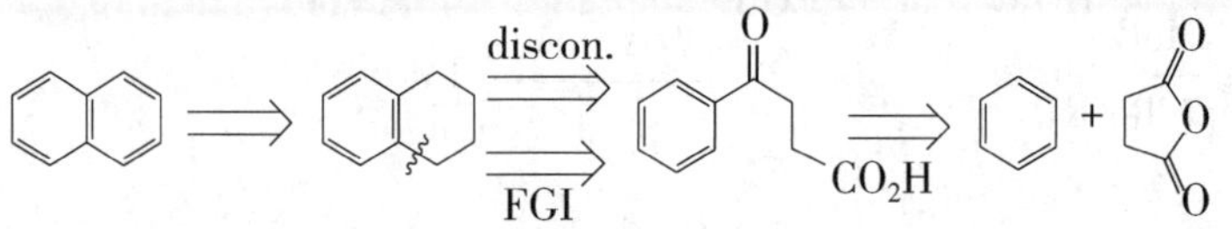

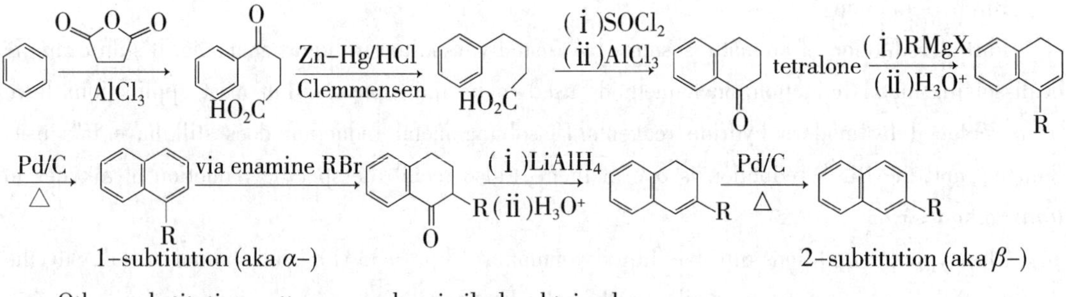

discon. O O

FGI CO_2H + O O

O O O O

$AlCl_3$ Zn-Hg/HCl Clemmensen (ⅰ)$SOCl_2$ (ⅱ)$AlCl_3$ tetralone (ⅰ)RMgX (ⅱ)H_3O^+

HO_2C HO_2C O R

Pd/C △ via enamine RBr (ⅰ)$LiAlH_4$ (ⅱ)H_3O^+ Pd/C △

R R O R R

1-subtitution (aka α-) 2-subtitution (aka β-)

Other substitution patterns can be similarly obtained.

8.2.5 芳环中位置阻断 Blocking Positions in Aromatic Rings

Functional groups that are introduced reversibly, or can be easily cleaved under mild condtions, can be used to access otherwise hard-to-make compounds[14].

$$\text{PhEt} \xrightarrow[H_2SO_4]{SO_3} \text{4-Et-}C_6H_4\text{-}SO_3H \xrightarrow[FeBr_3]{Br_2} \text{2-Br-4-}SO_3H\text{-ethylbenzene} \xrightarrow[H_2SO_4]{dil.} \text{2-bromoethylbenzene}$$

$$\text{PhNH}_2 \xrightarrow{Br_2} \text{2,4,6-tribromoaniline} \xrightarrow[(ii)\ H_3PO_2]{(i)\ NaNO_2/HCl} \text{1,3,5-tribromobenzene}$$

8.2.6 总结 Overall Summary

To devise a synthesis:

(i) Examine the TM; recognise functional groups and key structural features. In an exam you may be given a SM, if this is the case, check how it relates to the TM.

(ii) Use FG's present to help indicate disconnection points. Use latent polarities, umpolung and FGA to help if neccessary.

(iii) Consider FGI's appropriate to the TM; consider disconnections at branch points and heteroatoms. Be convergent—disconnect between FG's separated by a couple of carbon atoms.

(iv) Keep the number of steps as low as reasonably possible, but do use protecting groups where neccessary.

(v) Disconnect to good SM's.

(a) straight chain monofunctional compounds.

(b) branched monofunctional compounds containing six carbon atoms or fewer (for these purposes, including allyl, alkenyl and cycloalkyl compounds).

(c) simple mono-and disubstituted benzenes.

(d) common bifunctional compounds (acetoacetate esters, malonate derivatives etc.).

(e) hint: concerning regents & SM's... have you seen them before.

参 考 文 献
References

1. Smith A B, Dorsey B D. Strategies and Tactics in Organic Synthesis. New york: Academic Press, 1989. 369-414.

2. Smith M B. Organic Synthesis (Third Edition). Oxford: Academic Press, 2010. 1-76.

3. Wade L G. Organic Synthesis: The Disconnection Approach (Warren, Stuart). *J. Journal of Chemical Education*. 1984, 61(9): A248.

4. D'Angelo J, Smith M B. Hybrid Retrosynthesis. Boston: Elsevier, 2015. 35-44.

5. VanVeller B. A Decision Tree for Retrosynthetic Analysis. *J. Journal of Chemical Education*.

2021, 98 (8): 2726–2729.

6. Pinho e Melo M V D T, et al. Recent Advances on Functional Group Interconversion. *J. Current Organic Chemistry*. 2021, 25 (19): 2155.
7. Kennedy J F, et al. Protective Groups in Organic Synthesis (3rd Edition). *J. Carbohydrate Polymers*. 2001, 45 (1): 105–106.
8. Leslie A, et al. Functional Group Interconversion Reactions in Continuous Flow Reactors. *J. Current Organic Chemistry*. 2021.
9. Boon B A, et al. Using Ring Strain to Control 4π – Electrocyclization Reactions: Torquoselectivity in Ring Closing of Medium – Ring Dienes and Ring Opening of Bicyclic Cyclobutenes. *J. The Journal of Organic Chemistr*. 2017.
10. Eaborn C. Advanced Organic Chemistry. *J. Journal of Organometallic Chemistry*. 1993, 452 (1): C13.
11. Yeston J. Scaled-up sodium-free Birch reductions. *J. Science*. 2019.
12. Peerzada N. Classics in Total Synthesis. By K. C. Nicolaou and E. J. Sorensen. *J. Molecules*. 1998, 3 (2): 49.
13. Li C, et al. Synthesis and Applications of pi-Extended Naphthalene Diimides. *J. The Chemical Record*. 2016.
14. Nogi K, et al. Aromatic metamorphosis: conversion of an aromatic skeleton into a different ring system. *J. Chemical Communications*. 2017.

9 有机化合物的逆合成分析

Chapter 9 Retrosynthetic Analysis of the Organic Compounds

9.1 单官能团化合物逆合成分析 Retrosynthetic Analysis of the Compounds with One Functional Group

In the course of the retrosynthetic analysis of Target Molecule in Chapter 8, we met the first example of disconnection with the participation of one functional group. Participation of the hydrox yl group enabled disconnection of the C—C bond with the formation of two acceptable synthons, a neutral molecule and carbanion with an available reagent or synthetic equivalent. Now we start with the study of retrosynthesis by the problem - solving approach. This approach has characteristics of seminar work promoting knowledge of organic synthesis by retrosynthetic consideration of selected target molecules. They are either of commercial or scientific interest, and their retrosynthetic analysis has a certain didactic value[1].

A problem-solving approach to retrosynthesis is introduced with examples selected according to the functional group that participates in C—C bond disconnection or is interconverted. In the next few examples we approach the disconnection of compounds with one functional group of carbinols, alkenes, ketones and nitro compounds. The key retrosynthetic steps suggest important synthetic reactions, such as Diels-Alder, cyanhydrin, Wittig and Nef reactions. Basic principles for good disconnections are postulated. The importance of nitroalkanes as building blocks and precursors of primamines and ketones is exemplified. Retrosyntheses and syntheses of natural and commercial compounds of medium complexity are presented[2].

9.1.1 甲醇的切断 Disconnection of Carbinols

9.1.1.1 Cyanide

Cyanide is a good anion, and the cation is stabilized by a lone pair of electrons on oxygen. Example TM 9.1 suggests retrosynthesis then propose the synthesis of TM 9.1.

$(H_3C)_2C(OH)CN$

TM 9.1

Disconnection of any C—C bond from the carbinol C atom can be completed with participation of the hydroxyl group, avoiding the formation of an unstable cationic synthon. The stable cyanide anion is the preferred carbanionic synthon in the disconnection of TM 9. 1.

$(H_3C)_2C(OH)CN$ (**TM 9.1**) $\Rightarrow$ { H_3C–$C^+(CH_3)$–OH $\rightarrow$ H_3C–C(=O)–CH_3 ; ^-CN $\rightarrow$ $NaCN/H^+$ }

Note: there are many useful reagents for the cyanide ion, from inorganic salts (NaCN, KCN) to organic cyanides (R_3SiCN, NH_4CN). TM 9. 1 contains geminal hydroxyl and cyano groups. The proposed disconnection represents the retrocyanohydrin step. The cyanohydrin reaction is the standard route to α-hydroxy acids available on hydrolysis of cyano groups. The equivalent reagent of the carbocationic synthon is acetone.

Proposed synthesis route to TM 9. 1:

H_3C–C(=O)–CH_3 + HCN $\longrightarrow$ $(H_3C)_2C(OH)CN$

Example TM 9. 2 proposes retrosynthesis then give the synthesis of TM 9. 2:

C_6H_{11}–C(CH_3)(OH)CN

TM 9.2

Synthon TM 9. 2a is a neutral molecule, cyclohexyl methyl ketone. To the synthetic chemist acetophenone might spring to mind as an obvious next-generation target because of the structural relation to TM 9. 2a. Exhaustive hydrogenation of this aromatic ketone is expected to give 1-cyclohexylethanol TM 9. 2b, which can be oxidized to ketone by various protocols, e. g. , by pyridinium chlorochromate (PCC).

C_6H_{11}–C(CH_3)(OH)CN $\xRightarrow{dis}$ C_6H_{11}–C(=O)–CH_3 + $CN^-(+H^+)$

TM 9.2a

This is not a workable route, however, since the hydrogenation of acetophenone is not chemoselective and results in a mixture of products, 1-phenylethanol, ethyl benzene and ethyl cyclohexane, besides 1-cyclohexylethanol. The highest reported selectivity for TM 9.2b was 65%, obtained with rhodium nanoparticles entrapped in boehmite nanofibers（纳米软水铝石纤维）as catalyst.

CH_3 H_2,Cat. CH_3 OH pyridinium chlorochromate (PCC) CH_3 C=O

TM 9.2b TM 9.2a

The preferred retrosynthetic route leads over TM 9.2a to the next generation TM 9.2b by the addition of a double bond (FGA) at the proper position in the cyclohexane ring followed by retro-D.-A. disconnection:

H_3C C OH CN dis *retro*-cyanohydrin CH_3 C=O FGA CH_3 C=O dis *retro*- D.-A. + CH_3 C=O

TM9.2 TM 9.2a

9.1.1.2 末端炔

Aky-1-ne

Example TM 9.3 proposes good disconnection for TM 9.3.

OH C C≡CH H_3C dis OH C^+ CH_3 + O C CH_3 $\bar{C}$≡CH NaC≡CH

TM 9.3

Which C—C bond is preferably disconnected to the methyl, phenyl or ethynyl group depends on the stability of the carbanion, which appears as the synthon. All disconnections involve participation of a hydroxyl group. To evaluate the stability of the resulting carbanions, we consider the acidity of the C—H bond. For methane, pKa amounts to 42, for benzene 40 and for acetylene 25, revealing the acetylide anion as the most stable. The corresponding reagent, sodium acetylide, is easily available from acetylene and a strong base, e.g., sodium amide. The preferred disconnection of TM 9.3 leads to acetophenone and the acetylide anion. Participation of the OH group facilitates the disconnection of the neighboring C—C bond by the formation of the C=O bond in acetophenone with the departure of a proton.

Disconnection of acetophenone is denoted as *retro*-Friedel-Crafts (*retro*-F.-C.) since its synthesis is completed by the Friedel-Crafts reaction. It is important to note that there is no need to generate the phenyl anion in the synthetic direction since benzene is an acceptable reagent for highly reactive acetyl chloride activated by Lewis acid.

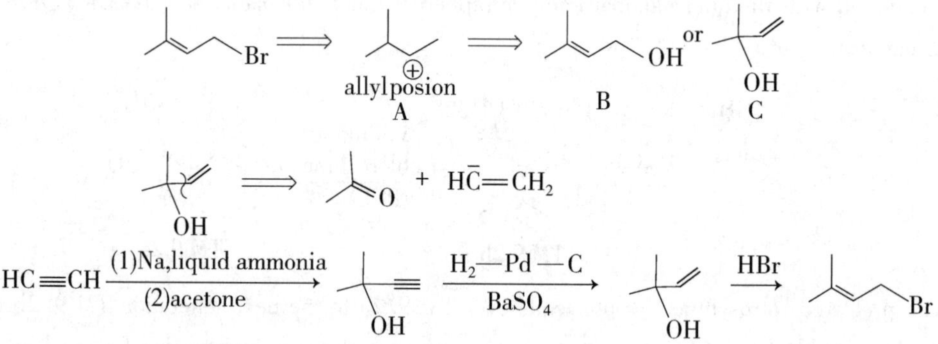

Br ⟹ allylposion A ⟹ OH B or OH C

OH ⟹ O + $H\bar{C}{=}CH_2$

$HC{\equiv}CH \xrightarrow[(2)\text{acetone}]{(1)\text{Na, liquid ammonia}}$ OH $\xrightarrow[BaSO_4]{H_2{-}Pd{-}C}$ OH $\xrightarrow{HBr}$ Br

9.1.1.3 Organometallic reagents

When all groups connected to the carbinol C atom give unstable carbanions, disconnection is guided by an additional principle met in the next example. ExampleTM 9.4 proposes possible disconnections of TM 9.4 and explains your choice.

Two possibilities for disconnection of TM 9.4 are shown in the Scheme 9-1.

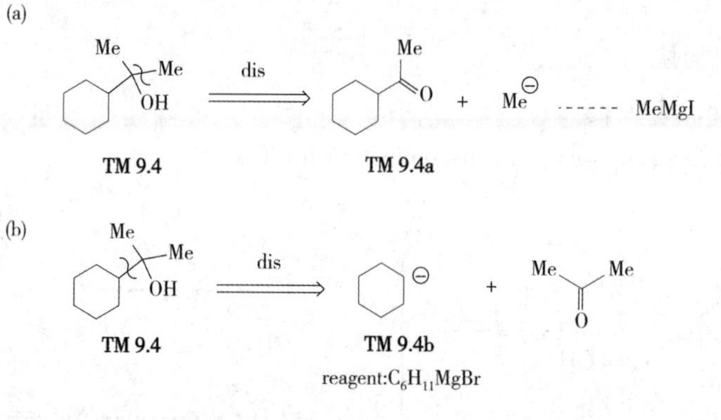

(a)

Me, Me, OH (TM 9.4) —dis⟹ Me, O (TM 9.4a) + $Me^{\ominus}$ ------ MeMgI

(b)

Me, Me, OH (TM 9.4) —dis⟹ $\ominus$ (TM 9.4b) + Me, Me, O

reagent: $C_6H_{11}MgBr$

Scheme 9-1 Disconnections of the methyl and cyclohexyl group.

With participation of the carbinolic hydroxyl group, disconnection affords anionic synthons, methyl and cyclohexyl carbanion. Both have proper synthetic equivalents in the corresponding Grignard reagents. To decide on the preferred disconnection, the second principle of retrosynthetic analysis helps; preferred disconnection leads to greater simplification of the target molecule. Maximum simplification of certain target molecules is usually achieved by disconnection, which results in two synthons of comparable size and complexity. Size is loosely defined by the number of C atoms in the skeleton and complexity by the number and relative position of functional groups. Greater simplification is obtained for TM 9.4 by disconnection (b) resulting in two synthons (C_6 and C_3), closer in their dimensions then synthons obtained in disconnection (a) (C_8 and C_1). Besides, by the second disconnection easily available acetone is obtained, different from methylcyclohexyl ketone 9.4a, which requires further retrosynthetic analysis. This is the moment to consider the availability of the TMs of the second-generation TM 9.4a and TM 9.4b. Assuming a Grignard reaction in the synthetic direction, cyclohexyl-bromide is needed. On the first glance this immediate precursor of Grignard reagent TM 9.4b is more easily available than cyclohexyl methyl ketone TM 9.4a. The two-step retrosynthetic analysis of TM 9.4b results in phenol, a commodity from the petrochemical industry. Its hydrogenation produces cyclohexanol, which is brominated under standard conditions. To enter the retrosynthesis of TM 9.4a, we need the retrosynthetic tool (functional group addition). Addition of the C=C bond at the proper position in the cyclohexane ring in TM 9.4a offers an unexpected opportunity. This FGA leads to a cyclohexene derivative amenable to *retro*-Diels-Alder (*retro*-D. -A.) disconnection to diene and dienophile. The retrosynthetic step and mechanism of the Diels-Alder reaction are discussed in Example TM 9.4. According to either of the two retrosynthetic sequences in Scheme 9-2, the synthesis of TM 9.4 can be completed from easily available stating materials by well-known, conceivable reactions. Discussing this example we get acquainted with three basic principles for good disconnection:

(i) disconnection should follow the correct mechanism.

(ii) disconnection should follow the maximal simplification of the target molecule.

(iii) by the sequence of disconnections we shall arrive at simple, easily available starting materials.

It is important that the principle of maximal simplification of the target molecule cannot be exactly defined or even quantified. Usefulness and easy understanding characterize this principle, like most rules in chemistry. In the former examples we applied a functional group addition to create target molecules that are more convenient for disconnection. This concept and other interconversions of functional groups are practiced in the examples that follow.

(a) Me, O, TM 9.4a ⟹ FGA ⟹ Me, O ⟹ dis ⟹ + Me, O

(b) MgBr, TM 9.4b ⟹ dis ⟹ Br ⟹ FGI ⟹ OH ⟹ OH

Scheme 9-2 Retrosynthetic analysis of (a) TM 9.4a and (b) TM 9.4b

Example TM 9.5 proposes the retrosynthesis then synthesis of TM 9.5:

CH_2OH, TM 9.5 — *retro*- D.-A. ? ⟹ + CH_2OH; FGI ⟹ CO_2Et ⟹ *retro*- D.-A. ⟹ + CO_2Et

Scheme 9-3 the Retro-Diels-Alder disconnection of TM 9.5

Example TM 9. 5 proposes the retrosynthesis then synthesis of TM 9. 5. This example requires more retrosynthetic imagination. We recognize the cyclohexene ring with the hydroxymethyl group in the β position of the C=C bond and consider a possible precursor for the cyclohexene ring. The benzene derivative can be excluded for good reason since partial and regioselective reduction to the substituted cyclohexene derivative represents a formidable task.

The second possibility conceives the construction of a cyclohexene ring from the proper building blocks and suggests retro-Diels-Alder disconnection of the cyclohexene derivative. Here we apply disconnection of two bonds resulting in two synthons and presented for TM 9. 5 in Scheme 9-3.

The *retro*-Diels-Alder step envisages homolytic disconnection of two bonds in the 2, 3-position related to the double bond of the cyclohexene ring. Two electrons of each σ bond and two electrons from the π bond move to the neighbor σ bonds (Scheme 9-3). Such disconnection corresponds to the pericyclic or concerted character of the Diels-Alder reaction[3].

The example of TM 9. 5 also serves to illustrate the importance of the order of retrosynthetic steps to propose a workable synthetic route (Scheme 9-4).

+ CO_2Et ⟶ CO_2Et $\xrightarrow[\text{THF},0℃]{LiAlH_4}$ CH_2OH

Scheme 9-4 proposal for synthesis of TM 9. 5

analysis

OH Ph ⟹ BrMg Ph + CHO $\overset{\text{FGI}}{\Longrightarrow}$ CH_2OH $\overset{\text{dis}}{\Longrightarrow}$ MgBr + CH_2O

Br $\xrightarrow[(2)CH_2O]{(1)Mg,Et_2O}$ CH_2OH $\xrightarrow[\text{pyridine}]{HCl,CrO_3}$ CHO (78%~92%) $\xrightarrow[\text{good yield}]{\text{BrMg Ph}}$ TM 9.6

analysis

OAc Ph Ph (TM 9.7) $\overset{\text{FGI}}{\Longrightarrow}$ OH Ph Ph (TM 9.7) $\overset{\text{dis}}{\Longrightarrow}$ Ph MgBr + H(C=O)OEt + BrMg Ph

$\Downarrow$ dis

Ph OH $\Longrightarrow$ $PhCH_2MgBr + HCHO$

$\Longrightarrow$ $PhMgBr + \overset{+}{C}H_2CH_2OH$

9.1.2 烯键的切断 Disconnection of Alkenes

In this section we shall closely inspect the retrosynthetic approach to simple alkenes, which includes disconnection of the C=C bond in two synthons, which have useful building blocks for synthetic equivalents.

Double disconnection of the C=C bond can be formally presented as in Scheme 9-5.

unacceptable synthon A unaccpetable synthon B

PPh_3 phopshonium yilde or wittig reagent

unacceptable equivalent for synthon A unacceptable equivalent for synthon B

Scheme 9-5 Formal presentation of double disconnection of the C=C bond

The presented disconnection looks ugly, mechanistically incorrect and energetically unacceptable, generating synthons with double charges on the terminal C atoms. Having in mind that any disconnection is an imaginative process, we can start searching for proper reagents for such "unacceptable synthons". Surprisingly, it turns out that they exist. The reagent for "unacceptable synthon A" is benzaldehyde, where an electronegative oxygen atom donates two electron pairs to the C=O bond, compensating two formal positive charges in this synthon. To discover an acceptable reagent for "unacceptable synthon B" with a double negative charge on the terminal C atom helps another imaginative consideration. For an effective reaction with the carbonyl group, carbanion is needed as the second reagent; so one negative charge in this synthon serves the purpose. The second negative charge is compensated by delocalization bound to an electropositive atom. The atom of choice is the P atom and as a highly effective synthetic equivalent the phosphonium ylide or Wittig reagent emerges.

Formation of the C=C bond enables the couple carbonyl compound-ylide as a nucleophile. An electron pair of carbanions in phosphonium ylide forms one of two C=C bonds with the carbonyl C atom; the second one is formed by an electron pair "hidden" in the P—C bond. Here a general scheme is presented for the reaction of carbonyl compounds with stabilized carbanions and ylides in the preparation of alkenes.

In the next example disconnection of an alkene reveals some typical "pitfalls" when proposing a synthetic route based on seemingly acceptable retrosynthetic analysis.

Example TM 9.8 proposes the retrosynthetic analysis and synthesis of TM 9.8 without use of the Wittig reagent in the formation of the C=C bond. The two possibilities in the retrosynthetic cnalysis of TM 9.8 are shown in the Scheme 9-6.

Ph **TM 9.8**

Conceiving the formation of the C=C bond by the elimination of water, two possible FGIs in the first retrosynthetic step lead to two alcohols, TM 9.8a and TM 9.8b, as the targets of the second generation.

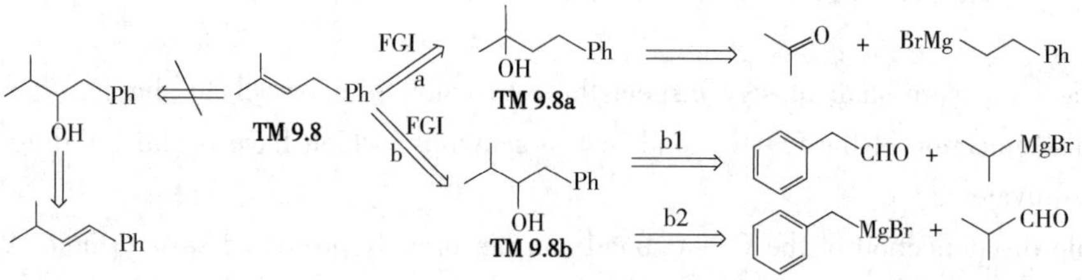

Scheme 9-6 Two possibilities in the retrosynthetic analysis of TM 9.8

FGI in step a opens a simple choice of which C—C bond in TM 9.8a to disconnect with the participation of the OH group. It results in the maximal simplification and two available reagents, acetone and the Grignard reagent, from 2-bromethylbenzene. FGI b, however, faces two possibilities for disconnection of the C—C bond in TM 9.8b, b1 and b2. Both lead to one mole of aldehyde and one mole of Grignard reagent. Disconnection b1 is unfavorable since Grignard reagent needs to stabilize carbanion on the sec C atom destabilized by hyperconjugation. Disconnection b2 leads to isobutyraldehyde, a large-scale industrial product available by the hydroformylation of propene catalyzed by Pt/Al_2O_3. Grignard reagent derived from benzyl bromide is not a good choice because the carbanion on the benzylic C-atom is destabilized by resonance. Since the radical-anion on the benzylic C atom is less destabilized, Grignard reagent has a radical character and is prone to polymerization. Besides this argumentation against retrosynthetic consideration b, it should be noted that elimination of water toward the benzylic C atom competes with elimination toward the *tert*-C atom of the isopropyl group affording the structural isomer of TM 9.8. The origin of this competition resides in the similar C—H acidity of the two C atoms (kinetic argumentation) and comparable stabilization of the C=C bond by conjugation with the aromatic ring and by hyperconjugation with methyl groups (thermodynamic argumentation). For these reasons, the retrosynthetic route to TM 9.8 over TM 9.8a is preferred[4].

In the frame of the next example TM 9.9, we consider in more detail the mechanism and steric aspects of the Wittig reaction:

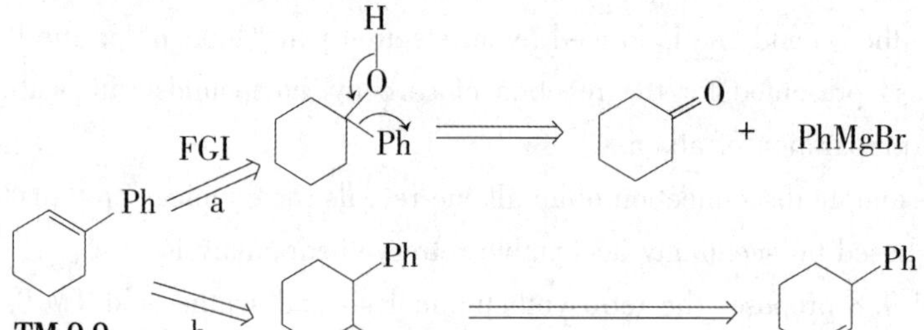

Thus:

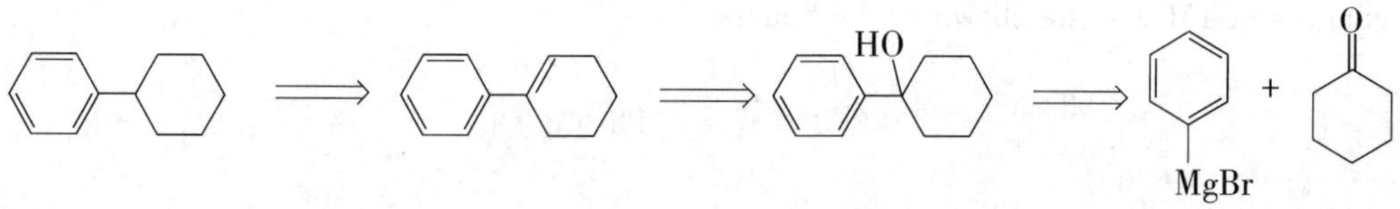

9.1.3 酮的切断 Disconnection of Ketones

9.1.3.1 Disconnection of dialkyl ketones

Retro-Friedel-Crafts and *retro*-Grignard reactions are useful in the synthesis of aryl alkyl ketones of diverse complexity. In dialkyl ketones we consider the carbonyl group as directing the preferred C—C bond disconnection.

Example TM 9.10 proposes the retrosynthetic analysis for TM 9.10 and suggest possiblereagents.

TM 9.10

Two plausible disconnections are presented in Scheme 9-10 (a) disconnection of the C—C bond between the α-C atom and carbonyl C atom and (b) disconnection of the C—C bond between the α- and β-C atoms to the carbonyl group. Both disconnections follow the principle of maximal simplification of the target molecule.

Synthons resulting from either disconnection have acceptable chemical species as reagents. We already met the disconnection of type a proposing retrosynthesis of TM 9.11. Disconnection b is based on a new concept, introduction of an activating group. Whereas carbocation has a logical reagent in 3-bromopropylbenezene TM 9.11c, carbanion cannot be selectively generated from pentan-2-one. Therefore, we conceive activating the carbethoxy group on the terminal α-C atom, enhancing its C—H acidity to pKa and thus the stability of carbanion in TM 9.11d.

β-Keto acid TM 9.11d is not available; therefore, the synthetic route according to disconnection a is preferred[2]. The reader should propose a synthetic route to TM 9.11 considering the most effective approach to TM 9.11a. The preferred disconnections of dialkyl ketone TM 9.11 are shown in the Scheme 9-7.

TM 9.11　TM 9.11a　TM 9.11b　TM 9.11c　TM 9.11d

Scheme 9-7 Preferred disconnections of dialkyl ketone TM 9.11

Example TM 9.11 completes the retrosynthesis and propose the synthesis of benzyl acetone

TM 9. 11.

TM 9.12

In Scheme 9 – 7 retrosynthesis is proposed following previous argumentation for TM 9. 11. Characteristic reaction conditions are given for the synthetic steps. Since p*K*a of ethanol is 18 and p*K*a of the methylenic group in acetoacetic ester is 11, deprotonation of TM 9. 11d by sodium ethoxide is completely chemoselective. C–alkylation on the addition of benzyl bromide is followed by acidification and heating to complete the hydrolysis and decarboxylation to TM 9. 11.

Note: Activation of the C—H bond by the carbethoxy group in β–keto esters and its elimination by hydrolysis and decarboxylation, the last two steps in Scheme 9–8, deserve comment.

Retrosynthetic analysis:

dis

Br

reagents

CO_2Et

Synthesis:

Me OEt NaOEt/EtOH 0℃ Me OEt + Br

EtOH/0℃to20℃ Me COOEt HCl/EtOH/Δ –EtOH $-CO_2$ Me

TM 9.12

Scheme 9–8 Retrosynthetic analysis and proposed synthesis of benzyl acetone

Easy and selective alkylation of the methylenic group in ethyl acetoacetate, the simple decarboxylation in the last steps and availability of this starting material make it the reagent of choice for acetonide carbanion. Ethyl acetoacetate is a commodity produced by the catalytic process discussed in connection with the retrosynthetic analysis of TM 9. 12.

Example TM 9. 13 proposes the retrosynthetic analysis of ketone TM 9. 13 taking into account the Z configuration of the double bond.

TM 9.13

This example shows how retrosynthetic consideration of the next generation target molecules is sometimes more demanding than the choice in thefirst step. The logical first retrosynthetic step of

TM 9. 13 is the disconnection of the C—C bond in the β position to the carbonyl group generating two stable synthons, carbanion on the α-C atom to the carbonyl group and carbocation on the α-C atom to the double bond, known as the allylic cation (Scheme 9-9).

Since the allyl cation is stabilized by resonance with the double bond, similar to the stabilization of the benzylic cation by an aromatic ring, the reactivity of the corresponding allyl halide 1-bromobut-2-ene TM 9. 13b is enhanced, resembling that of benzyl bromide. The anionic C_3 synthon we already met in the former example will appear in many of the disconnections that follow. Primary bromide TM 9. 13b is easily available by bromination of primary alcohol and this in turn by reduction of the corresponding ester, as anticipated by the first two FGIs in Scheme 9-10 and Scheme 9-11.

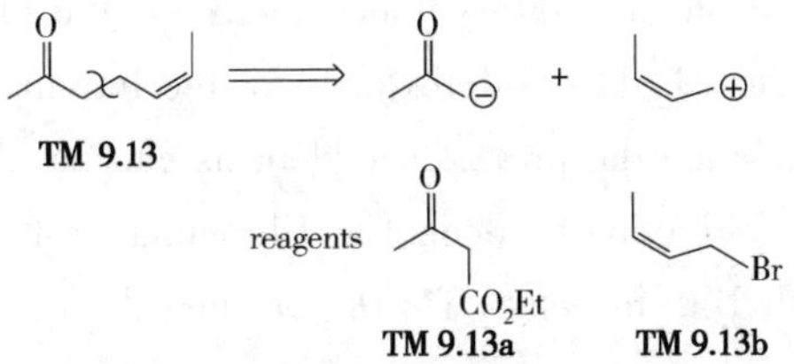

TM 9.13

reagents TM 9.13a TM 9.13b

Scheme 9-9 First disconnection step in the retrosynthesis of TM 9. 13

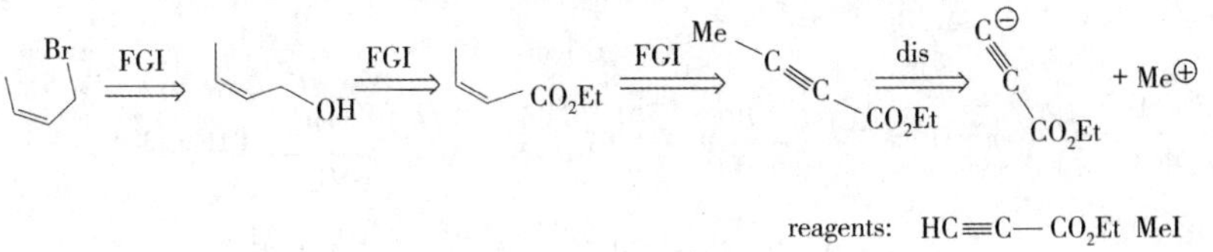

reagents: HC≡C—CO_2Et MeI

Scheme 9-10 Retrosynthetic analysis of TM 9. 13b

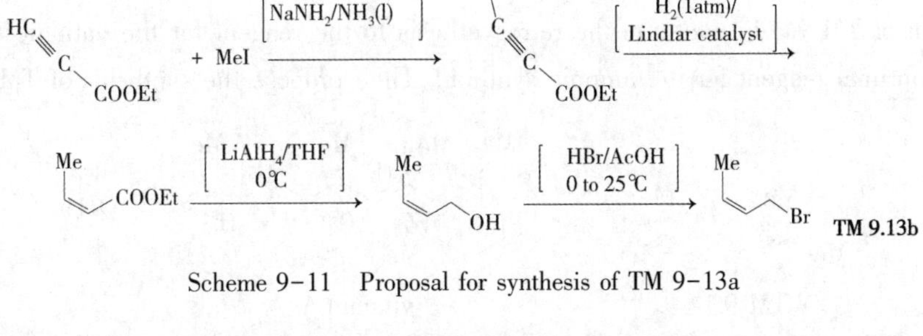

[$NaNH_2/NH_3$(l)] [H_2(1atm)/ Lindlar catalyst]

[$LiAlH_4$/THF 0℃] [HBr/AcOH 0 to 25℃] TM 9.13b

Scheme 9-11 Proposal for synthesis of TM 9-13a

The main synthetic issue on the route to TM 9. 13b represents the introduction of the C=C bond to the Z configuration since products of most C=C bond-forming reactions possess a thermodynamically stable E configuration. Here we need new knowledge; the triple bond can be selectively reduced to a double bond with Z configuration.

Note: Partial hydrogenation of the triple bond to the Z double bond is possible in the presence of Lindlar catalyst. This is a solid, heterogeneous catalyst based on Pd deposited on calcium carbonate and doped by various morphological forms of lead. Lead acts as a deactivator of palladium, and Pb(Ⅱ) acetate and Pb(Ⅱ) oxide also serve as "catalytic poisons". Since the addition of hydrogen on the triple bond occurs syn-selectively, the resulting alkene possesses a Z configuration of the double bond.

The second approach to the Z C═C bond offers the Wittig reaction with unstabilized ylide, but is a less attractive method for the large scale. Anticipating Z-selective hydrogenation of the triple bond, we propose a third FGI to the triple bond in ethyl butynoate and its disconnection to the methyl cation and anion of ethyl acrylate.

In summary, two interconversions of ally bromide afford allylic ester, followed by FGI of the Z double bond to triple bond and disconnection of C_4 alkyne to methyl halide and ethyl acrylate, both available reagents. Based on this retrosynthetic analysis, the plausible synthetic scheme for TM 9. 13b is proposed (Scheme 9-11). Methylation of the terminal acetylenic C atom requires deprotonation by strong bases since the pKa of acetylene is ca. 25. Z-selective hydrogenation of the triple bond is followed by reduction to alcohol and bromination to TM 9. 13b. In the last steps of synthesis, alkylation of ethyl acetoacetate TM 9. 13a then decarboxylation as in Scheme 9-12 affords TM 9. 13.

(1)EtO⁻; (2) Br; HO, OH; $NaNH_2$, MeI; HCl, H_2O; H_2-Pd/C, $BaSO_4$; TM9.13; CO_2Me

Scheme 9-12 Proposal for synthesis of TM 9. 13

Unsaturated ketone TM 9. 14 is the key intermediate in the industrial synthesis of β-carotene (β-胡萝卜素), a precursor of vitamin A (维生素 A). Start with the preferred disconnection of TM 9. 14, continue the retrosynthesis to the reagent for the cationic synthon and suggest the optimal reagent for the anionic synthon. Then propose the synthesis of TM 9. 14.

Me Me Me O
TM 9.14

Me Me Me Me (E) (E) OH (Z) (E) (E) Me
vitamin A

It is interesting to note that TM 9. 14 differs from TM 9. 14 by only one methyl group on the double bond; nevertheless, the preferred retrosynthetic analysis leads us to entirely different starting materials. As in Example TM 9. 14, the first disconnection of the C—C bond in the α, β-position bond to the carbonyl group is preferred.

The reagent for the allylic cation is the corresponding bromide TM 9. 14a, a commodity with wide application in the fragrance industry. Its industrial synthesis, along with the industrial production of ethyl acetoacetate, is presented in Scheme 9-13. Both building blocks are used in the proposed synthesis of TM 9.14. Starting from the petrochemicals acetylene and acetone, allylic carbinol is obtained by partial hydrogenation of the triple bond and brominated in the next step to TM 9.14a.

Note: Here comes the important point; during bromination in the acidic medium, the intermediary allylic carbocation formed on the *tert*-C atom is in equilibrium with the more reactive prim-carbocation, which is brominated. This synthesis of 3,3-dimethylallyl bromide TM 9.14a is the basis of multi-ton industrial production since this compound is used in the agrochemical, pharmaceutical and dyestuff fields. Two competitive methods are developed for the production of ethyl acetoacetate, both based on the dimerization of ketene. By the first method, ketene is produced by thermal breakdown of acetone at over 300℃, while the second "wet" method uses strong bases as catalysts for the elimination of hydrogen chloride from acetyl chloride. Spontaneous dimerization results in a relatively stable four-membered lactone, known as diketene on the market, which on alcoholysis affords ethyl acetoacetate.

In the last steps of the proposed synthesis of TM 9.14, the alkylation of TM 9.13b and decarboxylation are completed according to the previously described protocol for TM 9.11.

TM 9.14 ⟹ (DIS) Me₂C=CH–$CH_2^{\oplus}$ + $^{\ominus}H_2C$–CO–Me

reagents: TM 9.14a (Me₂C=CH–CH₂Br); TM 9.14b (EtOOC–CH₂–CO–Me)

Allyl bromide, industrial production:

MeCOMe + HC≡CH —($NaNH_2/NH_3(l)$, −70℃ to r.t.)→ $Me_2C(OH)$–C≡CH —(H_2/Lindar)→ $Me_2C(OH)$–CH=CH_2 —(HBr/AcOH/△)→ TM 9.14a

Acetoacetic ester, industrial production:

MeCOMe —(electric arc/△)→ $H_2C=C=O$ + CH_4

or

MeCOCl —(Et_3N/toluene/△)→ $H_2C=C=O$ + $Et_3N \cdot HCl$

ketene

$H_2C=C=O$ + $H_2C=C=O$ → diketene; diketene + EtOH → TM 9.14b

Proposal for synthesis:

Me₂C=CH–CH₂Br + EtOOC–CH₂–CO–Me —(NaOMe/MeOH)→ Me₂C=CH–CH₂–CH(COOEt)–CO–Me —((1)NaOH(aq)/△, (2)HCl(aq)/△)→ TM 9.14

Scheme 9-13 Production of intermediates TM 9.14a and TM 9.14b and proposal for the synthesis of TM 9.14

9.1.3.2 Disconnection of alkyl aryl ketones and diaryl ketones

As mentioned in the introduction to this section, *retro* – Friedel – Crafts is the basic disconnection of the Ar—CO bond in alkyl-aryl ketones. Hereafter, we will discuss an example where problems of regioselectivity in acylation and reactivity of the aromatic ring are tackled.

The retrosynthesis of TM 9.15 explains the preferred disconnection and solves the issue of regioselectivity in all synthetic steps. An obvious retrosynthetic step is *retro*-F. -C. disconnection, and the decision between two Ar—CO bonds is unambiguous in favor of the bond to the dimethoxyphenyl unit since this substrate is activated for F. – C. acylation. Alternative disconnection leads to nitrobenzene, an unreactive aromatic compound in F. -C. acylation. Its high inertness enables its use as the solvent for this reaction. The prevailing formation of *meta*, *para*-dimethoxy isomer TM 9.15 is directed by steric perturbation at *ortho*-positions by methoxy groups[3]. The proposed synthesis of TM 9.15 is shown in the Scheme 9-14, the complete disconnection scheme for TM 9.15 is shown in the Scheme 9-15.

OMe MeO NO_2 O TM 9.15

OMe MeO NO_2 O TM 9.15 *retro*-F. C. OMe MeO ⊖ ⊕ NO_2 O

reagent:*o*-dimethoxybenzene *p*-nitrobenzoyl chloride

Scheme 9-14 Proposed synthesis of TM 9.15

Cl NO_2 O TM 9.15a FGI Me NO_2 FGA Me

OMe MeO TM 9.15b FGI OH HO

Scheme 9-15 Complete disconnection scheme for TM 9.15

Note: Reagents for TM 9.15a and 9.15b are available aromatic compounds, products of the petrochemical industry. *Para*-nitrobenzoic acid is produced by nitration of toluene to *para*-isomer as the prevailing product, followed by oxidation of methyl to the carboxylic group. *Ortho* – dimethoxybenzene is produced from *ortho*-diphenol, which in turn is available by oxidation of phenol. One technological process uses hydrogen peroxide as oxidant, and annual production of ortho-diphenol reaches 20,000 tons/year, mainly intended for the production of pesticides and perfumes.

Now we can propose the complete synthesis of TM 9.15.

Note the parallel synthetic steps leading to key intermediates, which in the last step enter the F. -C. reaction affording TM 9. 15. This is a general characteristic of convergent syntheses where two large building blocks are prepared separately and in the last step coupled into a final target molecule. Convergent synthesis reflects retrosynthesis where maximal simplification of the target molecule is performed in the first retrosynthetic step[2].

HNO_3/H_2SO_4/△; (1)$KMnO_4$/△ (2)$SOCl_2$/△; H_2O_2/AcOH; Me_2SO_4 △; $ZnCl_2$/$CHCl_3$/△

TM 9.15

Scheme 9-16 Proposal for the complete synthesis of TM 9. 15

9. 1. 4 胺的切断 Disconnection of Amines

The amino group is usually introduced by substitution of a good leaving group with nitrogen nucleophiles or by reduction of the *imino* or *nitro* group. The second approach is preferred since ammonia and amines are weak nitrogen nucleophiles.

With the next examples we introduce a new synthetic strategy for target molecules with nitrogenfunctional groups. It is based on the use of building blocks with *nitro* groups available on the industrial scale by nitration of hydrocarbons. Reduction to the amino group or oxygenation to the *keto* group completes the approach to these functionalities.

Example TM 9. 16 considers the retrosynthesis of the anti-appetizer chlorphentermine TM 9. 16 and propose its synthesis[5].

TM 9. 16 Chlorphentermine (对氯苯丁胺)

The presence of the Me_2CNH_2 group suggests the use of the Me_2CHNO_2 building block. Hence, in the first retrosynthetic step we propose FGI to the nitro group in TM 9. 16a. This is an appealing solution since disconnection of the central C—C bond in TM 9. 16a results in stable synthons, benzyl cation and α-carbanion of 2-nitropropane.

In the above scheme, the corresponding reagents are immediately presented and an unambiguous and short synthesis can be proposed. Importantly, in the scheme the generation of carbanion before the addition of benzyl chloride is indicated since the methoxy anion behaves as a competitive nucleophile. On completed alkylation, TM 9. 16a is reduced chemoselectively. The retrosynthesis of chlorphentermine TM 9. 16 is shown in the Scheme 9-17.

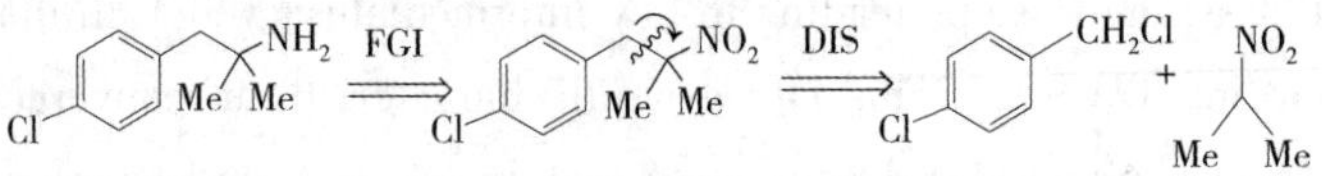

TM 9.16 TM 9.16a

Scheme 9-17 Retrosynthesis of chlorphentermine TM 9. 16

In the interconversion of the nitro group, nitroalkanes can be used as building blocks. The proposal for the synthesis of chlorphentermine TM 9. 16 is shown in the Scheme 9-18.

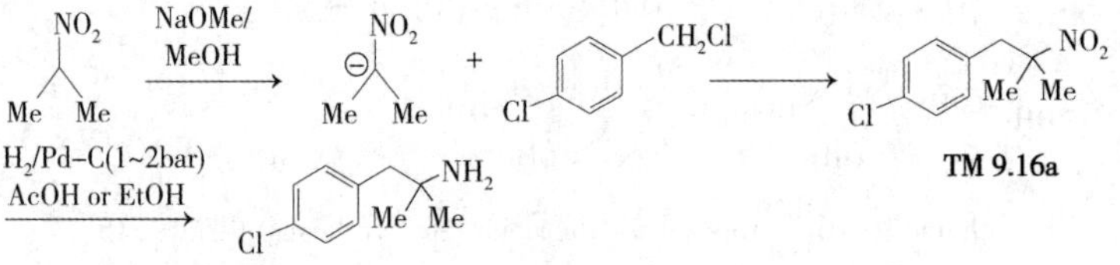

TM 9.16a

TM 9.16 Chlorphentermine

Scheme 9-18 Proposal for the synthesis of chlorphentermine TM 9. 16

Consider the retrosynthesis of diamine TM 9. 16 a structural congener of TM 9. 16 and propose a short synthesis:

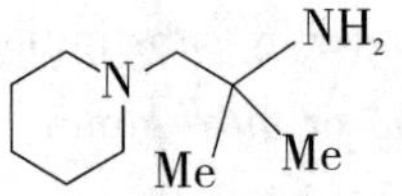

TM 9.17

Again we have an amino group on the *tert* - C atom where it cannot be introduced by direct nucleophilic substitution. Therefore, the FGI of the amino to nitro group springs to mind (Scheme 9-19).

Disconnection of TM 9. 16a requires new knowledge. This structure corresponds to the Mannich base since the *tert* - amino group is present in the β - position to the strong electron - withdrawing *nitro* group. The *retro* - Mannich type disconnection of two bonds leads to simple starting materials, piperidine, formaldehyde and 2 - nitropropane. In the same scheme are proposed reaction conditions for the synthesis of TM 9. 17.

Retrosynthetic analysis:

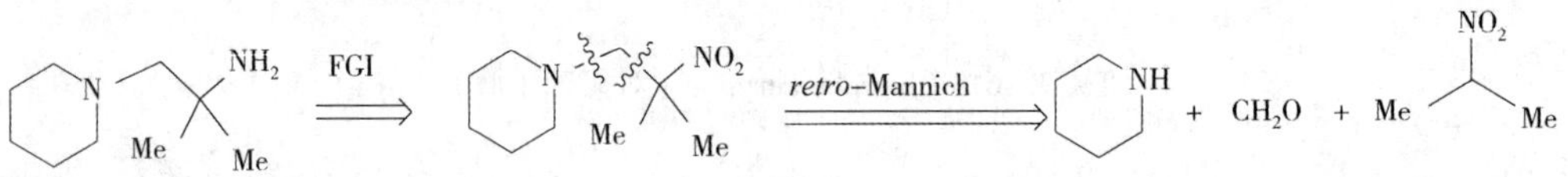

TM 9.17 TM 9.16a

Proposal for the synthesis:

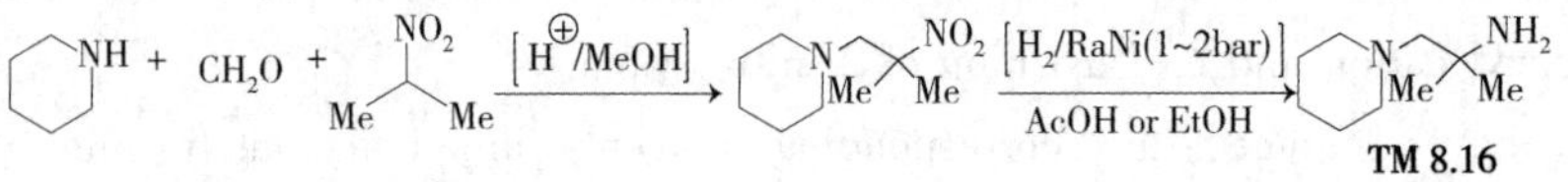

TM 8.16

Scheme 9-19 Retrosynthetic analysis and proposal for the synthesis of TM 9. 17

Complete the retrosynthetic analysis of C_8 diamine TM 9. 18, an important monomer for the industrial production of polyamides, and suggest its three - step synthesis starting from easy available commodities, nitromethane, acrylonitrile and isobutyraldehyde.

H_2N–CH₂CH₂CH₂–C(Me)(Me)–CH₂CH₂–NH_2

TM9.18

Obviously, both primary amino groups originate from the CN and NO_2 group in two building blocks. The next consideration requires more retrosynthetic skills. Since the CN group is incorporated as acrylonitrile, this C_3 building block is the origin of the amino group and three C atoms left from the quaternary one. Disconnection of the C—C(Me_2) bond requires the formation of carbanion on the *tert*-C atom of the Me_2C group and acrylonitrile as an electrophile. This reaction is known as the Michael addition of stable carbanions to enones as electrophiles and is discussed in more detail. The amino group on the right side and terminal C atom obviously originate from nitromethane. The remaining C_4 unit at the branching point then belongs to isobutyraldehyde. Having identified all building blocks in TM 9. 18, we can propose retrosynthetic Scheme 9-20.

The first two FGIs interconvert two amino groups into their precursors in TM 9. 18a; next FGA introduces the double C═C bond in TM 9. 18b, enabling retroaldol disconnection of nitromethane and TM 9. 18c. Cyano-aldehyde TM 9. 18c affords acrylonitrile and stabilized carbanion of isobutyraldehyde on *retro*-Michael disconnection.

According to this retrosynthetic consideration, the short synthesis of TM 9. 18 is proposed. The particular convenience of this three-step synthesis represents the contemporaneous reduction of the C═C bond and two nitrogen functionalities in the last step[6].

TM 9.18 (H_2N … NH_2) $\xrightarrow{2*FGI}$ TM 9.18a (CN … NO_2) $\xrightarrow{FGA}$ TM 9.18b (CN … NO_2)

$\xrightarrow{retro\text{-aldol}}$ TM 9.18c (CN … C(Me)(Me)CHO) + $H_2C^{\ominus}$–NO_2 (+$H^{\oplus}$)

TM 9.18c (H, Me, Me, CN, CHO) $\xrightarrow{retro\text{-Michael}}$ CN–CH=CH₂ + Me–$C^{\ominus}$(Me)–CHO

Scheme 9-20 Retrosynthetic analysis of TM 9. 18

NC–CH=CH₂ + Me–CH(Me)–CHO $\xrightarrow[\Delta]{NaOH/MeOH/}$ NC … C(Me)(Me)–CHO (TM 9.18c) + CH_3NO_2 $\xrightarrow{Et_3N/\Delta}$

NC … C(Me)(Me)–CH=CH–NO_2 (TM 9.18b) $\xrightarrow[AcOH\ or\ EtOH]{H_2/RaNi,Co\ or\ Ni}$ H_2N … C(Me)(Me) … NH_2 (TM 9.18)

Scheme 9-21 Proposal for the synthesis 1, 6-diamine TM 9. 18

Consider the synthesis of 1,3-disubstituted cyclohexane TM 9. 19 using C_5 diene as a strategic building block.

TM9.19

This information eliminates the cyclohexane derivative or its aromatic precursor as stating material. It also indicates the C_2 unit is the second building block and suggests the construction of a carbon skeleton by Diels-Alder reaction. Diene C_5 with a Me group is present in isoprene (2-methylbuta-1, 3-diene). C_2 dienophile should be activated by EWG, and the *nitro* group serves this purpose best. This analysis suggests amino-nitro FGI to TM 9. 19a as the target molecule of the next generation. By addition of the C=C bond in a strategic position on the cyclohexane ring, we arrive at key intermediate TM 9. 19b. This cyclohexene derivative is now prone to *retro*-D. -A. disconnection to isoprene and nitroethylene TM 9. 19c.

In spite of its relative instability nitroethylene (pure compound readily decomposes at r. t. , but is stable in benzene solution over months) is conveniently produced in kilo-quantities by dehydration of 2-nitroethanol according to the method outlined in the proposed synthesis of TM 9. 19.

Note that TM 9. 19a does not appear as an intermediate in the synthetic direction since both unsaturated functionalities in the product of the Diels-Alder reaction are contemporaneously reduced[5].

TM 9.19 $\xRightarrow{FGI}$ TM 9.19a $\xRightarrow{FGA}$ TM 9.19b $\xRightarrow{retro\text{-D.-A.}}$ izoprene + TM 9.19c

TM 9.18c $\xRightarrow{FGI}$ 2-nitroethanol $\xRightarrow{retro\text{-aldol}}$ $CH_3NO_2+CH_2O$

Scheme 9-22 Retrosynthetic analysis of TM 9. 19

$CH_3NO_2+CH_2O$ $\xrightarrow{KOH/MeOH/\Delta}$ 2-nitroethanol $\xrightarrow{P_2O_5 \text{ or polyphosphoric acid}/\Delta}$ nitroethylene + isoprene $\xrightarrow{toluene/\Delta/100℃ \text{ or} >140℃}$ Me-cyclohexene-NO_2 $\xrightarrow{H_2/Pd^0\text{-C, AcOH}}$ TM 9.19

Scheme 9-23 Proposal for the synthesis of TM 9. 19

Unsubstituted imines are unstable and cannot usually be made in good yield, but primary amines can still be made in a one-step reductive amination in which the imine is not isolated.

$$R^1R^2C{=}O \xrightarrow{NH_3} [R^1R^2C{=}NH] \xrightarrow[\text{or } NaB(CN)H_3]{H_2,cat} R^1R^2CH{-}NH_2$$

$$RCOCl \xrightarrow{NH_3} RCONH_2 \xrightarrow{\times} RCH_2NH_2$$

Primary amines are not usually made by reduction of amides but by other reductive process which are minor variations on this scheme. We can reduce cyanides to obtain unbranched amine. This method is especially suitable for benzylic amines since aryl cyanides can be made from diazonium salt, and for the homologous amines since cyanide ion reacts easily with benzyl halides.

$$RBr \xrightarrow{KCN} RCN \xrightarrow[\text{or } LiAlH_4]{H_2, PtO_2, H^+} RCH_2NH_2$$

$$ArNH_2 \xrightarrow[(2)Cu(I)CN]{(1)HONO} ArCN \xrightarrow[H^+]{H_2, Pd\text{-}C} ArCH_2NH_2$$

$$PhCH_2Cl \xrightarrow{CN^-} PhCH_2CN \xrightarrow[AlCl_3\ Et_2O]{LiAlH_4} PhCH_2CH_2NH_2$$

Disconnection again requires a preliminary FGI.

For branched chain primary amines, oximes are good intermediates since they can be made easily from ketones and reduction cleaves the weak N—O bond as well as reducing the C—N bond. FGI is again required before disconnection.

$$R^2COR^1 \xrightarrow[NaOAc]{NH_2OH, HCl} R^1R^2C{=}N\text{-}OH \xrightarrow[\text{or } H_2, cat]{LiAlH_4} R^1R^2CH\text{-}NH_2$$

The synthesis of Fenfluramine (芬氟拉明), a drug acting on the central nervous system, illustrates two amine disconnections. The ethyl group can be removed by the amide method leaving the branched chain primary amine available from the ketone by the oxime method.

The following reactions are important to understand the building blocks of *N*-atom target molecules:

$$(CH_3)_2C{=}NOH \xrightarrow[\text{or } Na/EtOH(Sol)]{LiAlH_4} (CH_3)_2CH\text{-}NH_2 \qquad Ph\text{—}NH\text{-}COCH_2CH_3 \xrightarrow{LiAlH_4} Ph\text{—}NH\text{-}CH_2CH_2CH_3$$

$$R\text{—}CN \xrightarrow[\text{or } H_2/Pd]{LiAlH_4} R\text{—}CH_2NH_2 \qquad R\text{—}NO_2 \xrightarrow[\text{or } H_2/Pd \text{ or } Na/EtOH(Sol)]{LiAlH_4} R\text{—}NH_2$$

analysis:

$$\underset{\text{Fenfluramine}}{m\text{-}F_3C\text{-}C_6H_4CH_2CH(CH_3)NHEt} \xRightarrow{FGI} ArCH_2CH(CH_3)NHCOCH_3 \underset{\text{amide}}{\xRightarrow{C\text{-}N}} ArCH_2CH(CH_3)NH_2 \xRightarrow{FGI} ArCH_2C(CH_3){=}NOH \underset{\text{oxime}}{\xRightarrow{C\text{-}N}} ArCH_2COCH_3$$

synthesis:

Neither oxime nor amide need be isolated-the published synthesis uses different methods of reduction in the two cases, no doubt developed by experiment.

$$m\text{-}F_3C\text{-}C_6H_4CH_2COCH_3 \xrightarrow[(2)H_2, cat]{(1)NH_2OH} ArCH_2CH(CH_3)NH_2 \xrightarrow[(2)LiAlH_4]{(1)MeCOCl} m\text{-}F_3C\text{-}C_6H_4CH_2CH(CH_3)NHEt$$

Fenfluramine:

The alkylation and reduction of aliphatic nitro compounds is one route to t-AlkNH_2 and is discussed. Another route uses the Ritter reaction followed by hydrolysis of the amide.

$$R^1CH_2NO_2 \longrightarrow R^3C(R^1)(R^2)NO_2 \xrightarrow[cat]{H_2} R^3C(R^1)(R^2)NH_2$$

$$R^3C(R^1)(R^2)OH \xrightarrow[H^+]{MeCN} R^3C(R^1)(R^2)NHCOMe \xrightarrow{^-OH/H_2O} R^3C(R^1)(R^2)NH_2$$

We have already seen that aromatic amines are made by the reduction of nitro compounds and aliphatic nitro compounds can be used in the same way.

Azides can also be reduced to amines. The importance of this method is that the azide ion, N_3^-, acts as a reagent for NH_2^-, so that the disconnection is the normal one for C—X bonds. Other reagents for this synthon are discussed in the next section[3].

$$RBr \xrightarrow{NaN_3} RN_3 \xrightarrow[or\ H_2,\ cat]{LiAlH_4} RNH_2$$

Amines of type can therefore be made by reduction of azides which can be derived from epoxides and azide ions.

analysis:

$$RCH(OH)CH_2NH_2 \xrightarrow{FGI} RCH(OH)CH_2 \text{-}N_3 \xrightarrow{1,2\text{-dis}} R\text{-epoxide} + N_3^-$$

synthesis:

$$R\text{-epoxide} \xrightarrow[dioxane\ reflux]{NaN_3} RCH(OH)CH_2N_3 \xrightarrow[or\ LiAlH_4]{H_2,PtO_2} RCH(OH)CH_2NH_2$$

Examples for the synthon NH_2 group:

9.1.5 羧酸及其衍生物的切断

Disconnection of Carboxy Acid and Acid Derivatives

The aliphatic acid can be made by the cyanide method. The cyanide method is better for reactive allylic and benzylic halides and has the advantages that eaters can be formed directly from the cyanide if required.

Direct disconnection is possible at this oxidation level since CO_2, especially convenient as solid "dry ice", reacts once only with Grignard reagents or RLi (iv). This method complements hydrolysis of cyanides(v) since the disconnection is the same but the polarity is different. Hence the t-alkyl acid could not be made by cyanide method as displacement at the tertiary centre would be difficult. The Grignard method works well.

Acids can be converted into a range of derivatives often via the acid chloride so that FGI or C—X disconnections may be needed before and after the C—C disconnection. The acid bromide requires the

analysis: FGI

analysis: FGI

analysis:

+

synthesis: OH, OH, Catechins; Base, Me_2SO_4; OMe, OMe; CH_2O,HCl, AcOH,80%, Chloromethylation; MeO, MeO, Cl; NaCN, H_2O,EtOH, reaction

MeO, MeO, CN; $LiAlH_4$; MeO, MeO, NH_2; TM

KOH,H_2O; MeO, MeO, COOH; $SOCl_2$; MeO, MeO, COCl

Grignard disconnection as nucleophilic displacement of aryl halides is not a good reaction.

analysis: ⟹ NH HO ⟹ BrMg ⟹ FGI HO ⟹ O BrMg

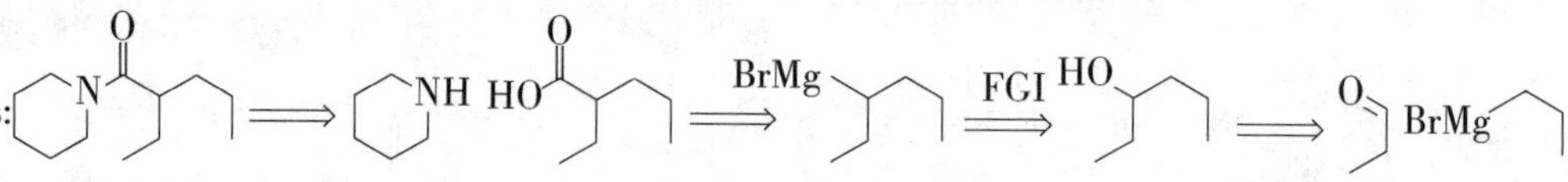

synthesis: Br —(1)Mg, Et_2O (2)EtCHO→ HO —(1)PBr_3 (2)Mg, Et_2O (3)CO_2→ HO —(1)SO_2Cl_2 (2) NH→

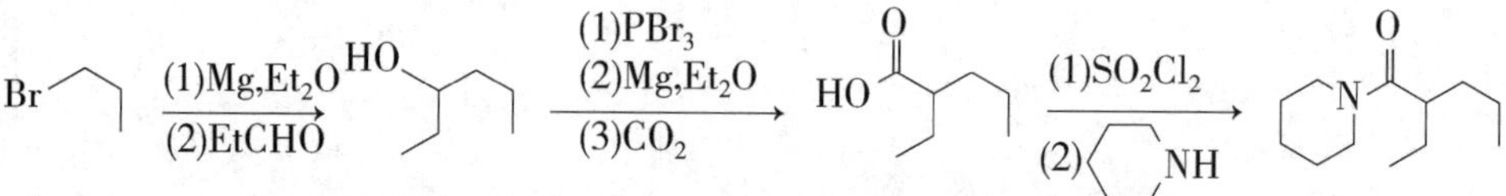

Diethyl malonate as building block usually prepares monocarboxylic acids:

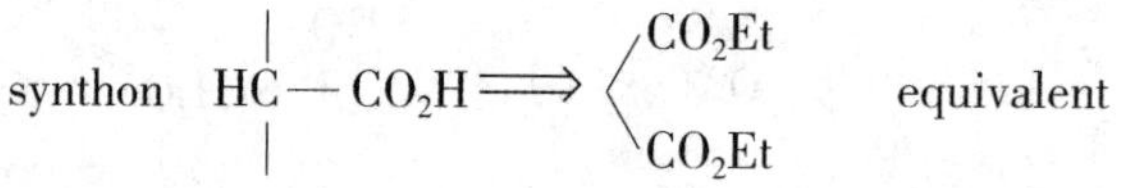

synthon $HC—CO_2H$ ⟹ CO_2Et CO_2Et equivalent

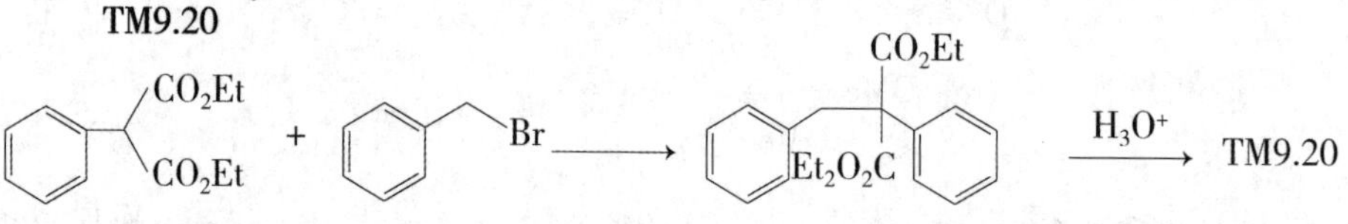

analysis: COOH ⟹ ⊕ ⊖ COOH

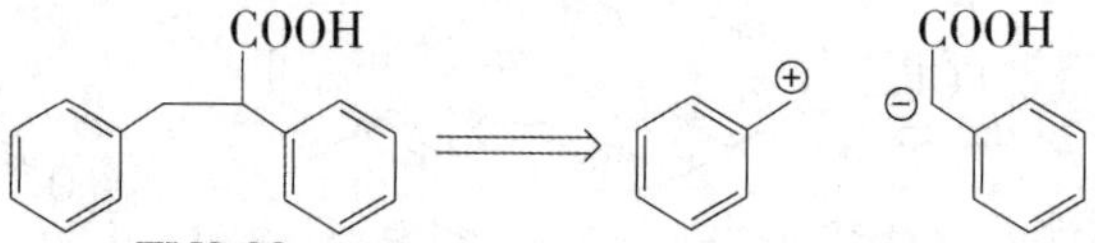

TM9.20

synthesis: CO_2Et CO_2Et + Br → CO_2Et Et_2O_2C —H_3O^+→ TM9.20

analysis: COOH ⟹ $\bar{C}H_2COOH$ + $\overset{+}{}$ Br ⟹ OH ⟹ COOH

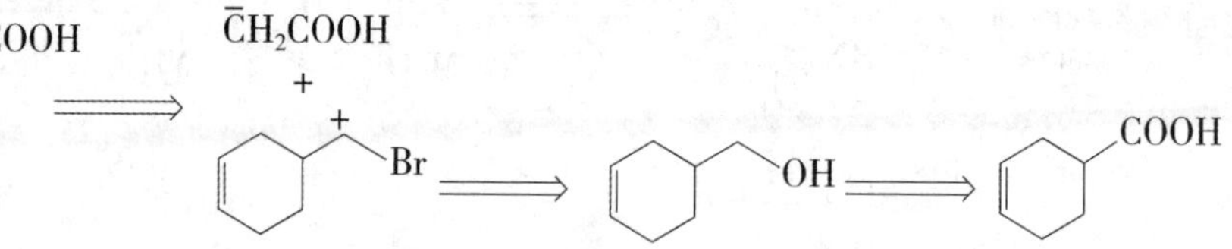

TM 9.21

synthesis: CO_2Et → CO_2Et —$LiAlH_4$→ CH_2OH → CH_2Br

CH_2Br + CO_2Et CO_2Et —EtONa→ CO_2Et CO_2Et —dilu.H^+→ CO_2H

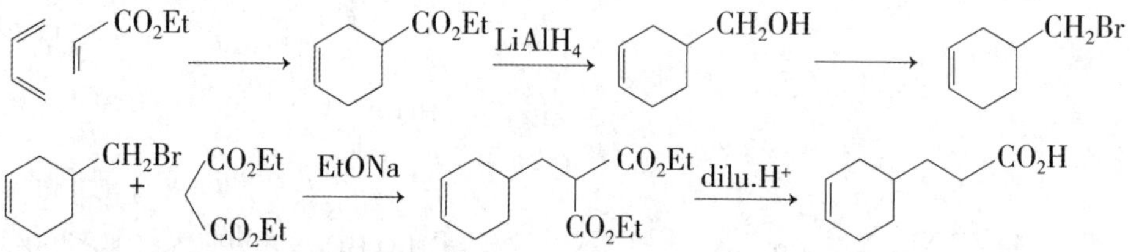

9.1.6 醚、卤代烷、硫化物的切断 Disconnection of Ethers, Alkyl Halides, and Sulphides

9.1.6.1 Alkyl halides

In one way, alkyl halides are easier than C—C disconnections. Reagents are available for both electrophilic (e.g. RBr) and nucleophilic (e.g. RMgBr) carbon whereas heteroatoms are almost always added as nucleophiles. It applies to C—X disconnection in aliphatic compounds gives a nucleophile XH and an electrophilic carbon species usually represented by an alkyl halide, tosylate, or mesylate. These compounds can all be made from alcohols and as alcohols can be made by C—C bond formation we shall treat the alcohol as the central functional group[7].

Table 9–1 Aliphatic compounds derived from alcohols

Reagents	Product	Class
ROH, base	ROR′	Ethers
RSH, base	RSR′	Sluphides
1.$(NH_2)_2CS$ 2.HO^-/H_2O	RSH	Thiols
Hal^-	RHal	Alkyl halides
Nu	RNu	Other derivatives

ROH ⇌ RX (X=halide, OTs, OMs)

$$RX \Longrightarrow XH + R^+ = RBr \text{ or ROTs or ROMs}$$

$$RBr \xleftarrow[\text{or HBr}]{PBr_3} ROH \xrightarrow[\text{pyr}]{TsCl} ROTs; \quad ROH \xrightarrow[Et_3N]{MsCl} ROMs$$

Conditions must be chosen to suit the structure of the molecules. Methyl and primary alkyl derivatives react by the S_N2 mechanism so powerful nucleophiles and non－polar solvents are effective. The nitro compound and the azide examples of the "other derivatives" in Table 9–1 are easily made from the corresponding bromides by S_N2 reactions as they are both primary alkyl compounds.

$$\text{CH}_2\text{=CHCH}_2\text{CH}_2\text{Br} \xrightarrow[\text{urea, DMSO}]{NaNO_2} \text{CH}_2\text{=CHCH}_2\text{CH}_2\text{NO}_2$$

$$\text{PhCH}_2\text{CH}_2\text{CH}_2\text{Br} \xrightarrow{NaN_3} \text{PhCH}_2\text{CH}_2\text{CH}_2\text{N}_3$$

Tertiary compounds react even more easily by the S_N1 mechanism via stable carbonium ions generated directly from alcohols, alkyl halides, or even alkenes. Powerful nucleophiles are no help here but polar solvents and catalysis (usually acid or Lewis acid) help by making the OH a better leaving group.

$$R^1R^2C{=}\text{(alkene)}R^3 \xrightarrow{H^+} R^1R^2R^3C^+$$

$$R^1R^2R^3C\text{–OH} \longrightarrow R^1R^2R^3C^+ \xrightarrow{X^-} R^1R^2R^3C\text{–X}$$

$$R^1R^2R^3C\text{–Br} \xrightarrow{\text{polar solvent}} R^1R^2R^3C^+$$

Compound can obviously be made by a Friedel–Crafts reaction from benzene and the tertiary chloride, which comes from the alcohol.

Br $\xrightarrow{Nu^-}$ Nu

Ar Br $\xrightarrow{Nu^-}$ Ar Nu

These interconversions are rather elementary in concept but are essential to synthetic planning. Compounds of the type R^1–X–R^2 offer a choice for the first disconnection and are more interesting.

9.1.6.2 醚和硫化物 Ethers and sulphides

We can often choose our first disconnection because of the reactivity (or lack of it) of one side of the target molecule. The oxygen atom in the wallflower perfume compound has a reactive side (Me, by S_N2) and an unreactive (Ar) side so disconnection is easy[8].

analysis:

Me–C₆H₄–O⌇Me $\xRightarrow[\text{ether}]{\text{C–O}}$ Me–C₆H₄–O⁻ + MeY

Dimethyl sulphate is used for methylation of phenols in alkaline solution where the phenol is ionized. Since the mechanism is S_N2, the more powerfully nucleophilic anion is an advantage.

synthesis:

Me–C₆H₄–OH $\xrightarrow[\text{NaOH}]{(MeO)_2SO_2}$ Me–C₆H₄–O⌇Me

85%

The gardenia perfume compound can be disconnected on either side as both involve primary alkyl halides. The benzyl halide is more reactive but the decisive point in favor of route (b) is that route (a) might well lead to elimination.

analysis:

Ph⌇b O⌇a (isoamyl) $\xRightarrow{a}$ Ph OH + X (isoamyl)

$\xRightarrow{b}$ Ph X + HO (isoamyl)

This is S_N2 again, so base catalysis helps.

synthesis:

HO (isoamyl) $\xrightarrow{\text{base}}$ -O (isoamyl) $\xrightarrow{\text{Ph Cl}}$ Ph O (isoamyl)

85%

If there is no obvious preference, it is more helpful to write both fragments as alcohols and decide later which to convert into an electrophile. Baldwin needed the ether to study the rearrangement of its carbanion. Both sides are reactive so we write the two alcohols. Baldwin does not reveal how he actually made the ether–both routes look good, although the one shown is less ambigious.

analysis: Ph⁀O⁀⁀= $\xrightarrow[\text{enter}]{\text{C—O}}$ Ph⁀OH + HO⁀⁀=

synthesis: HO⁀⁀= $\xrightarrow[\text{PhCH}_2\text{Br}]{\text{base}}$ Ph⁀O⁀⁀=

The same principles apply to sulphide (R^1SR^2) synthesis. The reaction is even easier by S_N2 as thiols ionise at a lower pKa than alcohols, the anion is softer than RO^- and thus more nucleophilic towards sp^3 carbon.

$$R^1\text{—S}\not\text{—}R^2 \Longrightarrow R^1S^- + R^2Y$$

The chlorbenside (氯杀螨) is first disconnected on the alkyl rather than the aryl side.

Chlorbenside:

analysis: C—S sulphide ⟹ Cl, S, S⁻, +, Cl, Cl, Cl, Cl

synthesis: SH, Cl, +, Cl, Cl, NaOEt, EtOH, S, Cl, Cl

9.2 双官能团化合物逆合成分析 Retrosynthetic Analysis with Participation of Two Functional Groups

We now leave disconnections of bonds between carbon and other atoms and turn to the more challenging C—C disconnections. These are more challenging because organic molecules contain many C—C bonds and we must learn which to disconnect. In this section, we discuss examples where two functional groups control selection of the preferred C—C bond for disconnection by their electronic effects. Most heteroatoms (O, N, S, Hal) in organic molecules are more electronegative, whereas P, Si and all metals are more electropositive than carbon. The polarization C-heteroatom bond dictates the direction of the imaginative flow of the electrons in the disconnected C—C bond at a distance from the functional group.

For the following discussion it is important to recall that any functional group in organic molecules rises the charge separation along the carbon chain. The through-bond distance of two functional groups in bifunctional molecules defines the position of the C—C bond and electron flow in disconnection. In this context, the terms match and mismatch of charges in the target molecules are presented, and the principal modes of logical disconnections of the 1,3- and 1,5-dioxygenated pattern in target molecules are discussed. Examples are given for asymmetric synthesis of chiral target molecules with these two deoxygenated patterns.

Disconnection of 1,3-CO and 1,5-CO patterns is supported by the matching distribution of charges in two synthons. One of them comprises a stabilized carbanion and the other the carbocation or C atom with high positive charge, both with recognizable and available reagents, as we saw in Chapter 8. Generally, patterns with an odd distance of functional groups support the preferred disconnection, different from patterns with an even distance of functional groups, 1,2-, 1,4- or 1,6-. The reason rests in non-matching distribution of partial charges on two C atoms on the bond to be disconnected so that disconnection always generates one "unacceptable" or illogical synthon.

Destabilized synthons with charges opposite to the natural polarization of the bond often have acceptable though not easily recognizable reagents or synthetic equivalents. After all, syntheses of numerous compounds with even distances of functional groups are successfully completed by formation of C—C bonds. They are important intermediates in the syntheses of complex natural compounds. In the sections, the retrosynthetic approach to such compounds with 1,2-,1,4- or 1,6-CO pattern are also discussed.

In addition, specific issues with illogical disconnections of target molecules with an even number of atoms separating two oxygen functionalities are addressed. The Mannich reaction, Michael addition and Robinson annulation are presented as synthetic approaches to this target molecule based on illogical disconnections.

9.2.1 1,3-二氧化模式 1,3-Dioxygenated Pattern (1,3-CO)

In this section we consider disconnection with participation of two oxygen functionalities, called the dioxygenated pattern. The rationale behind this selection rests in the fact that oxygen functionalities can be conveniently transformed into functional groups with other heteroatoms. Schematic structures of the 1,3-dioxygenated pattern are presented as the following framework (Fig. 9-1).

OH OR   OH O   O   O O

Fig. 9-1 Structures with a 1,3-dioxygenated pattern

9.2.1.1 1,3-羟基羰基化合物 1,3-Hydroxycarbonyl compounds

1,3-hydroxycarbonyl compounds or 1,3-dicarbonyl compounds with a 1,3-dioxygenated pattern are analyzed first in the section (Fig. 9-2).

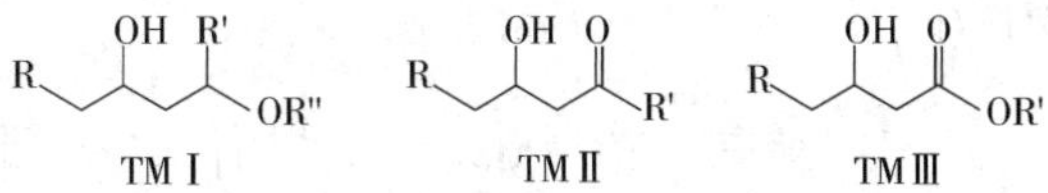

Fig. 9-2 Structures of 1, 3-Hydroxycarbonyl compounds

Maximal simplification of any of the three target molecules will be achieved by disconnection of one of two central or internal bonds (Scheme 9-24).

It is important to note that the oxidation level of the oxygenated functionality determines the position of disconnection. Changing the oxidation state of the oxygenated functionality, we can control the position of the C—C disconnection. As already seen, the OH group facilitates the disconnection of the C—C bond to the carbinolic C atom. The C—C bond to the OH group is preferably disconnected, and the disconnection in TM Ⅰ results in a carbonyl compound and unstabilized carbanion (plus proton). In TM Ⅱ and TM Ⅲ, instead, the emerging carbanion is stabilized by resonance with a second oxygen functionality. Disconnection is therefore directed by the participation of two functional groups. The carbonyl group in TM Ⅱ and carboxyl group in TM Ⅲ stabilize the carbanion by resonance, which is not possible for the *OR* group in TM Ⅰ. The synthetic reaction in which two carbonyl components are connected to a more complex molecule with a 1,3-hydroxycarbonyl pattern as in TM Ⅱ is the well-known aldol reaction, which takes place between the enolate and carbonyl form of aldehydes or ketones. The examples that follow serve to suggest the decision concerning which internal C—C bond in 1,3-dioxygenated compounds to disconnect.

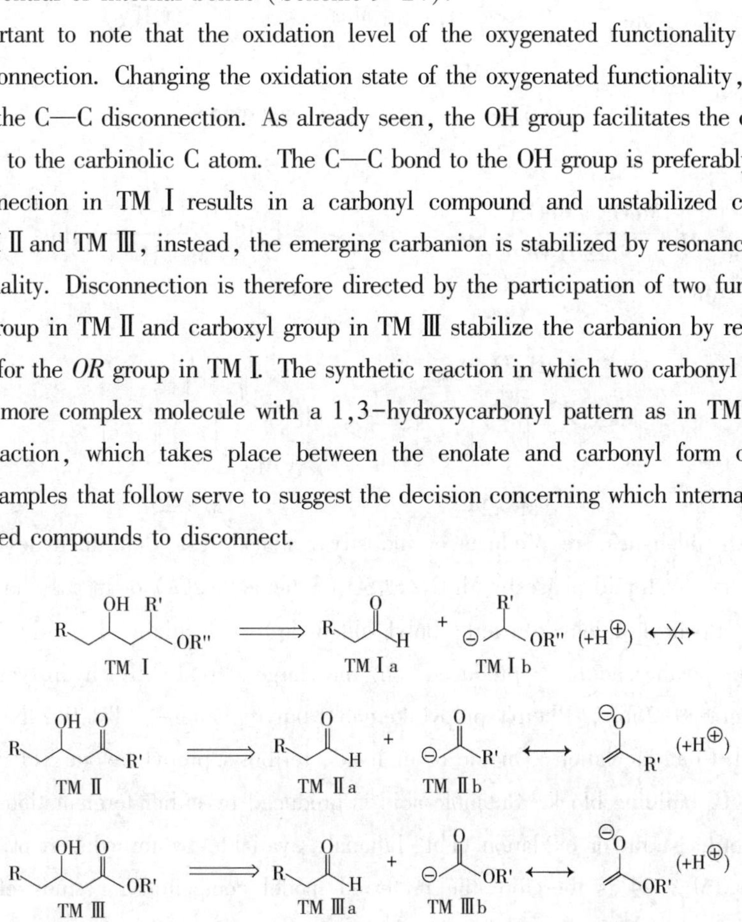

Scheme 9-24 Disconnections of a 1, 3-dioxygenated pattern in TM Ⅰ - Ⅱ

Propose the preferred disconnection for TM Ⅰ and then the synthesis from easily available startingmaterials. Suggest how to use TM Ⅰ for the synthesis of 2-ethylhexanol TM Ⅱ, a commodity produced in the amount of 2. 5 million tons/year.

Structural isomers TM 9. 24 and TM 9. 25 are equally acceptable for certain physico-chemical studies. Which one do you prefer for an easier synthetic approach?

Disconnections result in one molecule of aldehyde TM 9. 3a or TM 9. 4b and ketone TM 9. 24b or TM 9. 25a, presented in its enolic form according to the mechanism of disconnection.

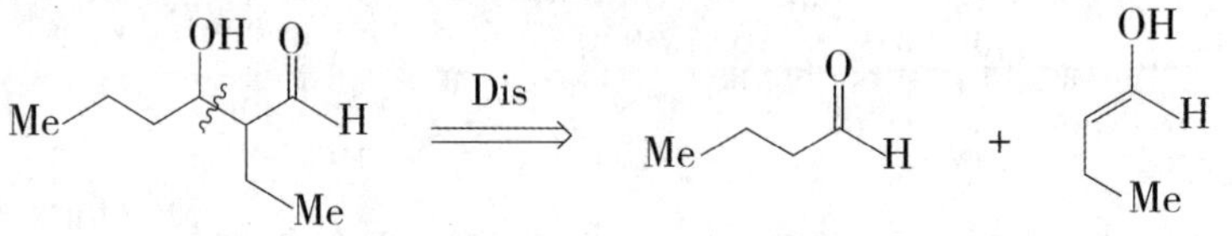

TM 9.22 Dis ⟹ TM 9.22a

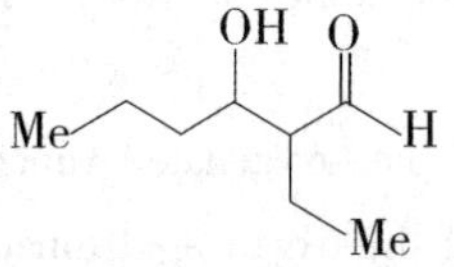

TM 9.22 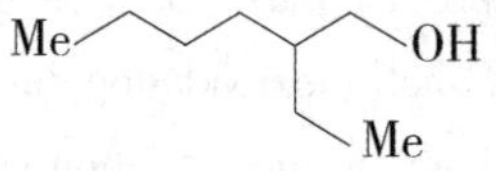 TM 9.23

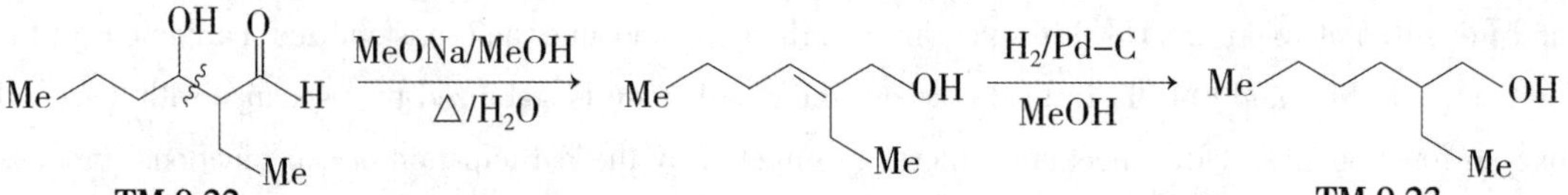

TM 9.22 MeONa/MeOH, Δ/H_2O → ; H_2/Pd–C, MeOH → TM 9.23

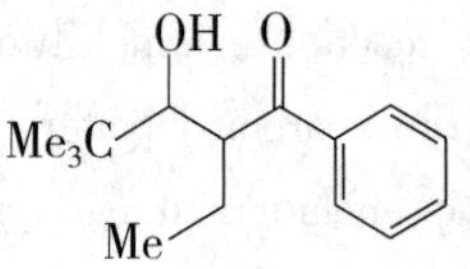

TM 9.24 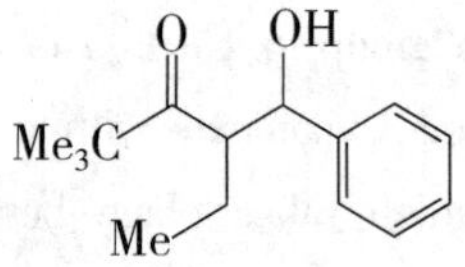 TM 9.25

Note: Both aldehydes are products of industry commodities. Benzaldehyde is produced by oxidation of toluene in liquid phase by MnO_2/H_2SO_4(Scheme 9-26a) or in gas phase at 400 ℃ by V_2O_5/K_2SO_4. Pivaloyl aldehyde (tert-butyl aldehyde) is available by reduction of pivaloyl chloride, while pivalic acid is produced on the large-scale by hydroxycarbonylation of isobutene (Scheme 9-26b). Phenyl propyl ketone (butyrophenone) TM 9.24b is available by common Friedel-Crafts acylation. On the other hand, tertbutyl propyl ketone TM 9.25a is not an easily available C_8 building block. Butanoic acid is produced by either fermentation of starch using the Bacillus subtilis strain or oxidation of n-butanal, available by formylation of propene. More easily available TM 9.24 is therefore the preferred model compound for study of the physico-chemical properties of 1,3-hydroxycarbonyl unit.

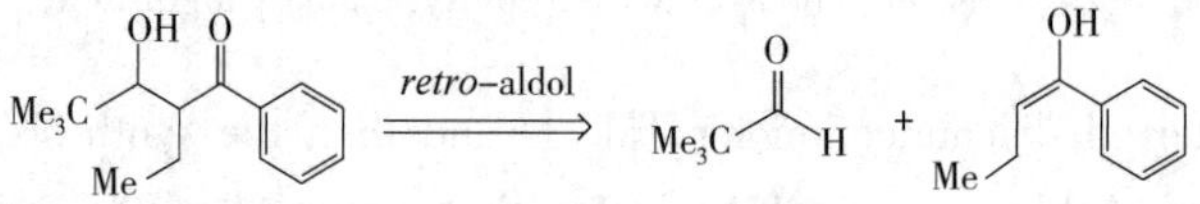

TM 9.24 *retro*-aldol ⟹ TM 9.24a + TM 9.24b

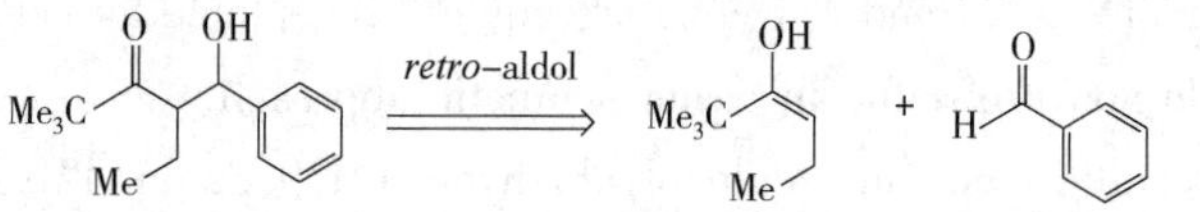

TM 9.25 *retro*-aldol ⟹ TM 9.25a + TM 9.25b

Scheme 9-25 Retro-aldol disconnection of TM 9.24 and TM 9.25

(a) toluene $\xrightarrow{MnO_2/H_2SO_4/\Delta}$ benzaldehyde TM 9.25b

(b) $H_2C{=}C(CH_3)_2$ + CO + H_2O $\xrightarrow{HF\ or\ H_3PO_4/\Delta}$ $Me_3C{-}CHO$ TM9.24a

Scheme 9-26 Technological methods for production of benzaldehyde and pivaloyl aldehyde

9.2.1.2 1,3-二羰基化合物 1,3-Dicarbonyl compounds

The second class of 1,3-dixygenated compounds is characterized by two carbonyl groups in the 1,3-position. Depending on the structure of group R, they belong to 1, 3- or β-diketones, 1,3-ketoaldehydes and 1,3-ketocarboxylic acids and their derivatives. Disconnection of either internal C—C bond in TM Ⅳ can be completed by participation of one carbonyl group (Scheme 9-27). Differently from the 1,3-hydroxycarbonyl pattern, disconnection of the 1,3-dicarbonyl pattern results in two logical synthons, an α-carbanion stabilized by conjugation and an acyl cation. The acyl cation has a number of acceptable reagents in reactive derivatives of carboxylic acids. Reactive enough with carbanions are carboxylic acid esters affording 1, 3-dicarbonyl compounds in the well-known ester condensation, also known as Claisen condensation. The next example serves as an introduction to the retrosynthetic analysis of this pattern.

Generic disconnection and the corresponding synthetic equivalents[9]:

Cl / OEt

Cl + EtO

Complete the retrosynthetic analysis of TM 9. 26 and suggest its synthesis:

COOEt

TM 9.26

DIS 1,3-CO a

DIS 1,3-CO b

TMⅣ

Scheme 9-27 Two possible disconnections of 1, 3-dicarbonyl pattern in TM Ⅳ

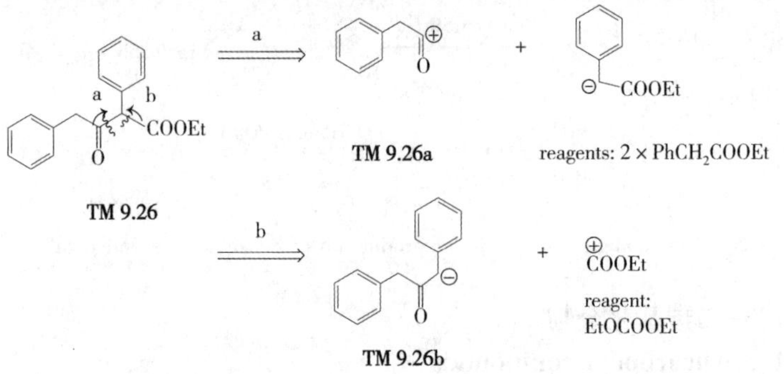

TM 9.26 ⟹(a) TM 9.26a + ⊖COOEt; reagents: 2 × $PhCH_2COOEt$

TM 9.26 ⟹(b) TM 9.26b + ⊕COOEt; reagent: EtOCOOEt

Scheme 9-28 Two possible disconnections of TM 9.26

The 1, 3-dicarbonyl pattern in TM 9.26 suggests obvious disconnection a leading to the maximal simplification of the molecule. Two C_7 synthons have the same reagent, an ester of phenylacetic acid (Scheme 9-28).

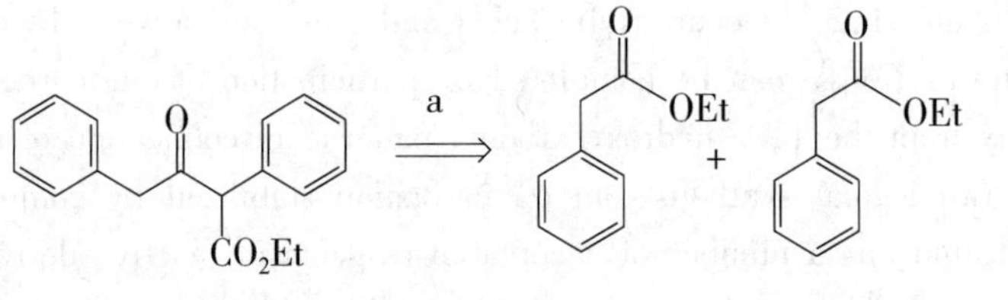

CO_2Et ⟹(a) OEt + OEt

One economic industrial production of phenylacetic acid is the oxidation of ethylbenzene by an aqueous solution of potassium bichromate in an autoclave at elevated temperature (Scheme 9-29).

It is surprising that the contradicting principle of maximal simplification disconnection b competes with disconnection a (Scheme 9-28). The carbanion of 1, 3-diphenylacetone and carbethoxy cation appear as synthons.

1, 3-Diphenylacetone is a commercial product, available from phenylacetic acid over the intermediary formed, β-keto acid, which under reaction conditions decarboxylates (脱羧酸盐) (Scheme 9-30).

On the other hand, the carbethoxy (乙氧羰基) cation, a seemingly illogical synthon, has an available synthetic equivalent in diethyl carbonate. Diethyl carbonate is produced by oxidative carbonylation of ethanol, promoted by various heterogeneous catalysts; one of the most effective is the mixed catalyst $CuCl_2/PdCl_2$ deposited on charcoal (Scheme 9-31).

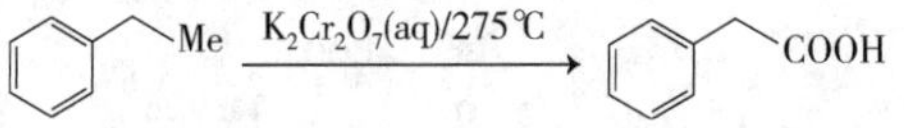

Me —$K_2Cr_2O_7$(aq)/275℃→ COOH

Scheme 9-29 Industrial method for production of phenylacetic acid

COOH —$(AcO)_2O$/KOAc, 140~150℃→ [COOH] —△/CO_2→

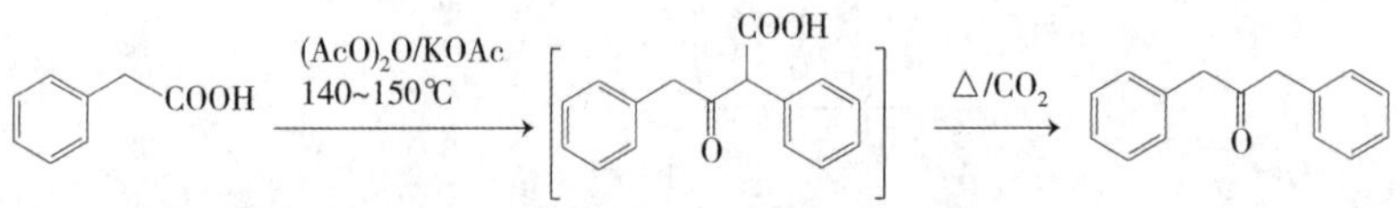

Scheme 9-30 Industrial production of diphenyl acetone

$$EtOH + CO \xrightarrow{CuCl_2/PdCl_2/\triangle} EtCOOEt$$

Scheme 9-31 Method for production of diethyl carbonate

Propose a synthetic route to TM 9. 27 and explain why the obvious disconnection according to maximal simplification of this target molecule is hampered by non-workable synthesis.

COOEt
COOEt
TM9.27

On first glance we recognize the 1,3-dicarbonyl pattern of diethyl malonate, which suggests the disconnection of the C—C bond to the aromatic ring. In view of the easy alkylation of diethyl malonate, one would expect a workable arylation of the diethylmalonate anion by aryl halides. However, this reaction represents a difficult synthetic task. Here we meet the important principle of reactivity in organic chemistry. The high stability of the malonate carbanion is due to its resonance stabilization by two geminal carbethoxy groups. Delocalization of the negative charge by resonance makes this carbanion a very soft nucleophile and therefore not reactive with hard electrophiles such as aryl halides, specifically bromobenzene TM 9. 27a (Scheme 9-32).

(a) COOEt COOEt ⟹ Br + COOEt COOEt

TM 9.27 **TM 9.27a**

(b) COOEt COOEt $\xrightarrow{\text{DIS 1,3-CO}}$ COOEt + EtOCOOEt

TM 9.27 **TM 9.27b**

Scheme 9-32 Two possible disconnections of TM 9. 27

As suggested by the discussion of HSAB, we invoke alternative disconnection (b) of TM 9. 27. On first glance, an illogical disconnection leads to the available staring material, ethyl phenylacetate and diethyl carbonate. In the synthetic direction, ester condensation is a well-known and workable process.

Example TM 9. 28 suggests the disconnection of TM 9. 28.

The first retrosynthetic step represents the logical disconnection of the exocyclic C—C bond with the formation of two large synthetic blocks (Scheme 9-33). next target molecule, TM 9. 28a, comprises a 1,3-dicarbonyl pattern with a keto group inside the cyclopentane ring and an exocyclic carboxylic group. Disconnection of the C—C bond in the ring is preferred, although we mentioned that ring-opening disconnections often result in more complex target molecules! *retro*-Ester condensation in TM 9. 28 results in easily available diethyl ester of adipic acid TM 9. 28b. The intramolecular variant of ester condensation is known as the Dieckmann reaction and preferred for construction of thermodynamically favored medium rings without steric strain, in particular cyclopentanone and cyclohexanone derivatives.

TM 9.28

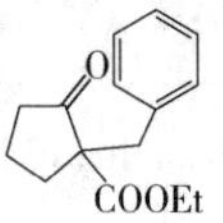

COOEt

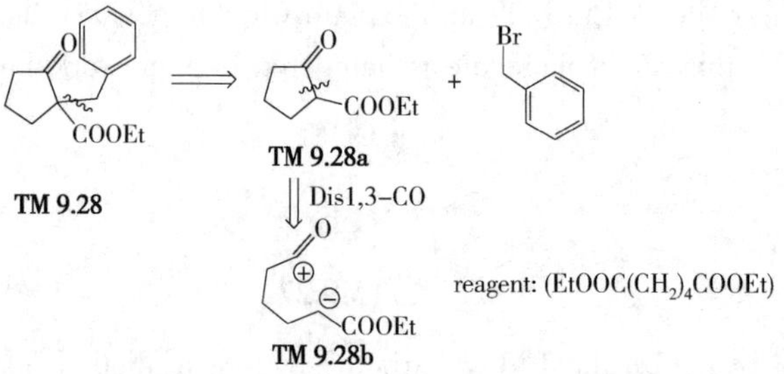

COOEt

TM 9.28a

TM 9.28

Dis1,3-CO

reagent: $(EtOOC(CH_2)_4COOEt)$

COOEt

TM 9.28b

Scheme 9-33 retro-Dieckmann disconnection of TM 9. 28

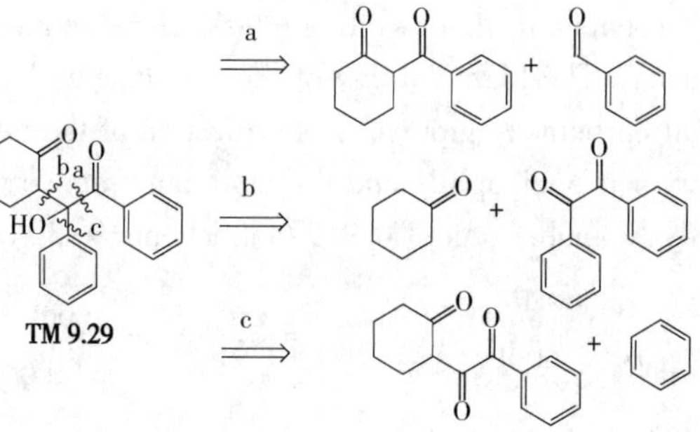

TM 9.29

Scheme 9-34 Overview of favorable and unfavorable disconnections

Disconnection approaches have more than one possibilities, the preferred disconnections often have acceptable easily recognizable reagents or synthetic equivalents. After all, syntheses of numerous compounds with even distances of functional groups are successfully completed by formation of C—C bonds. They are important intermediates in the syntheses of complex natural compounds[4].

1,3-dicarbonyl compound can be conveniently transformed from its derivatives via functional group interconversion.

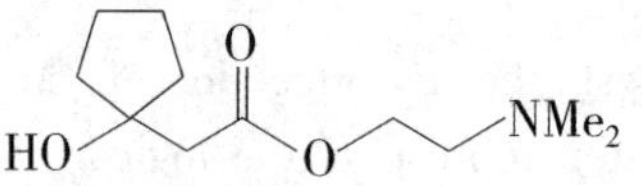

TM 9.30

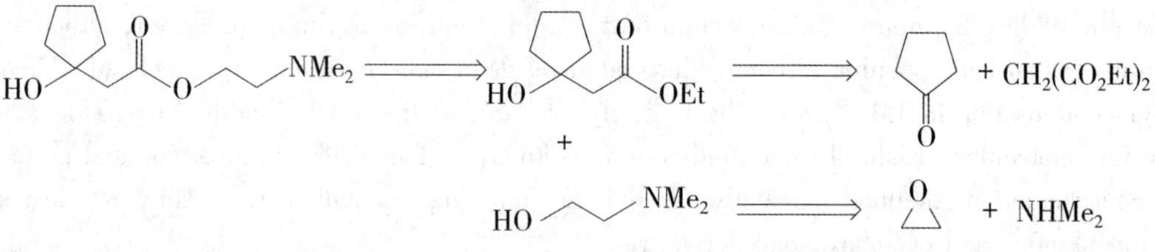

$+ CH_2(CO_2Et)_2$

$+ NHMe_2$

Aldehyde group in a 1,3-dicarbonyl pattern plays a role of protecting tert C—H bond, as seen the protection of C—H bond in Chapter 7.

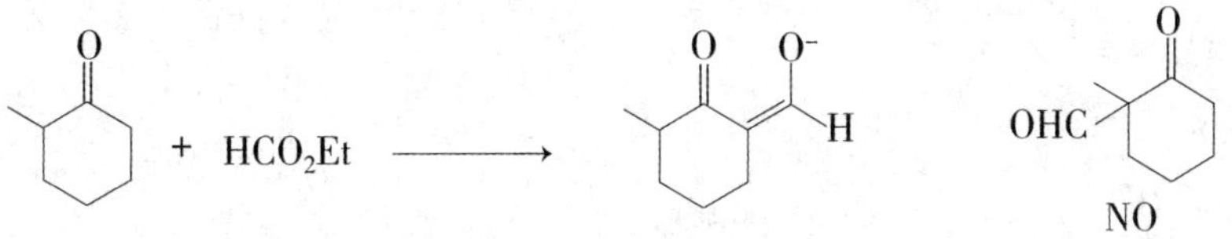

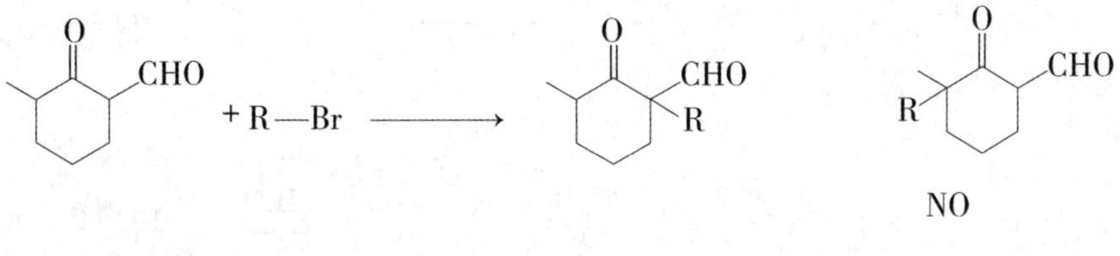

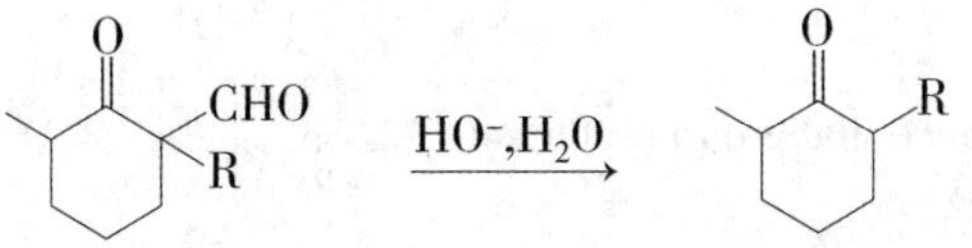

+ HCO_2Et ⟶

OHC

NO

CHO + R—Br ⟶ CHO R

CHO R

NO

CHO R $\xrightarrow{HO^-, H_2O}$ R

Propose the synthesis of TM 9.31. Aldehyde group as building block improves the selectivity of preparing TM 9.31.

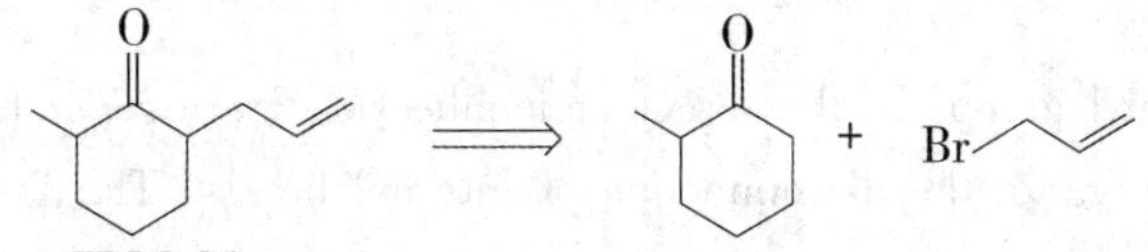

TM 9.31 ⟹ + Br

Base catalysis: Br ⟶ major minor

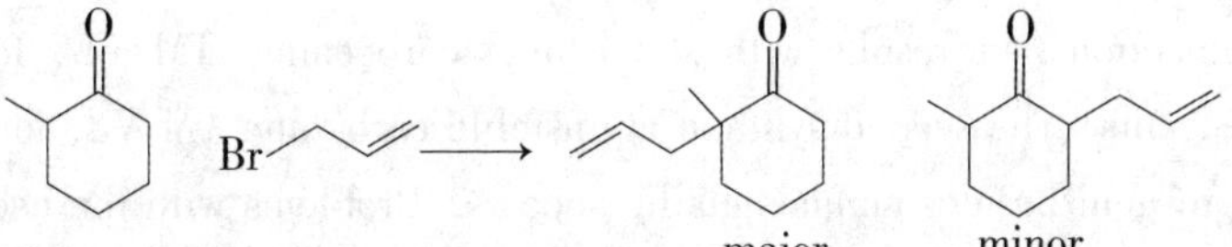

Aldehyde protection group: $\xrightarrow{Br}$ CHO $\xrightarrow{HO^-, H_2O}$

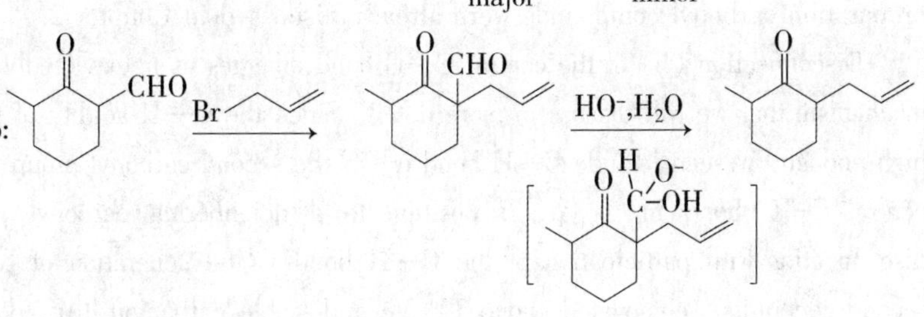

Consider disconnection with stable anion:

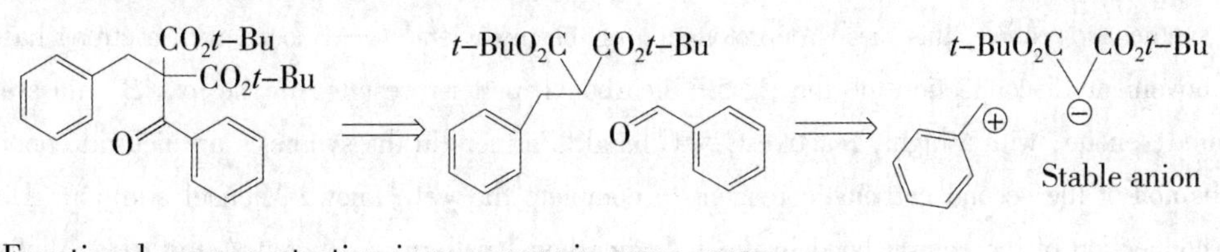

CO_2t-Bu CO_2t-Bu ⟹ t-BuO_2C CO_2t-Bu ⟹ t-BuO_2C CO_2t-Bu

Stable anion

Functional group protection is necessary in some cases:

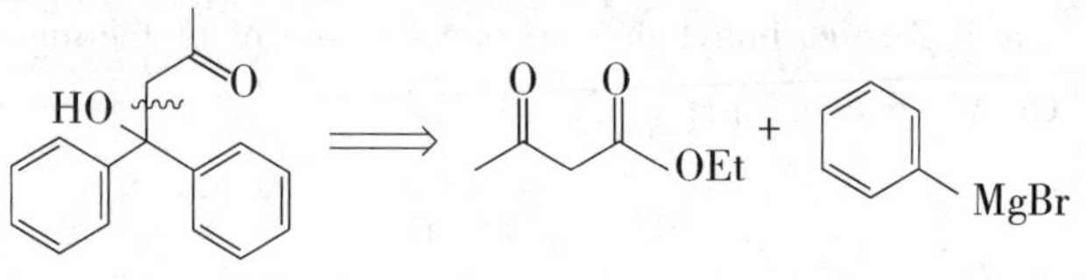

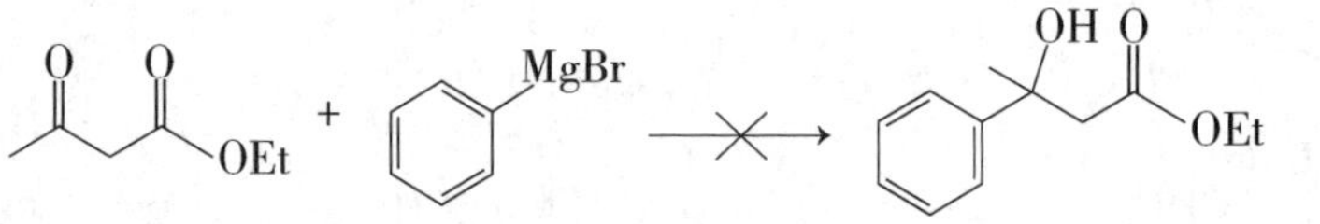

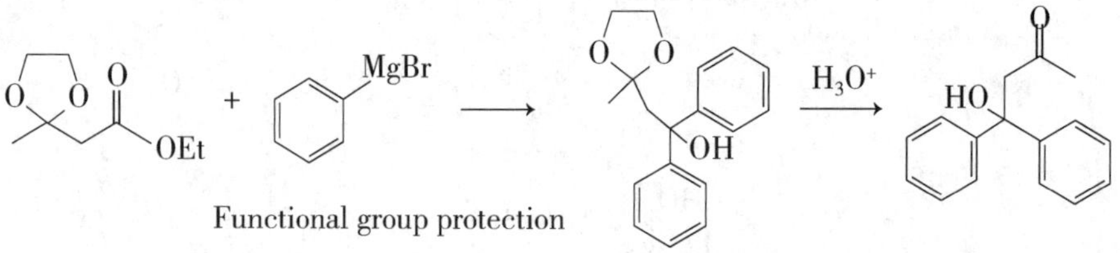

Functional group protection

9.2.2 1,5-二羰基模式 1,5-Dicarbonyl Pattern (1,5-CO)

Two carbonyl groups in the target molecule TM V are separated by five C atoms, and two possibilities emerge for the disconnection of internal bonds. The first one is disconnection of the C—C bond next to one of two carbonyl groups, and the second one is disconnection of one central C—C bond (Scheme 9-35).

Disconnection (a) results with stable acyl carbocation TM Vb, for which we introduced a series of reagents. The second synthon is unstable carbanion TM Va, for which we do not have an acceptable reagent besides organometallic species. Problems with the use of organometallics in the preparation carbonyl compounds were already discussed in Chapter 2.

Disconnection (b) of the central C—C bond emerges as unexpectedly convenient and involves a mechanism that we will discuss in more detail. Since the C—H acidity of the α-C atom in TM V is high enough, we can include C—H bond α- in the second carbonyl group in the disconnection of the central C—C bond in the α, β position to a neighboring carbonyl group. The net result of disconnection with participation of the C—H bond is the generation of two logical synthons of the second generation, conjugated enone TM Vc and stable carbanion TM Vd. Removal of the proton in disconnection (b) is used in similar schemes to complete the stoichiometry of the disconnection process. In the synthetic direction, this means deprotonation of the methylenic C—H bond with a strong base.

Convenient disconnection of the 1, 5 - dicarbonyl pattern results in the α, β - unsaturated compound, enone, with a highly reactive C═C bond. Partners in the synthesis are neutral enone and α-carbanion of the second carbonylic reagent to complete the well-known Michael addition. Hence, the disconnection of the central bond in the 1,5-dicarbonyl pattern is denoted as the *retro*-Michael.

Michael addition is a vinylogous analog of the aldol reaction, and most considerations of the aldol reaction in Sect. 9.2 apply here. The reaction is catalyzed by both acids and bases, but more efficiently by bases according to the mechanism presented in Scheme 9-36.

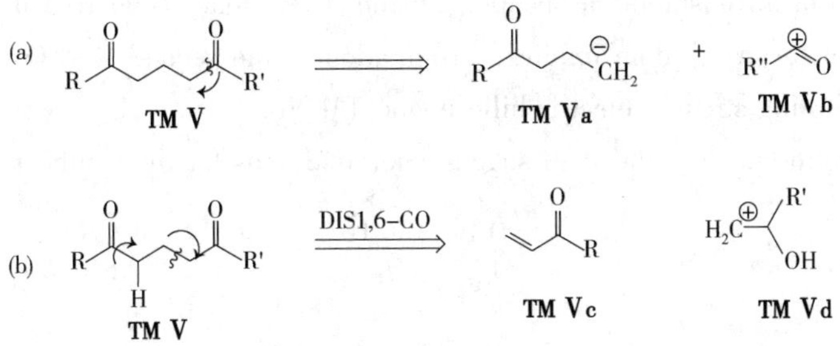

(a) TM V ⟹ TM Va + TM Vb

(b) TM V ⟹(DIS1,6-CO) TM Vc + TM Vd

Scheme 9-35 Disconnections of internal C—C bonds in TM V

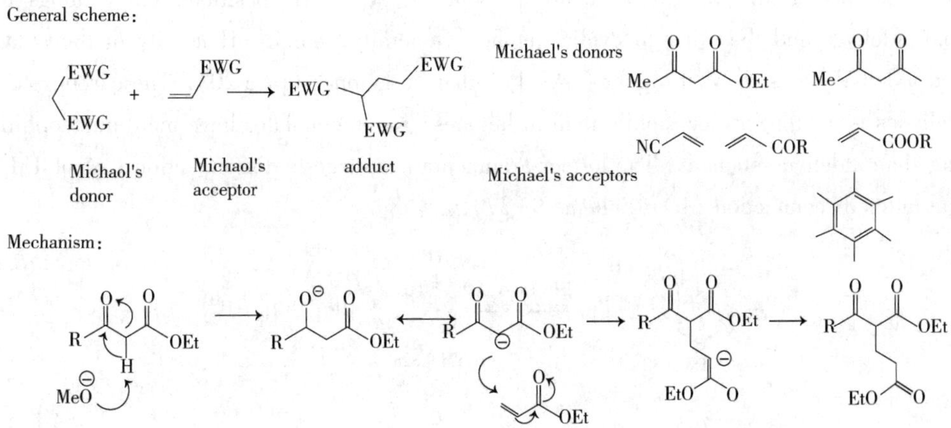

General scheme: Michaol's donor + Michaol's acceptor ⟶ adduct

Michael's donors; Michael's acceptors

Mechanism:

Scheme 9-36 General scheme and mechanism of Michael addition

Note: The vinylogy principle is important in retrosynthetic analysis. It states that two groups separated by one or more conjugated bonds exhibit the same reactivity as if they were directly connected. This is a consequence of the effect of thefirst group through conjugated double bonds. Fig. 9-3 presents a few molecules where the vinylogous position of the terminal methyl groups results in similar C—H acidity.

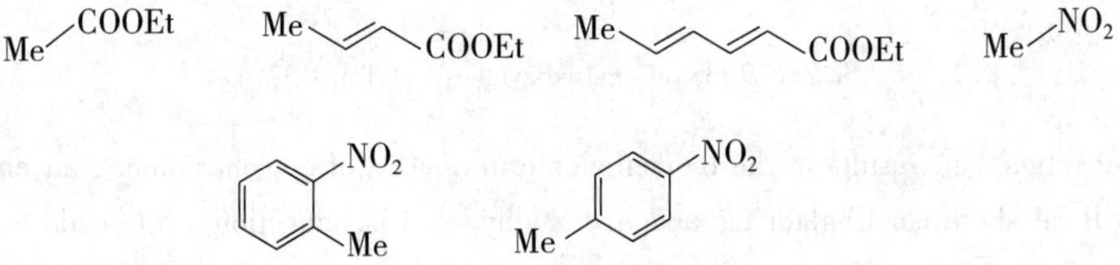

Fig. 9-3 Examples of molecules where the vinylogy principle gives similar C—H acidity to methyl groups

Selection of the preferred central C—C bond for disconnection in TM V in Scheme 9-35 depends on the electronic properties of the R and R' groups and should result in the most stable synthons. On heterolytic disconnection, the electron pair from the σ bond moves to the α-C atom in TM Vd where the carbanion is stabilized by the neighboring electron-withdrawing group. To avoid the formation of an unstable carbocation on the β-C atom, concerted deprotonation of the acid C—H bond in the α-C atom implies participation of the second C═O group in the 1,5-dicarbonyl unit affording strongly electrophilic enone TM Vc.

Propose the retrosynthesis and then suggest the conditions for the synthesis of TM 9. 32.

TM 9.32

Two carbonyl groups in TM 9. 32 are present in the 1, 6-position; one belongs to the terminal aldehyde and the other to cyclic ketone. Generally, the C—H acidity of the α-C—H bond in aldehydes is somewhat higher (p*K*a 17) than in ketones (p*K*a 20). The carbonyl C atom in aldehydes is a stronger electrophile than in ketones. Ketone enolates have more nucleophilic α-C atoms than aldehyde enolates. The former argumentation suggests disconnection (a) of TM 9. 32 and the latter disconnection (b) (Scheme 9-37).

(a) TM 9.32 —DIS 1,6-CO⟹ TM 9.32a + $H-C(=O)-\overset{\ominus}{C}H_2$ $(+H^{\oplus})$

(b) TM 9.32 —DIS 1,6-CO⟹ TM 9.32b + acrolein $(+H^{\oplus})$

Scheme 9-37 Alternative 1, 5-CO disconnections of TM 9. 32

cyclohexanone + acrolein —NaOMe, MeOH, Δ→ TM 9.32

Scheme 9-38 Proposed synthesis of TM 9. 32

Disconnection (a) results in the α-methylenic derivative of cyclohexanone, an enone whose synthesis will be shown in Chapter 2, and acetaldehyde. Disconnection (b) leads to the easily available raw materials cyclohexanone and acrolein. Acrolein is an industrial commodity produced by thermal oxygenation of propene with oxygen at 250℃. Cyclohexanone is commercially produced by co-catalyzed oxidation of cyclohexane or by controlled hydrogenation of phenol. The availability of both starting materials suggests a one-step synthesis of TM 9. 32 workable on the large scale (Scheme 9-38). The molar ratio of reactants is controlled to avoid double α, α'-alkylation of cyclohexanone.

Any disconnection will result in a synthon that has the consonant polarity（辅助极性）. Disconnection is preferred when it provides synthon with consonant polarity（辅助极性）and leads to the easily available raw materials.

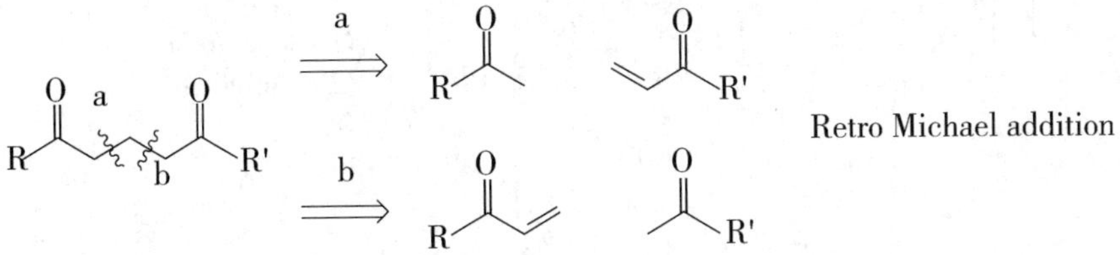

Retro Michael addition

In the next example we shall see that the preferred reagents in the synthetic steps depend on the correct order of retrosynthetic steps when alternative disconnections are conceivable.

Propose convenient synthetic routes to TM 9. 33 considering two alternative disconnections.

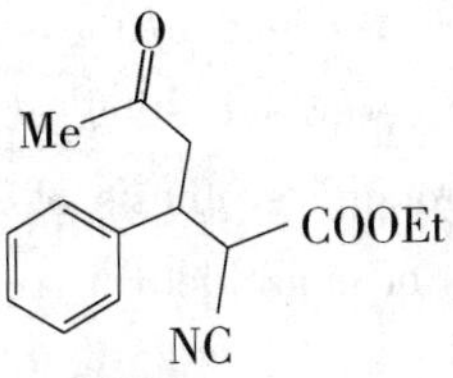

TM 9.33

First we observe that TM 9. 33 comprises a 1, 5-dicarbonyl pattern and select two central C—C bonds for disconnection（Scheme 9-39）. According to route（a）, the first retrosynthetic step, *retro*-Michael disconnection, affords an acetonide anion and target molecule of the second generation, building block TM 9. 33a. In the second step, retro-aldol disconnection of TM 9. 33a affords two easily available reagents, benzaldehyde and ethyl cyanoacetate.

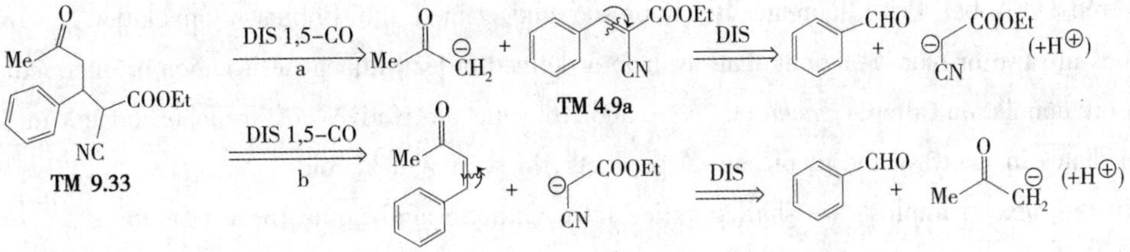

Scheme 9-39 Alternative 1, 5-CO disconnections of TM 9. 33

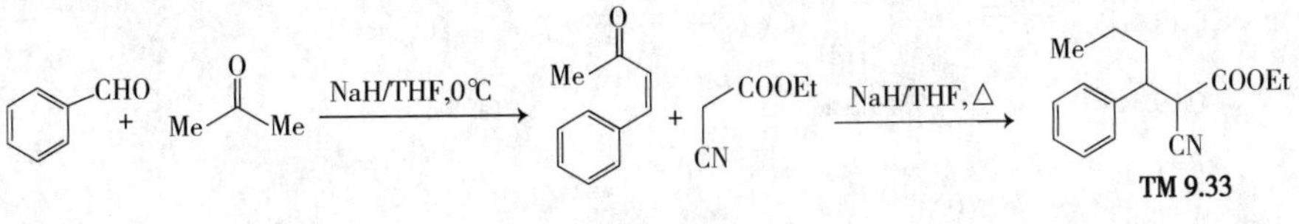

Scheme 9-40 Preferred synthetic route to TM 9. 33

$\xrightarrow[\text{EtO}^-]{\text{CHO}}$ Michael Addition

CO_2Et ⟹ CO_2Et (1)stable anion (2)obtainable

⟹ CO_2Et

Retro – Michael disconnection in the first step (b) results in enone TM 9.33b and one molecule of ethyl cyanoacetate. In the second step, disconnection of enone leads to benzaldehyde and acetonide anion[10].

Surprisingly, both disconnection routes lead to the same starting materials, benzaldehyde, ethyl cyanoacetate and acetone. We described ethyl acetoacetate as the preferred synthetic equivalent for acetone in Chapter 2. For the synthesis of TM 9.33b, a workable method is the condensation of acetone with an excess of benzaldehyde, a non–enolizable component to prevent polymerization of acetonide anion.

To decide whether disconnection (a) or (b) suggests a synthetically more convenient route, we consider the relative reactivities of single reactants. Benzaldehyde is more reactive with the acetonide anion than with enone TM 9.33b. Therefore, in the first step, aldol condensation is preferred, followed by Michael addition of enone to the carbanion of ethyl cyanoacetate. Consequently, retrosynthesis (b) suggests the preferred synthetic route to TM 9.33 (Scheme 9–40).

Michael addition – aldol condensation with cyclization is an example of a tandem reaction, discovered by Nobel Prize laureate R. Robinson and named the Robinson annulation[26]. Tandem reactions are two or more reactions that occur in a defined order without the isolation of intermediates. Robinson annulation affords *mono* –, *bi* – and tricyclic derivatives of cyclohexanone, important intermediates in many syntheses of natural products, in particular steroids.

In the next examples, we shall practice retrosynthetic analysis of these structures.

Propose the retrosynthetic analysis and then synthesis of TM 9.34.

TM 9.34 $\xRightarrow{\text{DIS}}$ TM 9.34a $\xRightarrow{\text{DIS 1,6-CO}}$ TM 9.34b + H_2C=CH–C(O)Me (+H)

⟹ + MeI

Scheme 9–41 Retrosynthetic analysis of TM 9.34

TM 9.34

In the bicycle target molecule we recognize a 1,5-CO pattern and enone unit. In the first retrosynthetic step disconnection of the C = C bond is preferred, leading to TM 9.34a, cyclohexane 1,3-dione derivative with a quaternary C-atom in the α position to both carbonyl groups (Scheme 9-41). *retro*-Michael disconnection of the new 1,5-CO pattern at the C—C bond to the ring affords a stable carbanion of 2-methyl-1,3-cyclohexandione TM 9.34b and methyl vinyl ketone. Final disconnection of the methyl group leads to 1,3-cyclohexandione.

Note: 1, 3 - Cyclohexandione (dihydroresorcinol) is available from 1, 3 - diphenol (resorcinol) by hydrogenation of its mono-sodium salt with one mole of hydrogen at temperatures below 50 ℃ catalyzed by Raney Ni in basic medium. Methyl vinyl ketone is a broadly used C_4 building block produced by gas - phase formylation, aldol condensation of acetone with formaldehyde catalyzed by metal oxides (Scheme 9-42).

Now we can propose a short synthetic route to TM 9.34 starting from the available building blocks (Scheme 9-43).

V_2O_5/P_2O_5 or Fe_2O_3/P_2O_5, >200℃; $-H_2O$

Scheme 9-42 Industrial production of methyl vinyl ketone

NaOH(aq), H_2/RaNi, 50℃; Me_2SO_4/Δ; NaOMe/MeOH

Scheme 9-43 Proposal for the synthesis of TM 9.34

On partial hydrogenation of resorcinol to 1,3-cyclohexanedione, C-methylation of enol is performed under standard conditions. In the next two steps, Robinson annulation is completed. Michael addition of methyl vinyl ketone affords an intermediate that spontaneously enters intramolecular aldol condensation to a stable six-membered ring in TM 9.34.

The next retrosynthetic analysis illustrates the application of the concept of an activating group.

Example TM 9.35 proposes the retrosynthesis and then synthesis of racemic natural product piperitone (薄荷酮) TM 9.35.

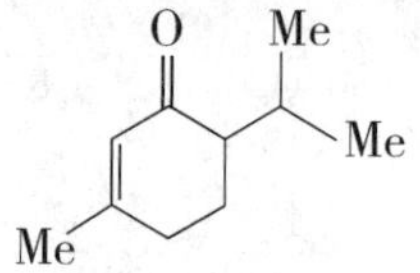

TM 9.35 Piperitone

Retro-Aldol type disconnection of cyclic enone in the first step is straightforward (Scheme 9-44). In the new TM 9.35a either of two central C—C bonds can be disconnected according to the retro-Michael. However, immediate disconnection of the C—C bond to the branching point, suggested by maximal simplification of the target molecule, results in methyl vinyl ketone and methyl isobutyl ketone, an inconvenient reagent for an unstabilized carbanion. This problem solves the FGA, the addition of an activating carbethoxy group at the branching point affording TM 9.35b.

Note: The concept of an activating group is related to the well-known concept of a protecting group and illustrated by two known examples in Fig. 9-4.

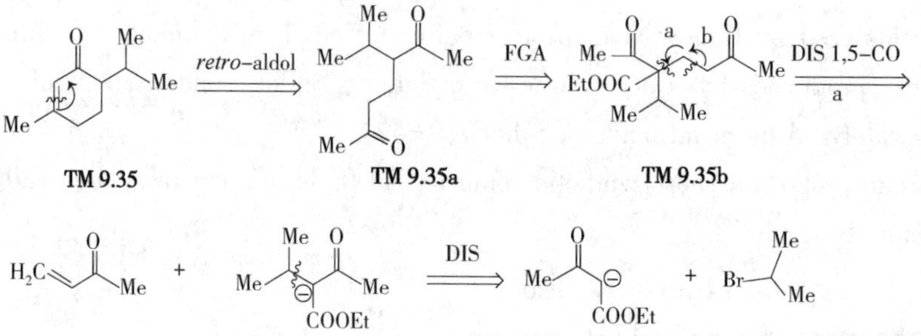

TM 9.35 — *retro*-aldol ⟹ TM 9.35a — FGA ⟹ TM 9.35b — DIS 1,5-CO (a) ⟹ H_2C=CH–CO–Me + Me–CO–C(⊖)(COOEt)–CHMe_2 — DIS ⟹ Me–CO–CH(⊖)–COOEt + Br–CHMe_2

Scheme 9-44 Proposal for retrosynthetic analysis of racemic piperitone TM 9.35

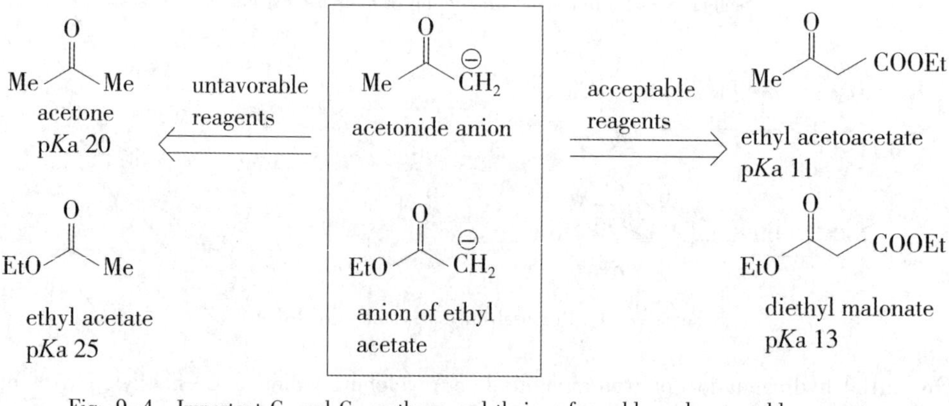

Unfavorable reagents	Synthon	Acceptable reagents
acetone p*Ka* 20	acetonide anion	ethyl acetoacetate p*Ka* 11
ethyl acetate p*Ka* 25	anion of ethyl acetate	diethyl malonate p*Ka* 13

Fig. 9-4 Important C_3 and C_2 synthons and their unfavorable and acceptable reagents

Higher C—H acidity in carbethoxy derivatives enables the formation of a stabilized carbanion prone to alkylation or Michael addition.

Now 1,5-disconnection (a) in Scheme 9-44 is preferred, resulting with α-isopropyl ethyl acetoacetate, an easily available TM 9.35c. *retro*-Michael disconnection (b) in TM 9.35b is not possible since no double bond can be formed to the quaternary C-atom.

Complete retrosynthesis of racemic piperitone is proposed in Scheme 9-44.

Synthesis of piperitone can be worked out from methyl vinyl ketone, ethyl acetoacetate and 2-bromopropane, where in all steps alkylation, Michael addition and an aldol condensation medium strong base are needed, e. g. , NaOEt/EtOH. For decarboxylation of intermediary TM 9. 35b, mildly acidic conditions areconvenient.

Suggest the retrosynthesis and then propose the synthesis of TM 9. 36, a model compound for the synthesis of steroids.

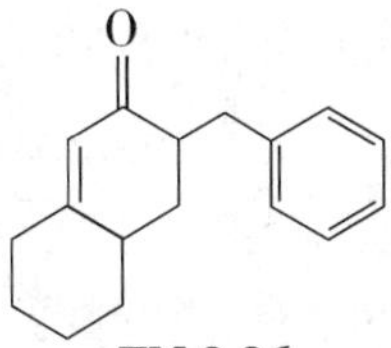

TM 9.36

On first sight, it is difficult to imagine *retro*-Mannich disconnection as the key step in the retrosynthesis of TM 9. 36. Starting the retrosynthetic analysis with the disconnection of the C=C bond in cyclic enone we arrive at a 1, 5-dicarbonyl pattern in TM 9. 36a (Scheme 9-45). Now two possible *retro*-Michael disconnections, a and b, lead to enones TM 9. 36c and TM 9. 36d, whereby greater simplification of the target structure is achieved by disconnection a.

To obtain the reagent for anionic synthon TM 9. 36b, we add an activating carbethoxy group to TM 9. 36e, and the new target molecule can now be disconnected to benzyl bromide and ethyl acetoacetate.

Enone TM 9. 36c can be disconnected by the sequence of steps to the Mannich base and retro-Mannich in the last step. Note that the Wittig reaction is less convenient for achieving TM 9. 36c because of the presence of the second carbonyl group, and a tandem reaction offers a practical alternative.

In the next example, a similar sequence of retrosynthetic steps suggests the synthesis of biologically important triene.

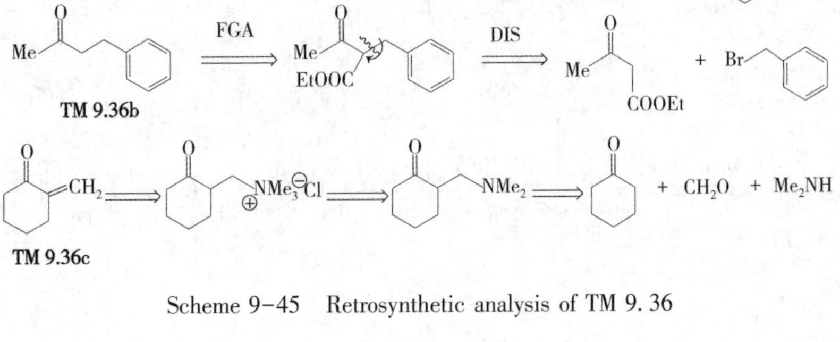

TM 9.36 *retro*-aldol ⟹ TM 9.36a ⟹(a) TM 9.36b + TM 9.36c

⟹(b) Me, H_2C + ⊖

TM 9.36b ⟹(FGA) Me, EtOOC ⟹(DIS) Me, COOEt + Br

TM 9.36c ⟹ $\overset{\oplus}{N}Me_3\overset{\ominus}{Cl}$ ⟹ NMe_2 ⟹ + CH_2O + Me_2NH

Scheme 9-45 Retrosynthetic analysis of TM 9. 36

Consideration of the basic mechanism of the Mannich reaction between the imine of prochiral aldehyde with methyl alkyl ketone serves as a useful introduction (Scheme 9-46).

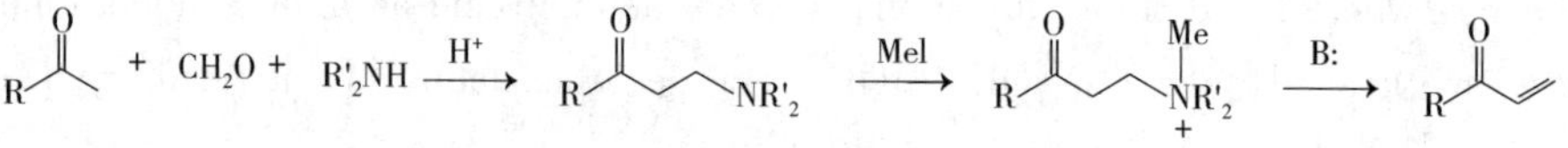

+ CH_2O + R'_2NH $\xrightarrow{H^+}$ R … NR'_2 $\xrightarrow{MeI}$ R … $\overset{+}{N}R'_2$ Me $\xrightarrow{B:}$ R

Scheme 9-46 Mannich reaction of prochiral aldehyde with methyl ketone

Examples:

analysis: Dis1,3-CO ⟹ Dis1,5-CO ⟹ + ⟹ + Br CO_2Et

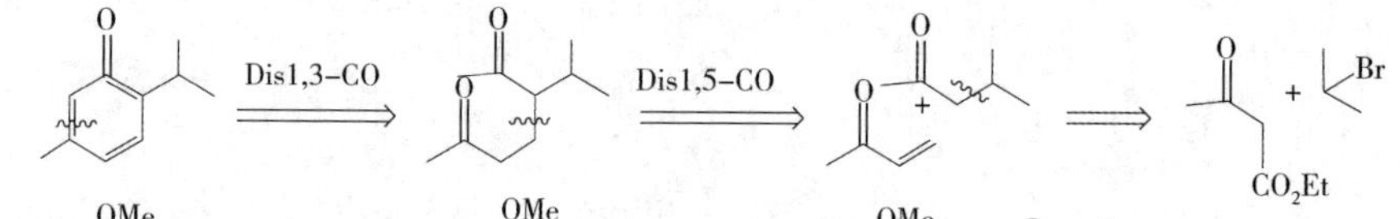

analysis: OMe Dis1,5-CO ⟹ OMe + Dis1,3-CO ⟹ OMe CHO + OEt +

synthesis: OMe CHO $\xrightarrow{EtO^-/EtOH}$ OMe $\xrightarrow{CO_2Et}$ OMe CO_2Et $\xrightarrow[H_2O]{H^+}$ TM

synthesis: CO_2Et $\xrightarrow[PhCH_2Br]{Base}$ CO_2Et Ph

(1) CH_2O, Me_2NH (2) △ → ; Base → Ph; Base △ → Ph

analysis: Dis1,3-CO ⟹ Dis1,5-CO ⟹ + EtO_2C Dis1,3-CO ⟹ + HCHO

Dis1,3-CO ⟹ Dis1,5-CO ⟹ + CO_2Et

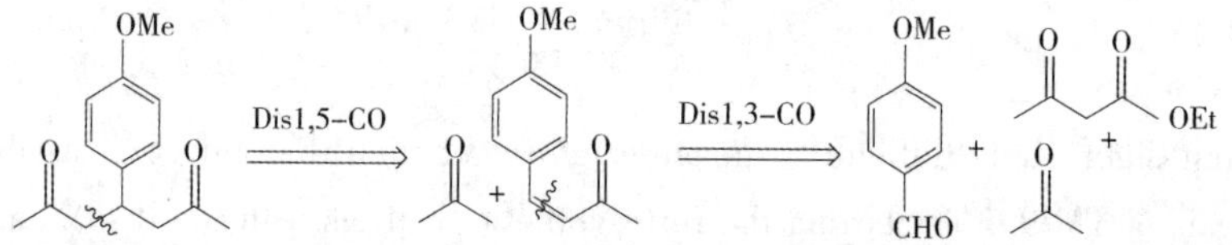

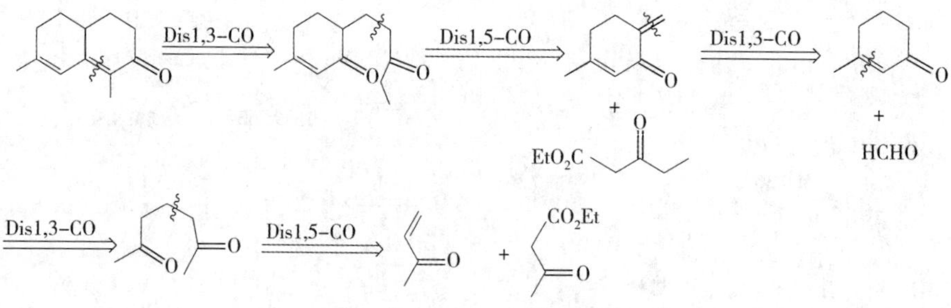

analysis: Ph CO_2Et Ph Dis1,3-CO ⟹ Ph CO_2Et Ph Dis1,5-CO ⟹ CO_2Et + Ph Ph Dis1,3-CO ⟹ PhCHO + Ph

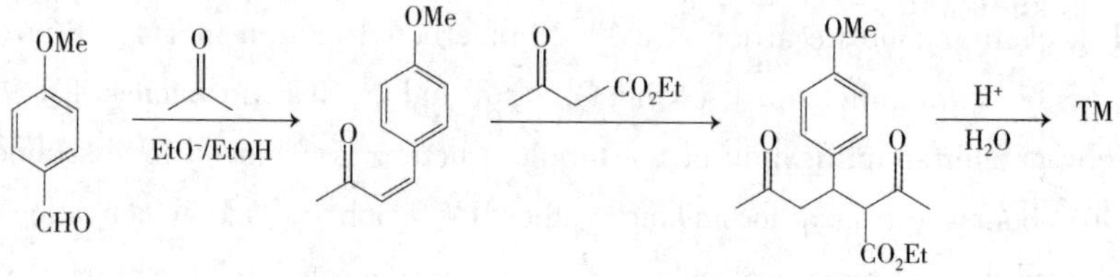

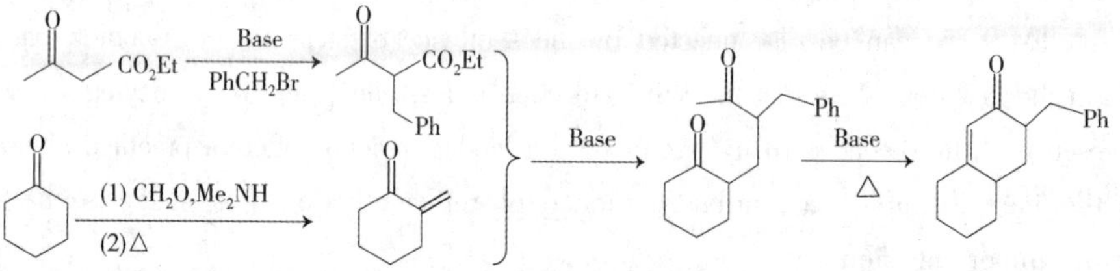

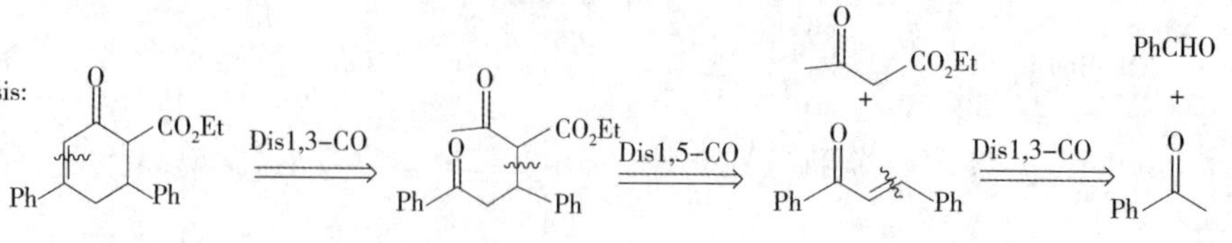

9.2.3 1,2-二羰基模式 1,2-Dioxygenated Pattern (1,2-CO)

In this section we analyze α-hydroxyl acids, α-hydroxy ketones, vicinal diol and the derivatives with a 1,2-dioxygenated pattern. Schematic structures of the 1,2-dioxygenated pattern are presented as TM Ⅰ-Ⅲ (Fig. 9-5).

HO O OH
Ⅰ

HO O
Ⅱ

HO OH
Ⅲ

Fig. 9-5 Featural molecular skeleton (特征分子骨架) with a 1,2-dioxygenated pattern

9.2.3.1 α-羟基酸 α-Hydroxy acids

α-Hydroxy acids belong to an important group of natural and synthetic compounds with a 1,2-dioxygenated pattern (Scheme 9-47).

Disconnection (a) corresponds to the synthesis of I, α-halogenation of carboxylic acid or its derivative followed by hydrolysis of halogen (Scheme 9-47). Disconnection (b) envisages building a carbon framework from C_n + C_1 synthons but looks unacceptable since illogical synthon-COOH with a negative charge on the carbonyl C atom appears. Let us, however, consider the next example[11].

(a) R OH O OH FGI ⟹ R O OH Br ⟹ R O OH

(b) R OH O OH DIS 1,2-CO ⟹ R O H + ⊖ O OH

Scheme 9-47 Retrosynthetic approaches to α-hydroxy carboxylic acids

Me OH OH O DIS 1,2-CO ⟹ O Me + ⊖ O O⊖

synthetic equivaient:CN

Scheme 9-48 Disconnection of TM 9.37 as a 1,2-CO pattern

Thus, HO O OH FGI ⟹ HO CN ⟹ O CN⁻

Suggest the retrosynthesis and then propose the synthesis of 2-hydroxy-2-phenylpropionic acid TM 9.37.

TM 9.37

analysis: ⟹ ⟹ $^{-}$CN

synthesis: Me + KCN —EtOH/H_2O, dil.H_2SO_4,r.t.→ HO (CN, Me) —EtOH/$H^{\oplus}$,Δ→ OH (Me, OH)

Scheme 9-49 Proposal for the synthesis of TM 9.37

Completing the disconnection of the internal C—C bond with participation of the hydroxy group, we generate the simple building block acetophenone and illogical synthon -COOH (Scheme 9-48).

Ketone is easily available so we start the search for the synthetic equivalent of the illogical anion, and as a matter of fact this is found in the cyanide anion. In the synthetic direction we use its salts as the source of the cyanide anion to complete the well-known cyanohydrin reaction. In the last step, the hydrolysis of nitrile affords the carboxylic group (Scheme 9-49).

Cyanide salts are available with either inorganic or organic cations, and a cyanide group can be conveniently prepared with radioactive ^{14}C-labeled atoms. The cyanohydrin reaction serves for the preparation of more complex ^{14}C-labeled carboxylic acids and their derivatives intended for pharmacological studies.

Consider the retrosynthesis and then propose the synthesis of ^{14}C-labeled TM 9.38.

TM 9.38

TM 9.38 ⟹(DIS1,2-CO) TM 9.38a + ^{14}COOH$^{\ominus}$ (+$H^{\oplus}$), reagent: $Na^{14}CN$ ⟹(DIS1,3-CO) + $H^{\oplus}$CHO, reagent: HCOOEt ⟹ TM 9.38b

TM 9.38b ⟹(FGA) TM 9.38c ⟹ Me(Me)CH$CH_2^{\oplus}$ + $^{\ominus}CH(COOEt)_2$
reagent: Me_2CHCH_2Br $CH_2(COOEt)_2$/NaOMe

Scheme 9-50 Complete retrosynthetic analysis of TM 9.38

The isotopically labeled α-hydroxy carboxylic group indicates the use of the ^{14}C cyanide ion in cyanhydrin synthesis and consequently 1, 2 - CO disconnection in the first retrosynthetic step (Scheme 9-50).

The target molecule of the second generation TM 9.38a comprises a 1, 3-CO pattern, and the ^{14}C-labeled cyanide anion indicates sodium cyanide as the second reagent. Disconnection of TM 9.38a leads to the carbanion of TM 9.38b and formyl carbocation, whose synthetic equivalent is ethyl formate. On first sight, the branched carboxylic acid TM 9.38b does not seem an easily available building block. On addition of the second activating carbethoxy group, the alkylated diethyl malonate unit appears as an available target molecule of the second-generation TM 9.38c. The last disconnection step leads to a stable diethyl malonate carbanion and synthon with carbocation on the primary C atom. Its synthetic equivalent is bromide, available in a few steps from isobutyraldehyde. An attempt to alkylate the diethyl malonate anion by isobutene in a Michael-type reaction fails because of the low reactivity of the terminal C atom on the double bond as an electrophile. Note that the hyperconjugative effect of methyl groups inverts the charge distribution in the $Me_2C{=}CH_2$ double bond as compared to enones.

Now we can propose the complete synthesis of ^{14}C-labeled TM 9.38 (Scheme 9-51).

Cyanide salts are also used in the Strecker synthesis of α-amino acids. This well-known name reaction generates compounds with a 1, 2-disubstituted pattern whose disconnection results in one illogical synthon. Example TM 9.38 presents some mechanistic details of this reaction, suggested by *retro*-Strecker disconnection.

H_2/Pd-C, AcOH; HBr/AcOH; + $CH_2(COOEt)_2$, NaOEt/EtOH; (1)dil.HCl/EtOH/△ (2)HCl/dry EtOH/△; + HCOOEt, NaOEt/EtOH; + $Na^{14}CN$, EtOH(aq); dil.HCl/EtOH/△

TM 9.38

Scheme 9-51 Proposal for the synthesis of ^{14}C-labeled TM 9.38

analysis:

Retrosynthetic analysis of some α-hydroxycarbonyl compounds can use the following reactions to provide simple starting materials.

Benzoin reaction: CHO $\xrightarrow{^-CN}$ O OH $\xrightarrow{[O]}$ O O

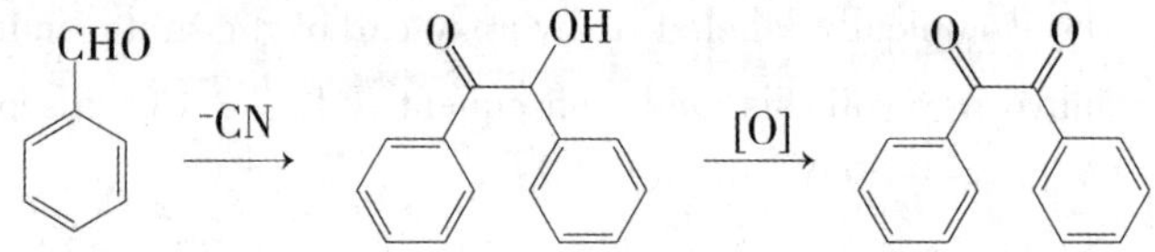

analysis:

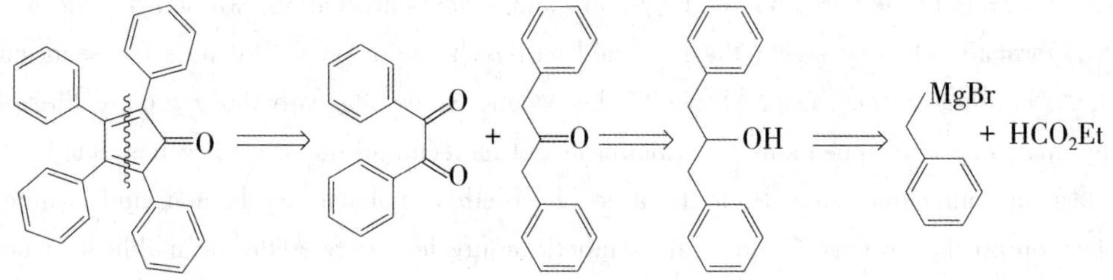

O $\Longrightarrow$ O O + O $\Longrightarrow$ OH $\Longrightarrow$ MgBr + HCO_2Et

α-Halogenated carboxylic acid as building blocks:

analysis: O CO_2H Cl $\Longrightarrow$ Br CO_2H + OH Cl $\Longrightarrow$ OH

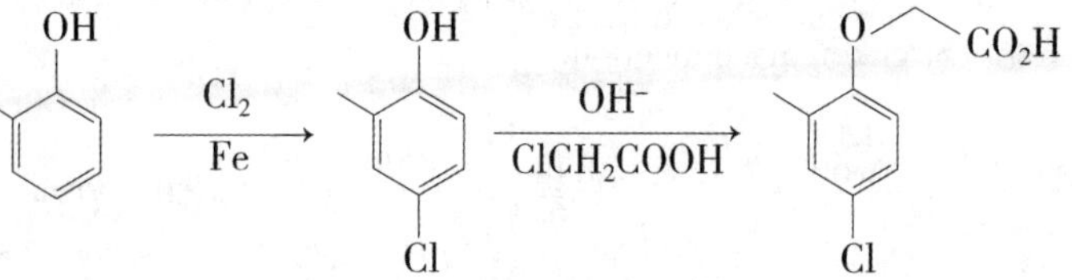

OH $\xrightarrow[Fe]{Cl_2}$ OH Cl $\xrightarrow[ClCH_2COOH]{OH^-}$ O CO_2H Cl

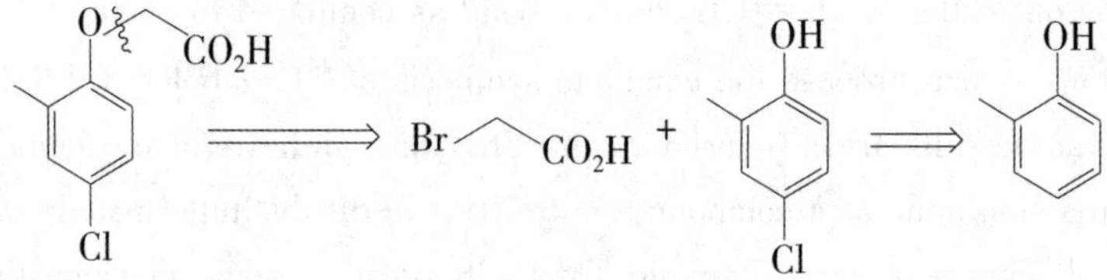

9.2.3.2 α-羟基酮 α-Hydroxy ketones

1,2-Dioxygenated patterns do not only appear in α-hydroxy carboxylic acids, but are also important functionalities in α-hydroxy ketones and their derivatives. In the former examples, we have seen negatively polarized α-C atoms in carbonyl compounds as nucleophilic centers[12].

Example TM 9.39 proposes the retrosynthesis of TM 9.39 and then its synthesis using a couple of reactions, alkynylation of ketone-hydration of the triple bond in the intermediary carbinol.

We described the alkynylation of the carbonyl group with the acetylide anion in Sect. 2.2. Before the initial retrosynthetic step, let us consider the mechanism of hydration of the triple bond activated by Hg(Ⅱ) ions (Scheme 9-52).

Note the orientation of the water molecule in the complex resulting in the addition of the OH group to the higher substituted C atom in accord with the Markovnikov rule. Having this reaction in mind, now we can approach the disconnection of TM 9.39 as in Scheme 9-53.

Such 1,2-CO disconnection results in ketone and an illogical synthon, the acetyl anion. Since we discovered sodium acetylide as the proper reagent for anionic C_2 synthons, a short and elegant synthetic scheme for TM 9.39 can be proposed (Scheme 9−54).

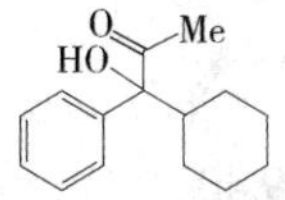

TM 9.39

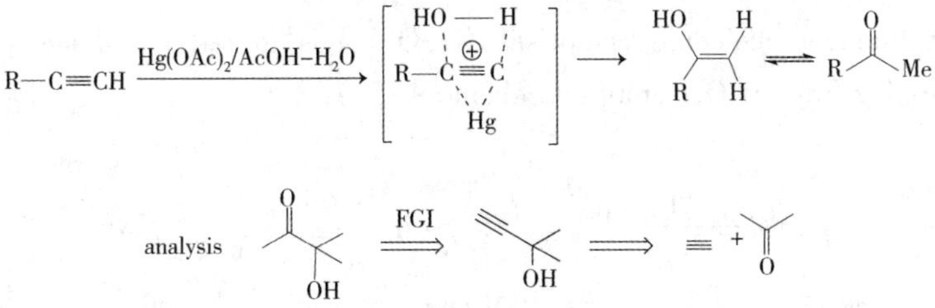

$R{-}C{\equiv}CH \xrightarrow{Hg(OAc)_2/AcOH{-}H_2O}$ [HO—H, R—C≡C ⊕, Hg] ⟶ HO, H, R, H (enol) ⇌ R(C=O)Me

analysis ⟹(FGI) ⟹ ≡ + (acetone)

Scheme 9−52 Hydration of alkynes catalyzed by Hg(Ⅱ) ions

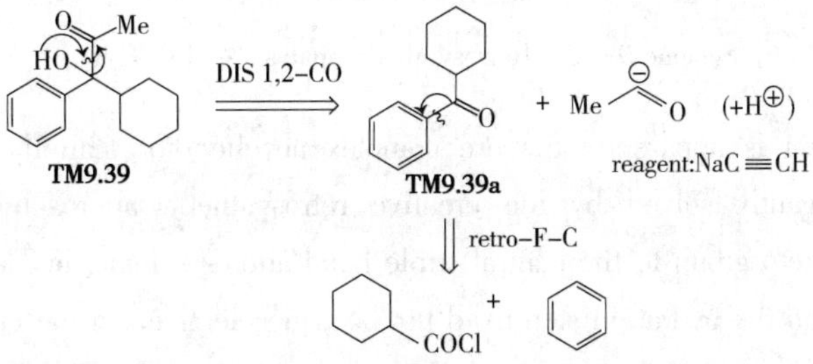

TM9.39 ⟹(DIS 1,2−CO) TM9.39a + Me(C=O)⊖ (+H⊕)

reagent: NaC≡CH

⇓ retro−F−C

COCl + (benzene)

Scheme 9−53 Retrosynthetic analysis involving the illogical disconnection of the 1,2−CO pattern in TM 9.39

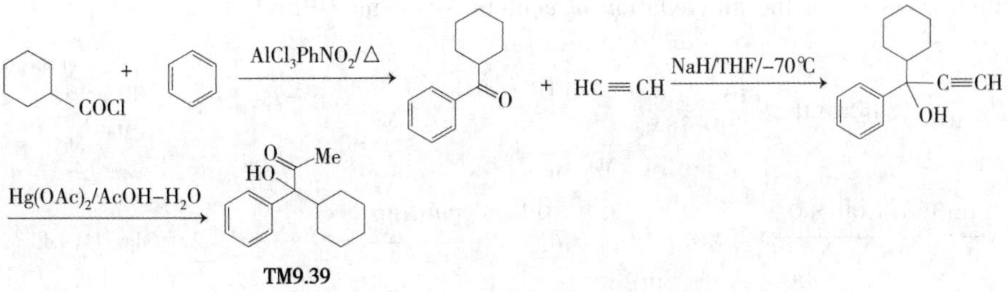

COCl + (benzene) $\xrightarrow{AlCl_3 PhNO_2/\Delta}$ (ketone) + HC≡CH $\xrightarrow{NaH/THF/-70℃}$ (C≡CH, OH) $\xrightarrow{Hg(OAc)_2/AcOH-H_2O}$ TM9.39

Scheme 9−54 Proposed synthesis of TM 9.39

The suggested retrosynthetic analysis of targeted α−hydroxy ketone might seem artificial, especially since it invokes a few less−known reactions, alkynylation−hydration of the triple bond. It is difficult to conceive of another practicable synthetic route to TM 9.39.

The next example reveals the power of retrosynthetic analysis where an α−hydroxycarbonyl pattern in the target molecule requires a sophisticated approach.

Propose the retrosynthetic analysis and then synthesis of cyclic ether TM 9.40.

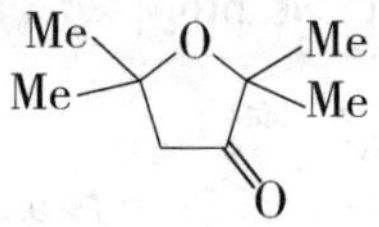

TM9.40

The target furanone molecule comprises C—O bonds to tert C - atoms, which suggests disconnection forming two tert OH groups (Scheme 9-55).

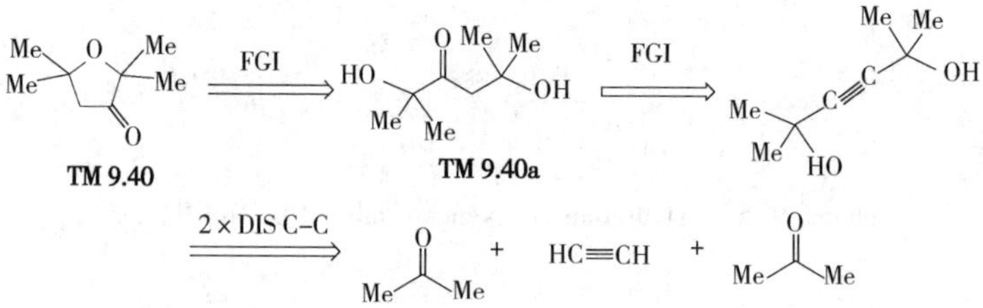

Scheme 9-55 Retrosynthetic analysis of TM 9.40

The new TM 9.40a is apparently a more complex acyclic diol than the target TM 9.40. An emerging issue is elegantly solved by the creative retrosynthetic approach to this intermediate. Interconversion of the keto group to the central triple bond affords symmetric TM 9.40b, and therefore in the synthetic direction the hydration step need not be regioselective. In the last retrosynthetic steps, two consecutive disconnections of the single C—C bonds by participation of OH groups afford acetylide anions and two moles of acetone, a surprisingly simple set of starting materials.

Synthesis of TM 9.40 is straightforward and needs two moles of a strong base to create the two alkyne anions needed for the alkynylation of acetone (Scheme 9-56).

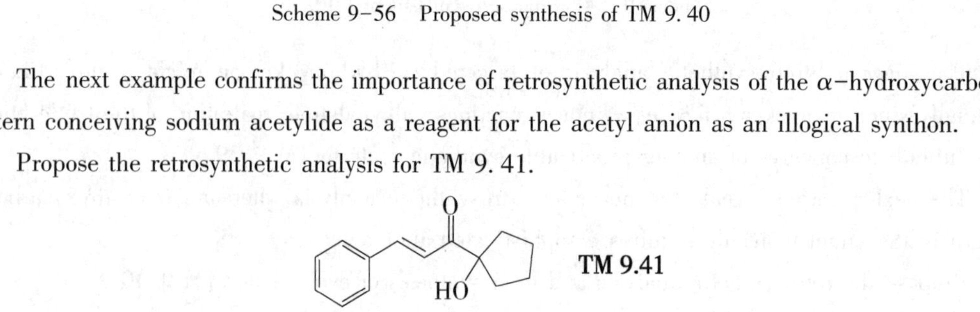

Scheme 9-56 Proposed synthesis of TM 9.40

The next example confirms the importance of retrosynthetic analysis of the α-hydroxycarbonyl pattern conceiving sodium acetylide as a reagent for the acetyl anion as an illogical synthon.

Propose the retrosynthetic analysis for TM 9.41.

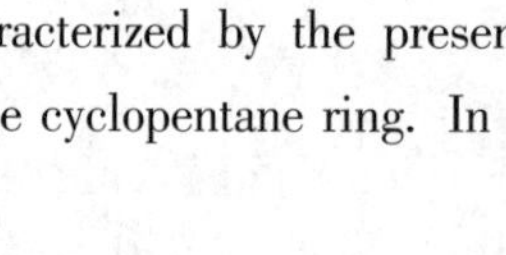

TM 9.41

The target molecule is characterized by the presence of enone and a 1,2-CO pattern that includes the tert OH group on the cyclopentane ring. In the first step, the maximal simplification of this structure is available by *retro*-aldol disconnection of the central C=C bond (Scheme 9-57).

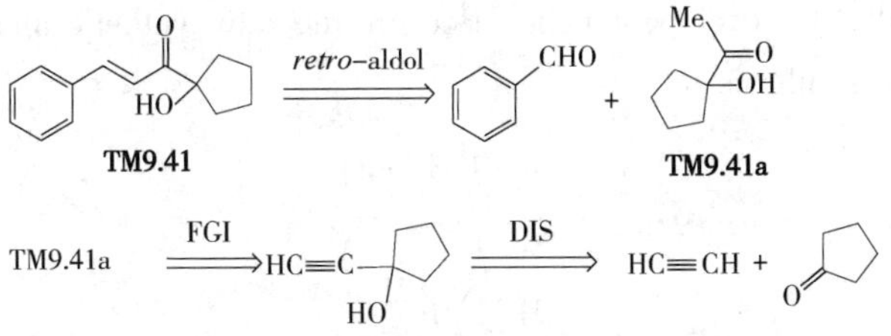

Scheme 9-57 Retrosynthetic analysis of TM 9.41

In the target molecule of the second - generation TM 9.41a, we follow a two - step retrosynthetic pattern from the former example. First, the FGI and then DIS of the central C—C bond end up with acetylene and cyclopentanone as convenient starting materials. Synthesis of TM 9.41 can be completed under the reaction conditions indicated in Scheme 9-57.

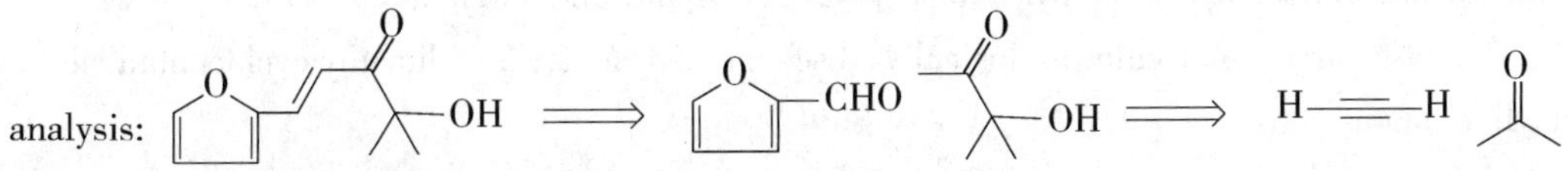

analysis:

9.2.3.3 邻二醇 Vicinal diol

In this section we analyze 1,2-diols or vicinal diols, with a1,2-dioxygenated pattern with two hydroxy groups (Scheme 9-58)[13].

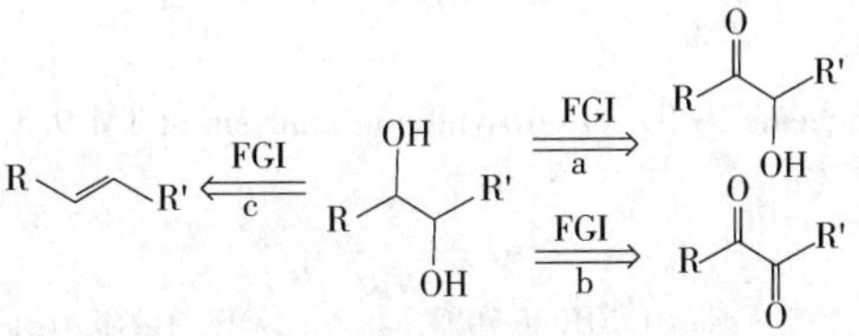

Scheme 9-58 Possible FGIs in the retrosynthetic analysis of 1,2-diols

In all three cases, FGIs are proposed since no disconnection of the central C—C bond gives a couple of synthons for which available reagents exist. In other words, 1,2-diols are not available by direct formation of the central, single C—C bond.

FGIs (a) and (b) involve retrosynthetic transformation of the hydroxy to carbonyl group; the third FGI (c) implies elimination of both OH groups to obtain alkene, the target molecules of the next generation. In the synthetic direction, this means that reduction of the C=O group(s) or dihydroxylation of the C=C bond is required for rich 1,2-diols.

Note there is an exceptional formation of C—C bonds in derivatives of vicinal diols. An interesting stereoselective method for the synthesis of anti-vicinal diols uses organometallic alkoxyallyl tins (γ-metallated enol-ethers) and aldehydes in the presence of $BF_3 \cdot Et_2O$ at -78℃. The reagents and reaction conditions limit this method to the laboratory scale.

In an early synthesis of camphor, the important intermediate TM 9.42 was prepared from easily available 2,2-dimethlycyclopentanone. Perform the retrosynthetic analysis of TM 9.42 and propose the three-step synthesis.

OH OH
Me
Me Me Me

TM 9.42

The 1,2-CO pattern in the target molecule appears as a vicinal diol. The starting 2,2-dimethylcyclopentanone indicates the formation of an OH group on the ring by alkylation of the keto group. It remains open, however, how the second carbinol OH group is introduced in the side chain. The retrosynthetic solution to this puzzle is hidden in *retro*-Grignard disconnection to TM 5.8a and interconversion of methyl ketone to an acetylenic unit (Scheme 9-59).

The last target molecule is logically disconnected to 2, 2-dimethylcyclopentanone and acetylide anions.

TM 9.42 $\xrightarrow{retro\text{-Grignard}}$ $Me^{\ominus}$ + TM 9.42a $\xrightarrow{FGI}$ (ethynyl carbinol) $\xrightarrow{DIS}$ 2,2-dimethylcyclopentanone + $^{\ominus}C{\equiv}CH$ $(+H^{\oplus})$

Scheme 9-59 Retrosynthetic analysis of TM 9.42

2,2-dimethylcyclopentanone + $HC{\equiv}CH$ $\xrightarrow{NaNH_2/NH_3(l)/-70℃}$ (ethynyl carbinol) $\xrightarrow{HgSO_4/H_2O/\triangle}$ (α-hydroxy ketone) $\xrightarrow{MeI/Mg/ether/0℃}$ TM 9.42

Scheme 9-60 Proposed synthesis of TM 9.42

Alkynylation of the cyclopentanone derivative by the acetylide anion leads to carbinol. Hg(Ⅱ)-catalyzed hydration in the next step affords α-hydroxy ketone, which in the last step is methylated by the Grignard reagent, completing the short synthesis of TM 9.42 (Scheme 9-60).

It is important to note that dihydroxylation of the tetrasubstituted C=C bond in the conceivable intermediate formed by the Wittig reaction with 2,2-dimethylcyclopentanone is not a workable route because of the steric perturbation of the gem-dimethyl group in the ring. Write and analyze this possible reaction route.

Perform retrosynthetic analysis and then propose the synthesis of TM 9. 43.

TM 9.43

The 1,2-dioxigenated pattern in TM 9. 43 comprises two single C—O bonds, one of them as an ether functionality. Having in mind the known property of three-member heterocycles as kinetically preferred and relatively thermodynamically unstable structures, we invoke the disconnection of the central C—O bond and epoxide TM 9. 43a as the target molecule of the next generation (Scheme 9-61).

Disconnection of the central C—O bond generates an unstable cation on the primary C atom beside the α-naphthyloxy anion as the second synthon. Here we have the case in which the cationic synthon has an acceptable reagent in the corresponding epoxide TM 9. 43a. Vicinally substituted 1-hydroxy-1-bromomethyl cyclohexane is the wrong choice since it spontaneously cyclizes to epoxide TM 9. 43a. Disconnection of the three-membered ring can be completed according to either a or b. In the first case, the resulting reagents are cyclohexanone and diazomethane, in the second methylenecyclohexane and peroxide. Both disconnections suggest synthetically feasible reactions.

Epoxides as building blocks:

MCBPA

MeO^-

OMe

OH

DIS 1,2-CO

a-naphthol

TM9.43

reagent: TM9.43a

a

+ CH_2N_2

b

+ H_2O_2

Scheme 9-61 Retrosynthetic analysis of TM 9. 43

Note α-Naphthol is available by oxygenation of tetraline (tetrahydronaphthalene) to 1-

tetralone followed by dehydrogenation (aromatization). The technological process is catalyzed by zeolites and oxygenation completed by oxygen. Intermediary 1-tetralone can be isolated or dehydrogenated to 1-naphthol by the same catalytic system.

The proposed synthesis of TM 9.43 in Scheme 9-62 characterizes the regioselective(区域选择性) ring opening of epoxide with an anion of α-naphthol under basic conditions. This bulky nucleophile(亲核试剂) approaches the less-substituted C atom of epoxide[14].

Cu_2Cl_2/Et_2O, 35℃ to r.t; CH_2N_2; HO; NaOMe/MeOH; OH; O; **TM 9.43**

Scheme 9-62 Proposal for the synthesis of TM 9.43

Widely used methods for the dihydroxylation of alkenes are discussed in the examples that follow.

Witting reaction: O; $Ph_3\overset{+}{P}\overset{-}{C}$; $KMnO_4$; OH OH

analysis: OHOH; O; MgBr

Consider the retrosynthesis of TM 9.44, a model structure for the pheromone of the bark beetle, and then propose its multistep synthesis.

O; O; **TM 9.44**

First, we observe the presence of intramolecular ketal, which is formed in the last step of synthesis. Therefore, the first FGI leads to TM 9.44a, a straight chain molecule with ketone and vicinal diol functionalities (Scheme 9-63).

By the next FGI, we introduce the terminal C=C bond interconverting diol in alkene TM 9.44b. On the route to the next target molecule convenient for disconnection with great simplification, the FGI of the C=C bond appears in the *retro*-Wittig manner. It does not greatly simplify the general structure, but provides TM 9.44c with a comfortable 1,5-CO pattern. retro-Michael disconnection results in the well-known starting materials acrolein and acetonide anion, for which ethyl acetoacetate is repeatedly accentuated as a proper reagent.

Although this retrosynthesis seems elegant and straightforward, it is not workable in a synthetic direction without certain modifications (Scheme 9-64).

A chemoselective Wittig reaction with one of the two carbonyl groups in TM 9.44c is not possible! It is also not possible to selectively protect only the keto group. The preferred solution is to introduce the aldehyde group only after protecting the keto group. The proposed synthesis solves this problem; still, seven steps are needed to TM 9.44. This route requires Michael addition on ethyl acrylate, decarboxylation and then esterification of the remaining carboxylic group, protection of the keto group and two further steps for transformation of the carboxylic ester to aldehyde. Only now is there convenient introduction of the seventh C-atom by the Wittig reaction. After dihydroxylation of the C=C bond, there is no need to remove the protecting group in a separate step since intramolecular transketalization occurs spontaneously.

TM 9.44 ⟹ (FGI) TM 9.44a (HO, OH) ⟹ (FGI) TM 9.44b ⟹ (*retro*-Wittig) TM 9.44c (CHO) + $\overset{\ominus}{C}H_2\overset{\oplus}{P}(Ph)_3$ ⟹ (Dis1,5-CO) CHO + ketone enolate$^{\ominus}$

reagent: $MeCOCH_2COOEt$/base

Scheme 9-63 Retrosynthetic analysis of TM 9.44

CO_2Et ketoester $\xrightarrow[\text{acrylate } CO_2Et]{NaOEt/EtOH}$ (CO_2Et, CO_2Et) $\xrightarrow[-CO_2]{(1)HCl/EtOH-H_2O\ (2)HCl/dry\ EtOH}$ (OEt)

$\xrightarrow[HO\quad OH]{\oplus H/THF/\triangle}$ ketal CO_2Et $\xrightarrow[(2)CrO_3/pyridine]{(1)LiAlH_4/THF}$ ketal CHO $\xrightarrow[THF/78℃]{Ph_3P{=}CH_2}$

ketal alkene $\xrightarrow[Cu(AcO)_2/AcOH-H_2O]{OsO_4/pyridine\ or}$ HO, OH (TM9.44a) $\xrightarrow{TsOH/toluene/\triangle}$ TM9.44

Scheme 9-64 Proposed synthesis of TM 9.44

Pinacol Rearrangement: acetone (=O) $\xrightarrow{Mg-Hg}$ (OH OH) $\xrightarrow{H^+}$ (R, O)

analysis: spiro ketone (O) ⟹ (OH OH) ⟹ cyclopentanone (=O)

Diels-Alder reaction:

analysis: (H, O, O, H, CO_2Me) ⟹ acetone (=O) + (HO, HO, H, H, CO_2Me) ⟹ (CO_2Me)

⟹ cyclopentadiene + CO_2Me (methyl acrylate)

Summary:

(a) HO, O, OH —FGI⟹ HO, CN ⟹ O, CN⁻

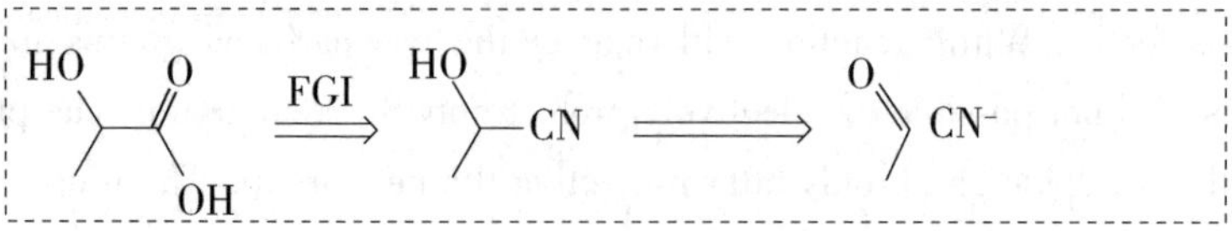

(b) HO, O —*Retro*-Acyloin Coupling⟹ O, OEt
—FGI⟹ O + ⁻≡R

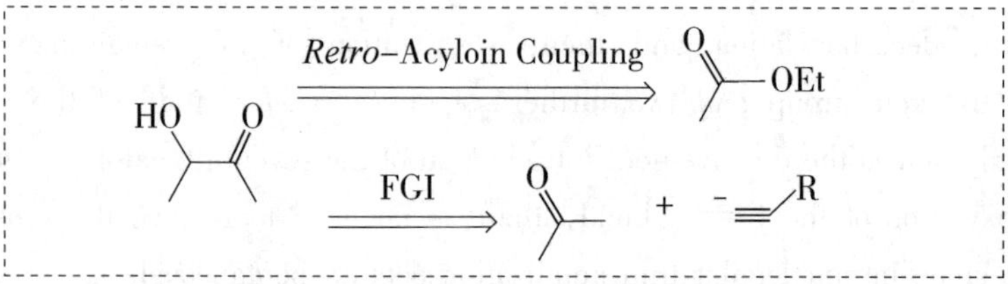

(c) HO, OH —FGI⟹ ⟹ OH
—FGI⟹ —*Retro*-Wittig⟹ O $\bar{C}H_2\overset{+}{P}Ph_3$
—*Retro*-Pinacol Coupling⟹ O

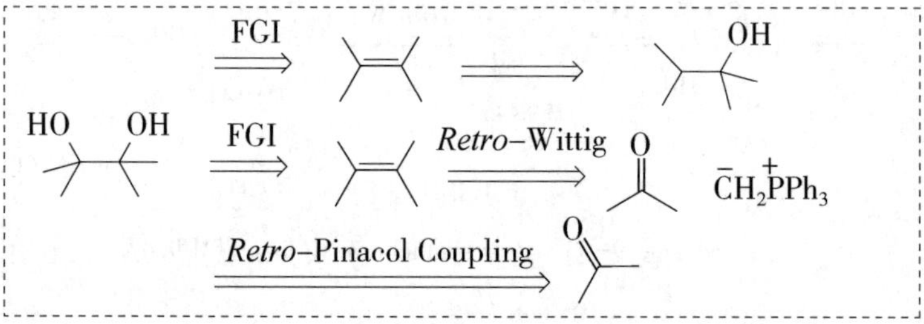

9.2.4 1,4-二羰基模式 1,4-Dioxygenated Pattern (1,4-CO)

In this section we analyze 1,4-dicarbonyl compounds or 1,4-hydroxy carbonyl compounds with a 1,4-dioxygenated pattern. Schematic structures of the 1,4-dioxygenated pattern are presented as TM I-II (Fig. 9-6).

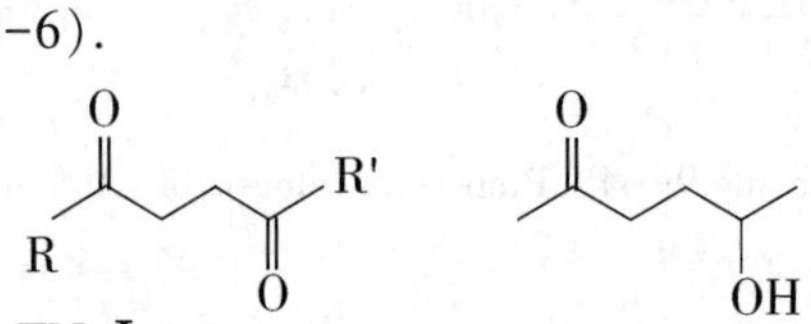

O, R', R, O — **TM Ⅰ**　　O, OH — **TM Ⅱ**

Fig. 9-6 Schematic structures of the 1,4-dioxygenated pattern

9.2.4.1 4-羟基羰基化合物 1,4-dicarbonyl compounds

As presented in Sect. 9.2, an even number of C atoms between oxygen functionalities results in a mismatch of partial charges on the atoms of the central C—C bond. Accordingly, disconnection of the central C—C bond in 1,4-dicarbonyl compounds results in an acceptable anionic synthon and illogical cationic synthon with a positive charge on the α-C atom (Scheme 9-65).

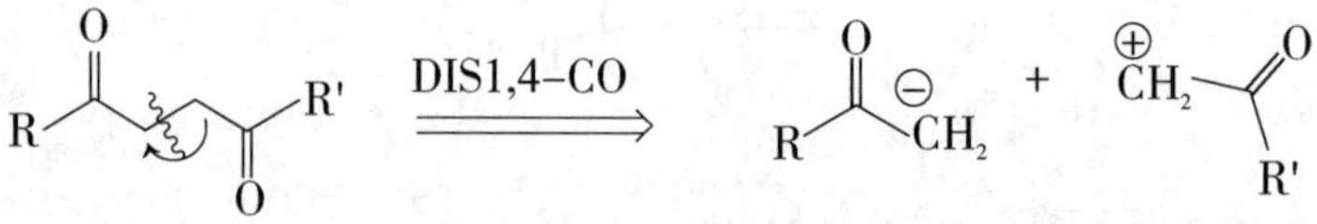

R, O, O, R' —DIS1,4-CO⟹ R, O, $\ominus CH_2$ + $\oplus CH_2$, O, R'

We have repeatedly introduced acceptable reagents for anionic synthons, while those for cationic synthons are discussed in Sect. 8.1. Inversion of polarity on α−C atoms in cationic synthons is achi eved by introduction of an σ−electron acceptor group.

Tricarbonyl, branch−chained TM 9.45 is an important intermediate in the synthesis of certain psychopharmaca. Perform the retrosynthetic analysis and then propose the synthesis of this compound.

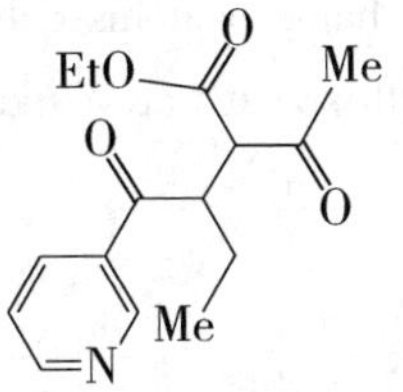

The decision to first disconnect the 1,4−dicarbonyl pattern is challenging. It results with in a carbanion of ethyl acetoacetate and cationic synthon charged on the α−C atom to the carbonyl group (Scheme 9−66). We therefore propose α−bromo ketone TM 9.45a as a new target molecule.

The next obvious target is TM 9.45b, the result of FGE, the elimination of bromine to pyridyl propyl ketone. retro−Friedel−Crafts disconnection of TM 9.45b is not acceptable for the good reasons we discussed. We also learn that the retro−Grignard disconnection of TM 9.45b presented in the above scheme results in tert alcohol if it starts from ester and not from the corresponding nitrile.

Another convenient carboxylic acid derivative for acyl cations in the Grignard reaction is Weinrib's amide, a less known reagent presented in Scheme 9−66. Weinrib's amide reacts with Grignard reagent, affording an intermediary chelate that decomposes to ketone TM 9.45b (Scheme 9−67).

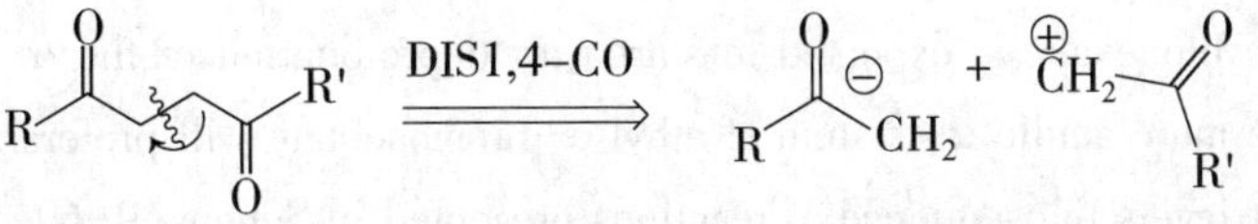

DIS1,4−CO

Scheme 9−65 General scheme of disconnection of 1,4−dicarbonyl compounds

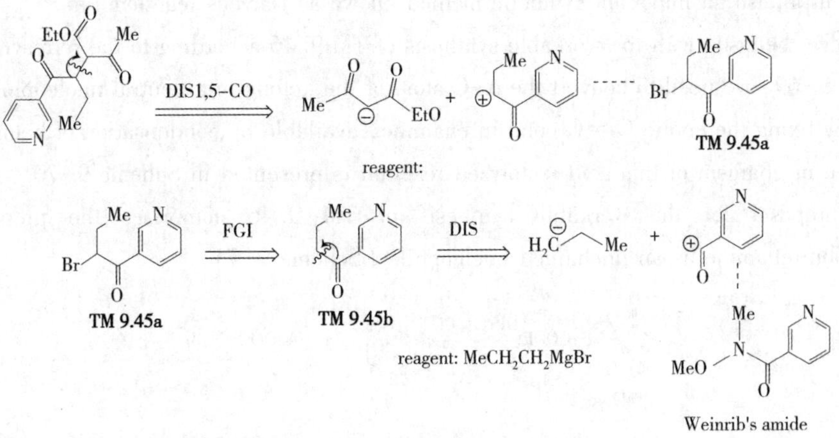

DIS1,5−CO

TM 9.45a

reagent:

FGI

DIS

TM 9.45a

TM 9.45b

reagent: $MeCH_2CH_2MgBr$

Weinrib's amide

Scheme 9−66 Complete retrosynthetic analysis of TM 9.45

The intermediary chelate is stable at low temperatures and decomposes on workup in aqueous medium. Ketone TM 9. 45b is brominated under standard conditions, with bromine in weak acidic medium, and used for alkylation of ethyl acetoacetate anions to TM 9. 45.

As an introduction to the next subject, an interesting phenomenon deserves comment. The acidity of the α–C—H bond in ketones (pKa about 20) is usually lower than that of the α–C—H bond in α–halogenated esters (pKa about 17) since halogen atoms enhance C—H acidity. We shall see in the next two examples what happens when both species are expected to react and a strong base is added.

NaOEt/EtOH
0℃to r.t.
TM 9.45a

Scheme 9–67 Proposal for the synthesis of TM 9. 45

Suggest an obvious disconnection of TM 9. 46 and evaluate the feasibility of the corresponding one–step synthesis.

COOEt

TM 9.46

An obvious 1, 4 – CO disconnection is presented in Scheme 9 – 68 using reagents for the conceived synthons.

Alkylation of cyclohexane is expected to start with deprotonation of the α–C atom by a strong base. However, the more acidic α–C atom of ethyl α–bromoacetate will preferably be deprotonated. This deprotonation triggers the sequence of reactions presented in Scheme 9–69. The formation of α, β–epoxy carboxylic esters from carbonyl compounds and α–halo esters is not just an undesired side reaction, but is also an important synthetic method known as Darzens reaction.

To solve the issue with the workable synthesis of TM 9. 46 according to the retrosynthetic step in Scheme 9–69, we need to convert the α–C atom of the ketone to a neutral nucleophile. This is achieved by fixing the enolic C=C bond in enamine, available by condensation of ketone and sec amine. The mechanism of this acid–catalyzed reaction is presented in Scheme 9–70.

The proposal for the workable synthesis of TM 9. 46 comprises the preparation of cyclohexanon–enamine as an uncharged nucleophile (Scheme 9–71).

DIS1,4–CO
COOEt ⟹ + EtOOC Br
TM9.46

Scheme 9–68 1,4–CO Disconnection of TM 9. 46

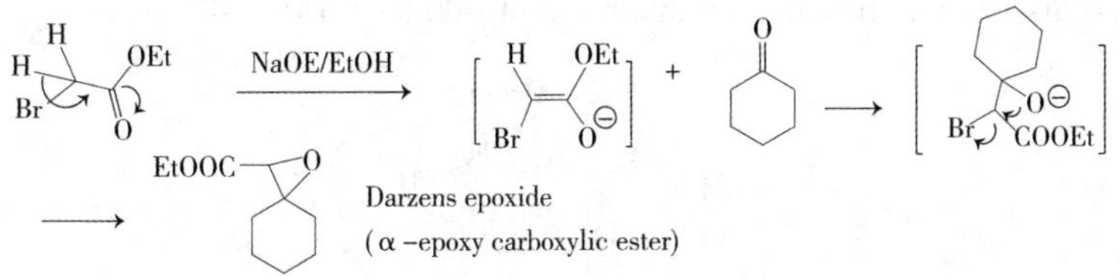

Scheme 9-69 Base-promoted reaction between cyclohexanone and enolate of ethyl α-bromoacetate

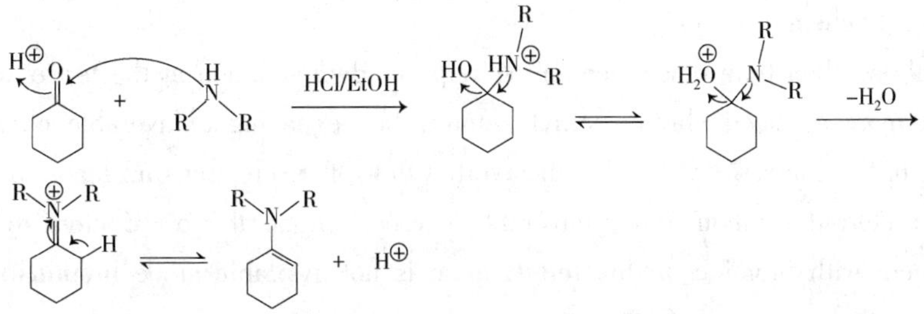

Scheme 9-70 Mechanism of the formation of cyclohexanon-enamine

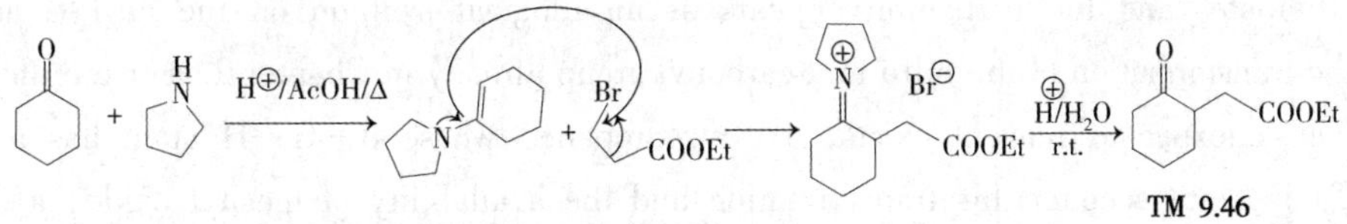

Scheme 9-71 Proposal for the synthesis of TM 9.46

Using this concept, complete the retrosynthetic analysis and suggest a workable synthesis of TM 9.47.

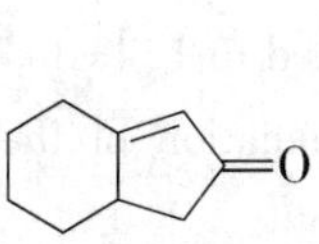

TM 9.47

Retro-aldol disconnection of α, β-unsaturated ketone leads to 1,4-diketone TM 9.47a. By disconnection of the central C—C bond, we generate two synthons, α-carbanion of cyclohexanone and α-carbocation in acetone. Synthetic equivalents for two synthons are the enamine of cyclohexanone TM 9.47b and α-chloroacetone TM 9.47c (Scheme 9-72). It should be taken into account that quaternization of the tert *N*-atom is reversible under the thermal conditions used for C-alkylation. Both characteristics of the enamine structure favor C-alkylation.

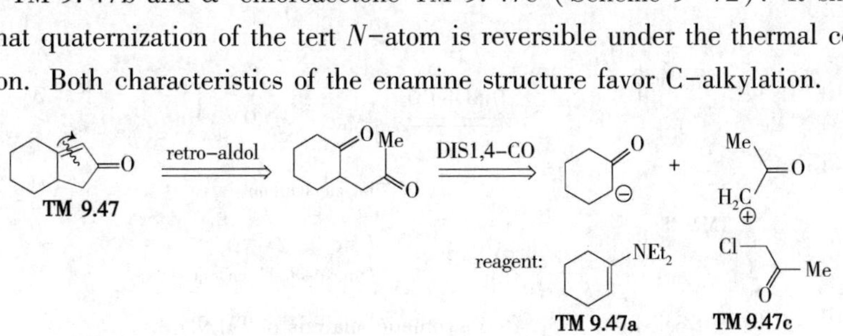

Scheme 9-72 Retrosynthetic analysis of TM 9.47

Propose the disconnection and then synthesis of diketone TM 9. 48.

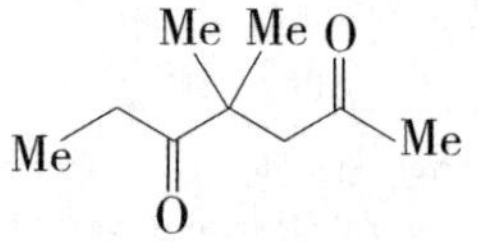

In this 1,4-dioxygenated pattern, we can expect that all disconnections of central bonds will result in one preferred and one illogical synthon. Let us first consider two variants of 1,4-CO disconnection a (Scheme 9-73).

An anionic synthon from disconnection a1 requires deprotonation at the tert α-C atom in the presence of a more acidic α-bromomethyl ketone, hence having unfavorable chemoselectivity. Disconnection of the same C—C bond with inverted flow of σ-electrons a2 leads to the acetonide anion as the preferred synthon and unfavorable carbocation on the α-C atom to the carbonyl group. A reagent with bromine on the tert C atom is not available since bromination preferably occurs on the methylenic sec α-C atom.

A surprising solution is offered by *retro*-Mannich type disconnection b where the acidic C—H group participates and the carbanion appears as an illogical synthon on the acyl C atom. We showed the transformation of the nitro to a carbonyl group already in Chapter 2. For the illogical C_3 synthon, an unexpected reagent exists, 1-nitropropane, whose α-C—H atom has a pKa of ca. 10. Taking into account this transformation and the availability of mesityl oxide, a dimer of acetone, as the second product of disconnection b, we can now propose a short synthesis of TM 9. 48 (Scheme 9-74).

By Michael-type addition of stable carbanion α- to the nitro group intermediary, 1,4-nitroketone is obtained, which is oxygenated to 1,4-diketone TM 9. 48 by $TiCl_3$ in acidic medium. An overview of the methods for the oxygenation of the nitro to keto group, known as the Nef reaction, is available.

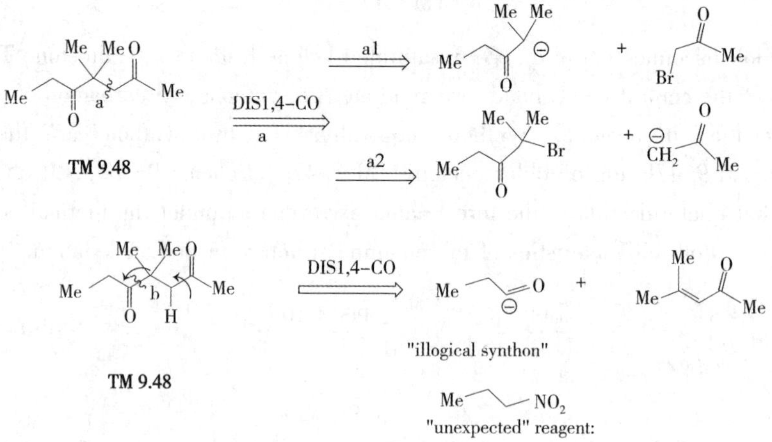

TM 9.48

DIS1,4-CO a; a1; a2

TM 9.48

DIS1,4-CO

"illogical synthon"

"unexpected" reagent: Me$\diagdown$$\diagup$$NO_2$

Scheme 9-73 Retrosynthetic analysis of TM 9. 48

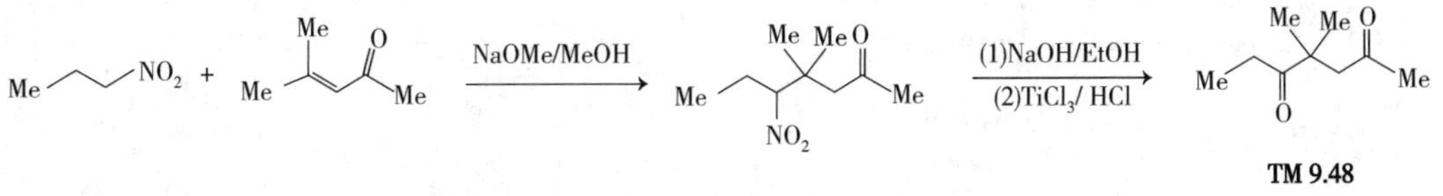

Scheme 9-74 Proposal for the synthesis of TM 9.48

9.2.4.2 1,4-羟基羰基化合物 1,4-hydroxy carbonyl compounds

1,4-Hydroxy carbonyl compounds seem available by selective reduction of one carbonyl group in 1,4-dicarbonyl compounds. Such chemoselectivity usually is not workable, and the retrosynthetic step where the hydroxy group in 1,4-hydroxycarbonyl compounds is interconverted to a keto group does not suggest any appealing solution to the synthetic problem. Therefore, we retrosynthetically consider the disconnection of 1,4-hydroxycarbonyl compounds to the available building blocks.

In the previous chapter, we saw α-halo carbonyl compounds as acceptable reagents for illogical synthons I with carbocation α- to the carbonyl group (Scheme 9-75).

Synthon Ⅱ, an analog of I with an oxygen atom at a lower oxidation state, is also conceivable. Here, the positive charge can be compensated intramolecularly by an electron pair from an oxygen atom, i.e., by the formation of epoxide. In other words, epoxides are ideal reagents for illogical synthons Ⅱ.

We saw the electrophilic character of epoxide toward the alkoxide ion. The carbanion is another charged nucleophile that opens a three-membered ring in reaction with epoxides. When such an anion is unstabilized, its synthetic equivalent is an organometallic compound affording α-alkylated or arylated alcohols as products (Scheme 9-76a).

When as the nucleophilic component α-carbanion stabilized by a carbonyl group or "masked" as enamine reacts, the reaction products are 1,4- or γ-hydroxycarbonyl compounds. The next example illustrates the utility of such retrosynthetic considerations.

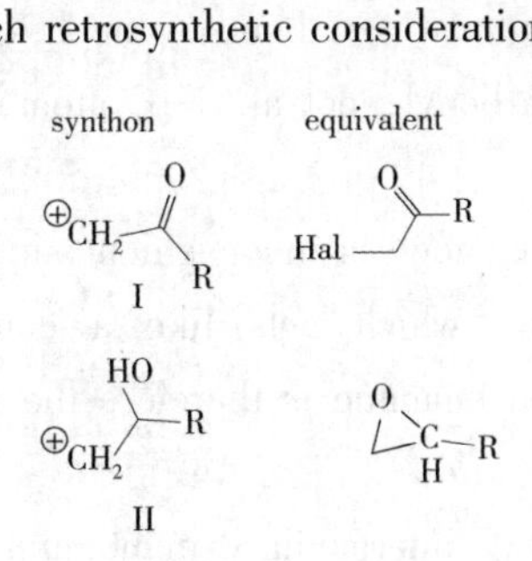

Scheme 9-75 Illogical carbocationic synthons and acceptable reagents

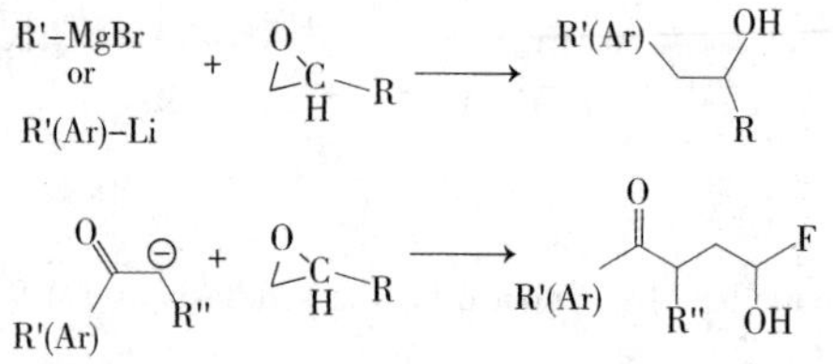

Scheme 9-76 Alkylation of epoxides by (a) unstable and (b) stabilized carbanions

Complete the retrosynthetic analysis and then propose the synthesis of TM 9.49:

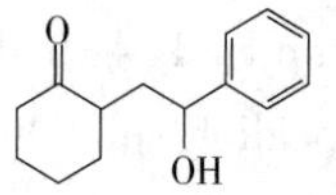

TM 9.49

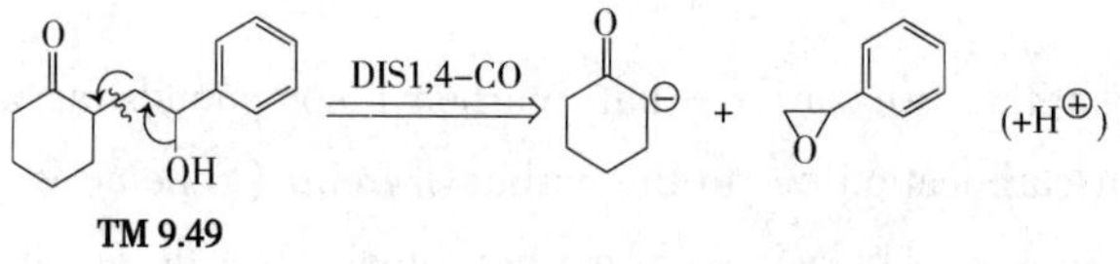

Scheme 9-77 Retrosynthetic analysis of TM 9.49

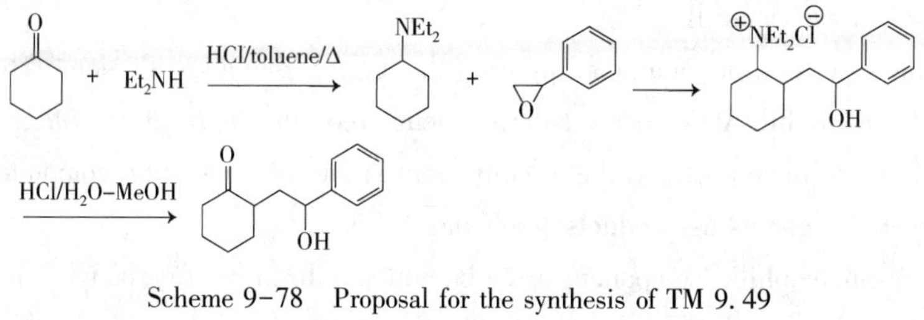

Scheme 9-78 Proposal for the synthesis of TM 9.49

Proper disconnection of the central C—C bond of the 1,4-hydroxycarbonyl pattern generates a carbanion on the α-C atom to carbonyl, not an α-C atom to the hydroxy group, and styrene epoxide (Scheme 9-77).

An anionic synthon for cyclohexanone is inconvenient since a strong anionic base is needed for the generation of a carbanion, which acts like a competitive nucleophile and attacks epoxide[15](环氧化物开环). Neutral enamine is therefore the preferred nucleophile (Scheme 9-78).

On the completed ring opening, intermediary immonium salt hydrolyzes easily to the *keto* group in TM 9.49.

Examples:

analysis: O CO_2Me ⟹ O + Br CO_2Me

synthesis a: O + Br CO_2Me $\xrightarrow{MeO^-}$ O O Me Br ⊖ CO_2Me strong base promotes Darzen's condensation

synthesis b: O R_2NH ⟶ NR_2 $\xrightarrow{Cl\ \ CO_2Me}$ $\overset{+}{N}R_2$ CO_2Me $\xrightarrow{H_3O^+}$ O CO_2Me

analysis: =O ⟹ O =O ⟹ O =O Cl—

synthesis: O $\xrightarrow{\text{N H}}$ N $\xrightarrow[(2)H_3O^+]{(1)Br}$ O =O $\xrightarrow{B:}$ =O

Because methylene anion of ethyl acetoacetate is stable enough, the target molecule gives synthons with definite charge.

Examples:

analysis: O EtO_2C O ⟹ O EtO_2C O

analysis: O O OEt OH $\overset{Dis}{⟹}$ O O OEt O

synthesis: O O OEt O $\xrightarrow{EtO^-}$ O O O

Oxygen functionalities can be conveniently transformed into functional groups with other heteroatoms.

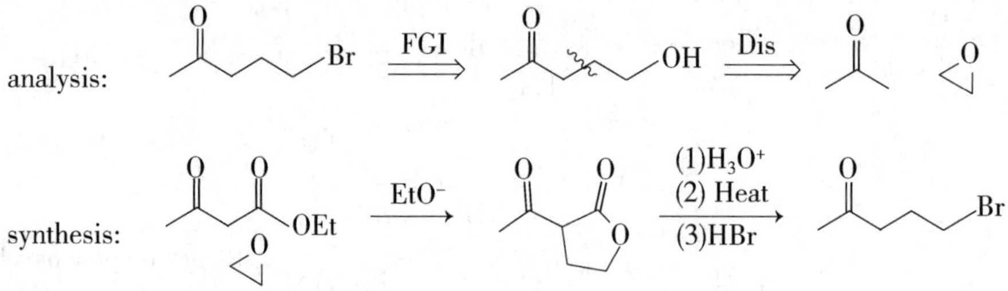

analysis: O, Br FGI ⟹ O, OH Dis ⟹ O + O

synthesis: O O OEt + O EtO^- ⟶ O O O (1) H_3O^+ (2) Heat (3) HBr ⟶ O Br

Cyanide is a good anion, and it attacks unsaturated ketone.

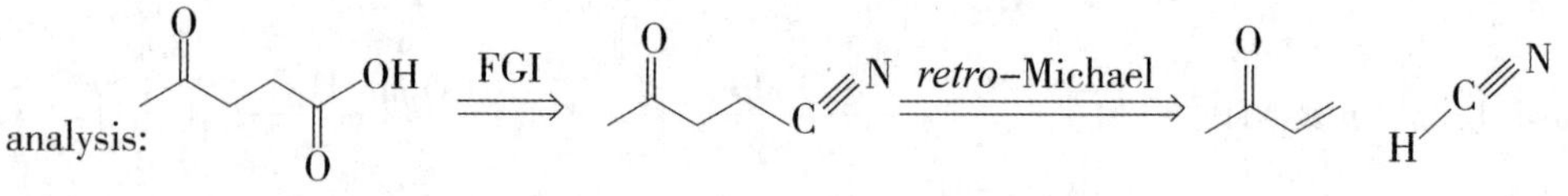

analysis: O OH O FGI ⟹ O C≡N *retro*-Michael ⟹ O + H—C≡N

9.2.5 1,6-二羰基模式 1,6-Dicarbonyl Pattern (1,6-CO)

The large through-bond distance between two oxygenated functional groups in a target molecule with a 1,6-CO pattern leaves the impression that no workable C—C bonding reaction can bind two building blocks since disconnection of any central C—C bond unavoidably results in one acceptable and one illogical synthon (Scheme 9-79).

Disconnection a is feasible only if protection of carbonyl group in carbanionic synthon is completed before the preparation of the Grignard reagent as a synthetic equivalent. Disconnection b indicates the synthetic route that starts with alkylation of the stabilized α-carbanion. Disconnection c does not offer any good solution for the synthon where the carbanion appears on the β-C atom. For the cationic synthon, instead, enone $RCOCH{=}CH_2$ is a convenient reagent.

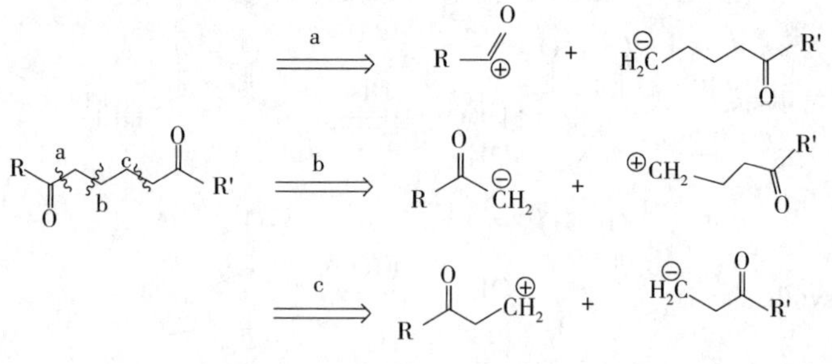

R, a, b, c, O, R' ⟹(a) R—C(=O)⊕ + ⊖H_2C … C(=O)R'
⟹(b) RC(=O)⊖CH_2 + ⊕CH_2 … C(=O)R'
⟹(c) RC(=O)C⊕H_2 + ⊖H_2CC(=O)R'

Scheme 9-79 Overview of disconnections of the central C—C bonds in the 1,6-CO pattern

The preferred retrosynthetic solution for 1,6-dicarbonyl compounds invokes a completely different concept of reconnection (RCN). This retrosynthetic step connects two functional groups into a cyclic, easily available structure, a target molecule of the next generation. Reconnection of the 1,6-dicarbonyl pattern results in a cyclohexene ring. Scheme 9-80 presents the synthetic step, the oxidative split of the C=C bond on the route to 1,6-dicarbonyl compounds and the corresponding retrosynthetic step, reconnection of this acyclic structure.

This scheme indicates that reconnection of the 1, 6-dicarbonyl compound to a cyclic alkene corresponds to the ozonolysis of the cyclic alkene in the synthetic direction. It is important to observe that no real synthetic reaction corresponds to the reconnection. Recognition of the derivatives of cyclohexene as target molecules of the next generation for the 1, 6-dicarbonyl pattern is of evident retrosynthetic utility. This concept is discussed in the next examples.

Propose the retrosynthetic analysis and then synthesis of TM 9. 50:

Me N HOOC Me COOH

TM 9.50

R R O_3/DCM/-75℃ R O R O RCN1,6-CO R R

Scheme 9-80 Correlation between the synthetic step and reconnection of the 1, 6-dicarbonyl compound

Me N HOOC Me COOH **TM 9.50** FGI Me N OHC Me CHO RCN1,6-CO Me N Me **TM 9.50a**

DISC-N O Me + HN Me *retro*-D-A + O Me

Scheme 9-81 Retrosynthetic analysis of TM 9. 50

On recognition of the 1, 6 - dicarbonyl pattern in dicarboxylic acid TM 9. 50, two retrosynthetic steps lead to TM 9. 50a (Scheme 9-81). The first step is interconversion of 1, 6-carboxylic to aldehyde groups that on reconnection afford cyclohexene derivative TM 9. 50a.

Further retrosynthetic steps lead to *N*-methylaniline, butadiene and methyl vinyl ketone, three available building blocks.

In Scheme 9-82, oxidation of dialdehyde, a product of the ozonolysis of TM 9. 50a, is indicated as an additional step on the route to TM 9. 50. There is an important aspect of ozonolysis: under oxidative conditions, aldehydes are spontaneously oxidized to carboxylic acids.

Note: Ozonolysis is a specific reaction that requires an ozonizator to produce ozone in the electric arc. Equipment is available in laboratory and industrial construction revealing the usability of this reaction on the technological scale. The mechanism of ozonolysis of the C=C bond is presented (Scheme 9-83).

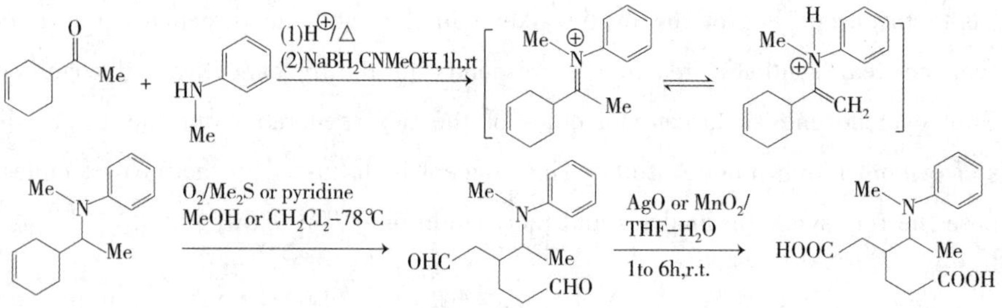

Scheme 9-82 Proposal for the synthesis of TM 9. 50

In the next example, an issue with the synthesis of 1, 3-disubstituted benzene is hidden. Introduction of the second substituent into an aromatic ring is often a non-trivial problem in the synthesis of complex target molecules. There are probably other acceptable synthetic approaches to TM 9. 51, and the reader is encouraged to consider the alternatives.

Propose the retrosynthetic analysis and synthesis of TM 9. 51.

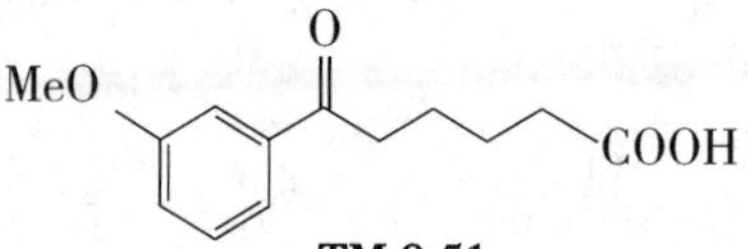

Scheme 9-83 The mechanism of ozonolysis of the C=C bond.

The 1,6-CO pattern suggests the reconnection of the side chain to the cyclohexene ring as the first retrosynthetic step (Scheme 9-80).

The presence of 1-phenylcyclohexene in TM 9. 51a suggests the FGI of the C=C bond to OH group TM 9. 51b followed by *retro*-Grignard disconnection. Grignard reagent TM 9. 51c is available from 3-bromoanisole. An approach to this *meta* substituted benzene derivative requires a few steps[16]. Both substituents are of the first order characterized by positive Hammett σ-values, and both direct the next substituent in the *ortho/para* position. In other words, none of them can orient a second substituent into the *meta*-position. Preparation of 3-bromoanisole follows the route where the first substituent, the meta-directing nitro group, after introduction of the second substituent, is transformed into a substituent of the first order present in the target molecule (Scheme 9-84).

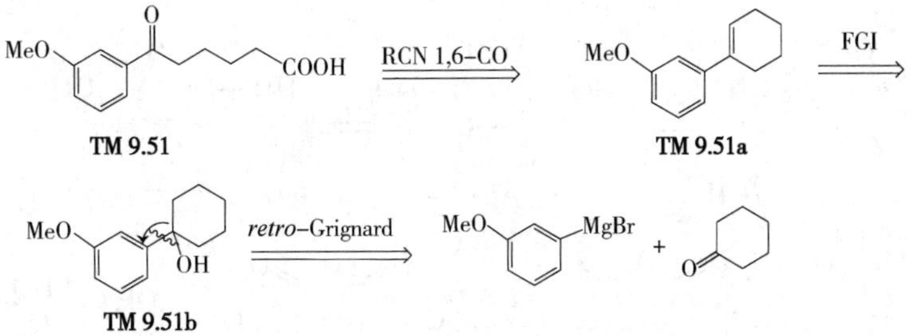

MeO, O, COOH, RCN 1,6-CO, MeO, FGI
TM 9.51 TM 9.51a
MeO, OH, *retro*-Grignard, MeO, MgBr, +, O
TM 9.51b

Scheme 9-84 Proposal for the retrosynthetic analysis of TM 9. 51

The complete synthesis of TM 9. 51 from *meta*-bromophenol and cyclohexanone as available raw materials is presented in Scheme 9-85.

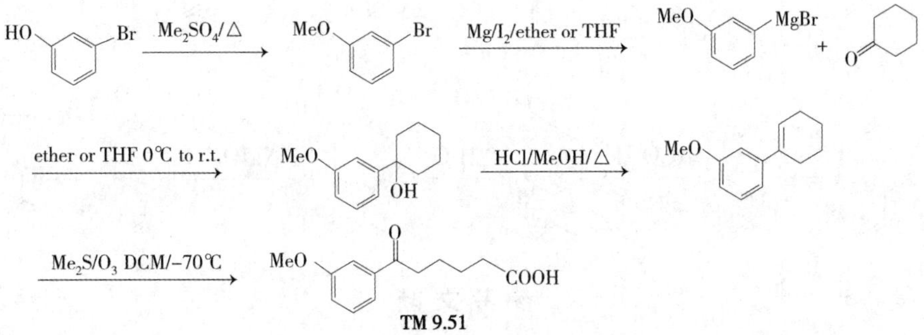

HO, Br, Me_2SO_4/Δ, MeO, Br, Mg/I_2/ether or THF, MeO, MgBr, +, O
ether or THF 0℃ to r.t., MeO, OH, HCl/MeOH/△, MeO
Me_2S/O_3 DCM/-70℃, MeO, O, COOH
TM 9.51

Scheme 9-85 Proposal for the synthesis of TM 9. 51

How would you approach the retrosynthesis of unsaturated ω-hydroxy ester TM 9. 51?

The terminal hydroxyl and carboxyl groups are the highest and the lowest oxidation states of functional groups with oxygen atoms.

In conclusion, retrosynthetic analysis of the 1,6-CO pattern reveals the value of reconnection when no productive route to the C_6 carbon chain is available from the $C_m + C_n$ building blocks:

Examples:

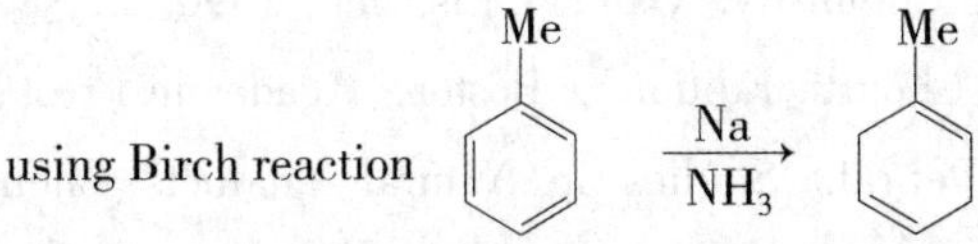

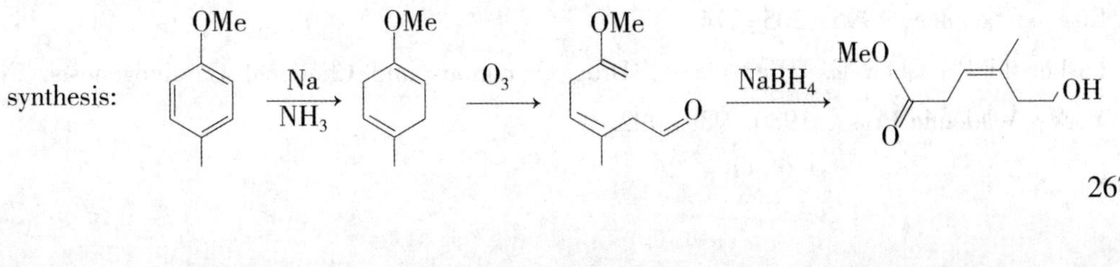

using Birch reaction: Me → (Na / NH_3) → Me

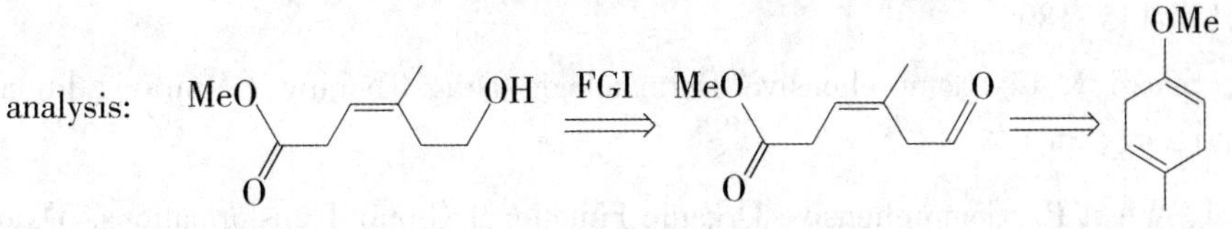

analysis: MeO, O, OH ⟹ (FGI) MeO, O, O ⟹ OMe

synthesis: OMe → (Na / NH_3) → OMe → (O_3) → OMe, O → ($NaBH_4$) → MeO, O, OH

analysis: ⟹ ⟹

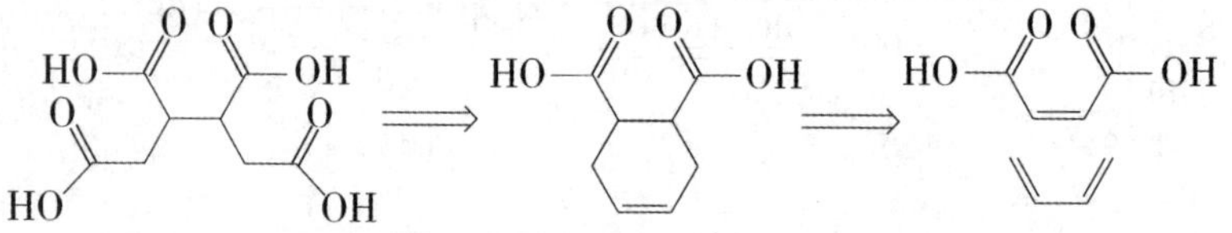

synthesis: → → $\xrightarrow[2.H_3O^+]{1.OH^-}$

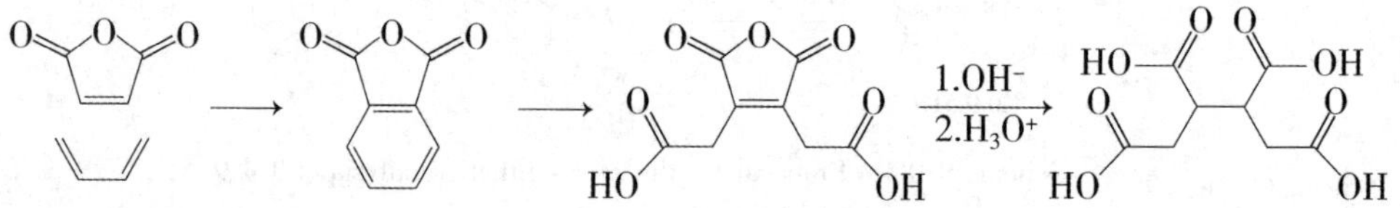

analysis: CHO ⟹ CHO ⟹ ⟹

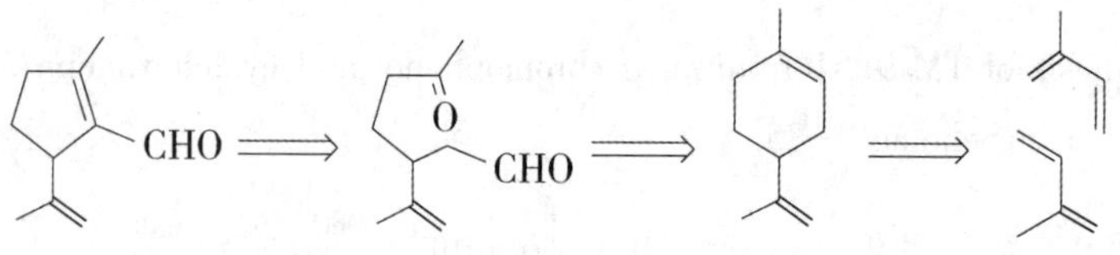

synthesis: → $\xrightarrow{RCO_3H}$ $\xrightarrow{H_2O}$ $\xrightarrow{NaIO_4}$ → CHO

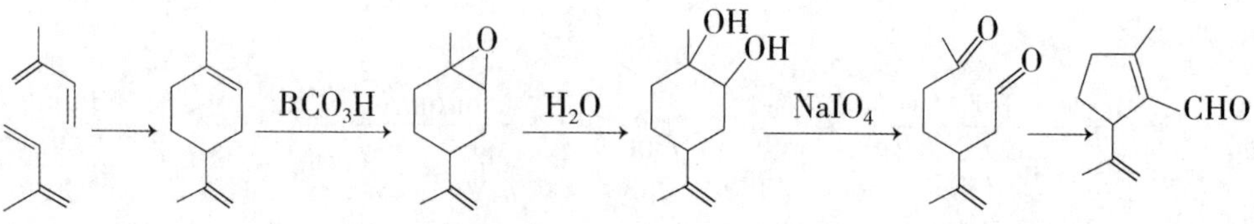

参 考 文 献
References

1. D' Angelo J, Smith M B. Hybrid Retrosynthesis. Boston: Elsevier, 2015. 19–25.
2. D' Angelo J, Smith M B. Hybrid Retrosynthesis. Boston: Elsevier, 2015. 1–17.
3. D' Angelo J, Smith M B. Hybrid Retrosynthesis. Boston: Elsevier, 2015. 45–62.
4. Smith M B. Organic Synthesis (Third Edition). Oxford: Academic Press, 2010. 999–1159.
5. Parashar R K. Reaction Mechanisms in Organic Synthesis. Amsterdam: Elsevier, 2008. 1–50.
6. D' Angelo J, Smith M B. Hybrid Retrosynthesis. Boston: Elsevier, 2015. xiii–xvii.
7. Smith M B. Organic Synthesis (Fourth Edition). Boston: Academic Press, 2017. 419–482.
8. Serratosa F. Studies in Organic Chemistry. Oxford: Elsevier, 1990. 62–87.
9. Smith M B. Organic Synthesis (Fourth Edition). Boston: Academic Press, 2017. 863–915.
10. Tschersich R, Bisterfeld C, et al. Studies in Natural Products Chemistry. Amsterdam: Elsevier, 2018. 145–180.
11. Lewis A B, Suozzi K C. Comprehensive Dermatologic Drug Therapy (Fourth Edition). Elsevier. 2021: 585–591.
12. Ebenezer W J, Wight P. Comprehensive Organic Functional Group Transformations. Oxford: Elsevier Science, 1995. 205–276.
13. Laskowski P, Chen C. Microsomes, Drug Oxidations and Chemical Carcinogenesis. New York: Academic Press, 1980. 989–992.

14. Yaragorla S, Koteshwar Rao R. Green Sustainable Process for Chemical and Environmental Engineering and Science. Oxford: Elsevier, 2020. 255-273.
15. Tang M, et al. Direct Thioamination of Cyclohexanones via Difunctionalization with Thiophenol and Aniline. *J. Advanced Synthesis and Catalysis*. 2022, 364 (13): 2205-2210.
16. Smith M B. Organic Synthesis (Fourth Edition). Boston: Academic Press, 2017. 547-603.

第四部分
有机合成技术
Part Ⅳ Organic Synthesis Techniques

10 具体的合成方法和技术

Chapter 10 Specific Synthetic Methods and Techniques

Organic synthetic chemistry is progressing thanks to the development of new and specific experimental methods. Some innovative synthetic methods in organic chemistry are concisely presented. Among them, the largest interest has been attracted by multicomponent reactions, parallel syntheses and combinatorial chemistry, mechanochemically promoted organic reactions, organic reactions promoted by microwave irradiation, syntheses in ionic liquids, digitization and validation of a chemical synthesis literature database in the ChemPU and modern automated biomolecule synthesis. In the next sections we briefly describe the basic characteristics of these methods and offer examples of their successful application. This chapter is more informative then educative and is expected to expand the imaginative capacity of synthetic organic chemists.

10.1 多组分反应 Multicomponent Reactions

Multicomponent reactions (MCRs) have long been known in laboratory praxis, but not recognized as a general concept. In the last decades their industrial utility has been recognized and multicomponent syntheses developed for large-scale production. Before consideration of specific examples, it is important to emphasize that in MCR you should not react all reagents at the same time since this is a highly unfavorable process entropically. What actually happens in MCR is a well-defined sequence of bimolecular reactions where an intermediate formed in the first step reacts with the third component in the second step and in rare cases with the fourth component in the third step. Intermediates of the whole process usually are not known and regularly are not isolated[1].

10.1.1 多组分反应的一般概念 General Concept of Multicomponent Reactions

Multicomponent reactions represent a flexible tool for the synthesis of a large number of target molecules from three or more starting molecules. They are one-step, one-pot reactions, economic regarding resources and today considered to be close to what is defined as "ideal synthesis"[2].

Their general synthetic value was recognized when I. Ugi and collaborators reported on some importantvariants of four-component reactions and their application in the production of known drugs. It is interesting that three multicomponent reactions were broadly used over 150 years, known as name reactions according to their inventors: Strecker synthesis of amino acids, Hantsch synthesis of 1, 4-dihydropyrimidines and particularly the important Mannich reaction. Hantsch four-component synthesis of 1, 4-dihydropyrimidines entered laboratory praxis long ago, but only recently became an industrial method. As an example, synthesis of nifedipine **3**, a broadly used drug in the therapy of hypertension, is presented in Scheme 10-1.

ortho-Nitrobenzaldehyde **1**, ammonia and two moles of ethyl acetoacetate **2** combine on heating in an aprotic solvent to fully substituted 1, 4-dihydropyridine **3**.

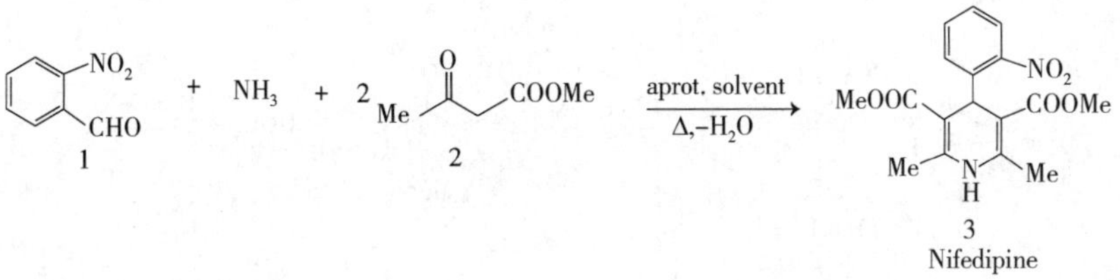

NO_2 CHO 1 + NH_3 + 2 Me COOMe (O) 2 —aprot. solvent, Δ, $-H_2O$→ MeOOC COOMe NO_2 Me N H Me 3 Nifedipine

Scheme 10-1 Four-component Hantsch reaction in the synthesis of nifedipine 3

CHO CHO 4 + $MeNH_2$ 5 + COOMe =O COOMe 6 ⟶ COOMe NMe =O COOMe 7 Tropinone

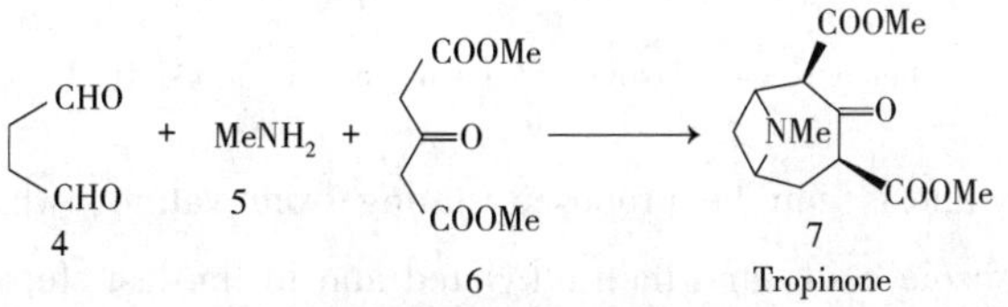

Scheme 10-2 Synthesis of tropinone 7 by the three-component Mannich reaction

We already described the application and mechanism of the Mannich reaction in Chapter 2. An elegant application of this reaction is Robinson synthesis of the bicyclic alkaloid tropinone 7 starting from simple building blocks 4~6 and completed with great atom economy (Scheme 10-2).

10.1.2 Ugi 多组分反应 Ugi Multicomponent Reactions

Ugi reactions are characterized by the use of isocyanides as a strong electrophilic component. In the next example, formaldehyde, dimethylamine and 2,5-dimethylphenylisocyanide are combined in one step to xylocaine 11, an important local anesthetic (Scheme 10-3). The next example explains why most multicomponent reactions are not easily amenable to rational retrosynthetic analysis.

CH_2O + Et_2NH + 10 ⟶ 11 Xylocaine

Scheme 10-3 Three-component Ugi reaction in the synthesis of xylocaine 11

Recognition of TM 10. 1 as an anilide of N-alkylated and N-acylated α-amino acid valine suggests "obvious" retrosynthetic analysis. Retrosynthesis leads over three consecutive disconnections of C—N bonds to valine and the three additional building blocks (Scheme 10-4).

TM 6.1 ⟹ (DIS C—N) ⟹ (DIS C-N) ⟹ (DIS C—N) Valine

Scheme 10-4 "Obvious" retrosynthesis of TM 10. 1

A corresponding synthesis can be proposed stating from valine, which is in the first step acylated by activated benzoic acid, and then alkylated and in the last step coupled with aniline to amide TM 10. 1. A completely different approach offers an Ugi four-component reaction, whose mechanism is presented in Scheme 10-5.

I II III TM 10.1

Scheme 10-5 Ugi four-component synthesis of TM 10. 1

On mixing isobutyraldehyde, ethylamine, phenyl isocyanide and benzoic acid, all steps in the abovereaction occur in "one-pot". Intermediates Ⅰ ~ Ⅲ are not isolated, and the presence of some of them was confirmed by spectroscopy.

Immonium ion I is formed in the benzoic acid-catalyzed condensation of aldehyde and amine. This intermediate reacts with the strongly electrophilic C-atom of phenyl isocyanide forming a C—C bond in nitrilium intermediate Ⅱ. The formal triple bond in Ⅱ is highly susceptible to nucleophilic attack of weak nucleophiles such as the carboxylate anion forming intermediary acyl-imine III, an analog of an anhydride. This unstable species spontaneously rearranges into stable diamide TM 10. 1. Such a complex sequence of bond-forming and bond-breaking events is nearly impossible to conceive retrosynthetically.

10. 2 平行合成与组合化学 Parallel Synthesis and Combinatorial Chemistry

Parallel synthesis and combinatorial chemistry are two closely related concepts, result of intentions to automatically perform more synthetic reactions. Parallel synthesis enables preparation of a set of defined compounds in a number of physically separated reaction vessels or micro-departments. Combinatorial chemistry instead uses a combinatorial process for preparation of a large number of compounds from a defined set of building blocks. Combinatorial chemical synthesis generates a large number of compounds, so-called libraries, at the same time and in predictable mode[3].

Productive methods in organic synthesis, also called technologies, are born in the pharmaceutical industry, then developed in academic institutions and applied in broad synthetic fields. The pharmaceutical industry was the first to suffer from slow organic syntheses as the "narrow throat" in the process of the discovery of a new drug. It was estimated that synthetic chemists are able to prepare one preparation-one compound by the "classic method", approximately 50 ~ 100 compounds/year, with a cost $5000 ~ 7000 per compound. An automated combinatorial synthesis can generate up to 100, 000 compounds/year per chemist, with an average price of $5 ~ 7 per compound. The many new compounds in the pharmaceutical industry substantially enhance the chance to identify lead compounds (LCs) on the route to a new drug entity (NDE). The availability of a large number of compounds prompted high-throughput screening (HTS) of biological activity. There are numerous examples of successful parallel synthesis of heterocyclic compounds, e. g., for indoles and benzofuranes. Here we refrain from a detailed discussion of combinatorial syntheses combined with high throughput screening of biological profiles. From the large number of monographs and review articles on this topic here, we select citing only a few.

Parallel synthesis does not require sophisticated equipment and software support. It is therefore the method of choice for the preparation of a dozen compounds with well-defined purity for biological testing in academia and industry. Some examples of parallel synthesis are presented to support this. Typically they are performed in 1 ~ 30 mL reactors with shaking or magnetic stirring and cooling/heating, which allows working temperatures between −78℃ and 250 ℃[4].

Using parallel synthesis, a series of 4 (3H)−pyrimidinone derivatives Ⅳ was prepared. According to previous knowledge from these compounds, anti−HIV activity was expected.

From Ⅳ as the starting compound, where R′ = H, Y = OH in R″ and R‴ = Me, a virtual library for parallel synthesis was generated (Fig. 10−1), and then, by combinatorial synthesis, a library of 522 compounds V was prepared (Scheme 10−6).

The synthetic protocol characterizes the use of solid support with a carbonate unit and flexible linker (Merrifield resin) to construct library V. All compounds are screened on binding to the non−nucleoside binding pocket (NNBP). In vitro binding constants >5 kcal/mol are regarded as indicators of potential in vivo anti−HIV activity.

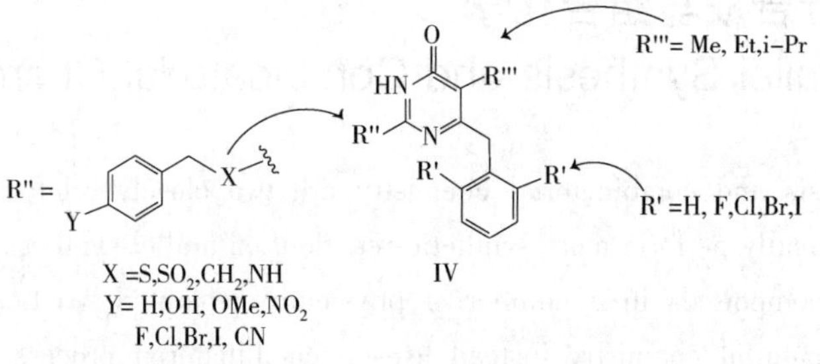

R‴= Me, Et,i−Pr

R″ =

X =S,SO$_2$,CH$_2$,NH
Y= H,OH, OMe,NO$_2$
F,Cl,Br,I, CN

R′ =H, F,Cl,Br,I

IV

Fig. 10−1 Fragments used for generation of a virtual library of 4(3*H*)−pyrimidinones IV

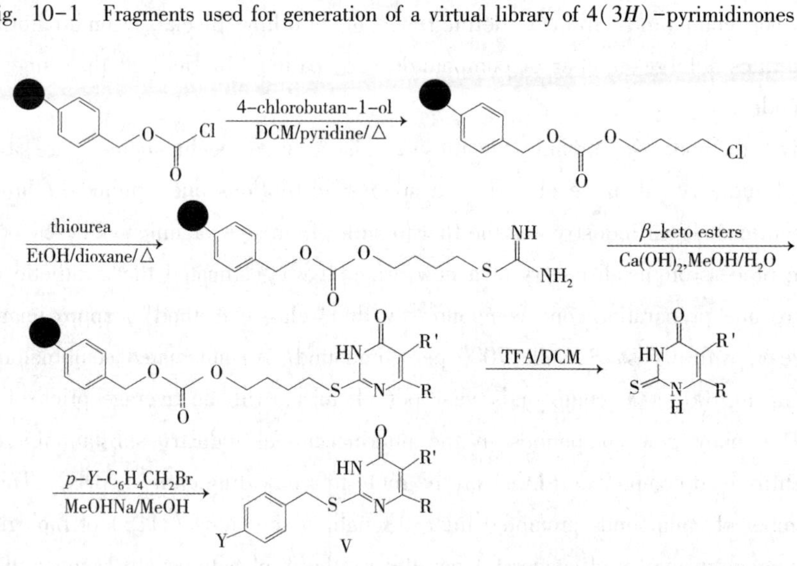

4−chlorobutan−1−ol
DCM/pyridine/△

thiourea
EtOH/dioxane/△

β−keto esters
Ca(OH)$_2$,MeOH/H$_2$O

TFA/DCM

p−*Y*−C$_6$H$_4$CH$_2$Br
MeOHNa/MeOH

V

Scheme 10−6 Preparation of the library of 2−*thio*−3(4*H*)−pyrimidinones V by parallel synthesis

On cyclization of 1,2− and 1,3−hydroxyalkylazides in the presence of Lewis acid, the library of dihydrooxazolines VI is prepared by one−step parallel syntheses (Scheme 10−7).

This method characterizes the use of polymer−bound phopshine as a scavenger of the excess of hydroxyl azide in the reaction solution. The library of ca. 60 compounds is prepared with ca. 40%

average yield and ca. 90% purity of isolated compounds meeting the criteria for high-throughput screening of their biological activity.

Scheme 10-7 Parallel synthesis of dihydrooxazoline derivatives

10.3 有机合成中的机械化学 Mechanochemistry in Organic Synthesis

According to the accepted definition, mechanochemical reactions are those performed by direct absorption of mechanic energy. Chemical reactions promoted by mechanical force are ever more frequently explored in organic synthesis. It was demonstrated that mechanochemical reactions can be performed in the absence of any solvent and therefore meet the challenges of modern chemistry that require economic use of energy and auxiliary materials and fulfill the requirements for environmentally acceptable processes[5].

Mechanochemical reactions are performed in planetary mills, moving reactors where the collision of metal balls of a defined radius transfers mechanical energy to the reacting species, prompting their chemical reactivity. Since no solvent is used, isolation of the products is regularly simplified. Mechanochemical reactions have been traditionally explored in the synthesis of organometallic compounds, porous materials, salts and cocrystals. More recently, this technology has found application in the production of active pharmaceutical compounds, known as active pharmaceutical ingredients, APIs[6].

The next two examples illustrate the successful application of mechanochemical conditions in organic synthesis.

The Suzuki-Miyaura reaction is today the most valuable method for coupling aryl components into biaryl structures. While traditional protocols for this reaction require large quantities of expensive nonpolar aprotic solvents and often proceed with low yields, the mechanochemical variant of this reaction affords biaryls 3 in high yields (Scheme 10-8).

R	Hal	yield/%
$COCH_3$	Br	94
$COCH_3$	I	97
OCH_3	Br	93
OCH_3	I	30

Scheme 10-8 Example of the Suzuki-Miyaura reaction in the mechanochemical variant

Interestingly, when in substrates 2 Hal =Cl, they proved completely nonreactive under standard conditions in apolar organic solvents, revealing insufficient Pd-catalyzed activation of the aryl-Cl bond. This energetically demanding step is surmounted in the mechanochemical variant, optimizing some parameters such as rotation speed and ball diameter.

4 + 5 → (solvent free, planetary mill) → 6

Conditions:
(i) 400rpm,60min
(ii)800rpm,60min

R	(i)yield/%	(ii) yield/%
H	31	94
4-Me_2N	2	68
2,3-OMe	6	94
4-Cl	10	99
4-OH	77	96

Scheme 10-9 Knoevenagel reaction in the mechanochemical variant

This example illustrates non - catalytic Knoevenagel condensation completed under mechanochemical conditions (Scheme 10-9). Aromatic aldehydes 4 react with malononitrile, affording geminal dicyano derivatives 6 in high yield. Interestingly, using 15-mm-diameter steel balls, the yields increased substantially when the rotation was doubled from 400 rpm to 800 rpm.

10.4 微波辐射促进的有机合成 Organic Synthesis Promoted by Microwave Radiation

Microwaves act as a high-frequency electric field; hence, they warm up all materials possessing a mobile electric charge, such as polar molecules in an apolar solvent. Polar solvents are also warmed up by microwaves since their molecules are forced to rotate in the electric field and lose energy by collisions[7].

In the last decades, microwaves were occasionally used in chemical laboratories and soon found application in organic synthesis. Microwave radiation enables heating of reactants without heating of the reaction vessel by heating a sample through its volume and not over the walls of the reactor. It allows uniform heating and great energy savings. Moreover, different compounds transform microwave radiation into energy to different extents, enabling faster and selective heating of the components of the reaction medium. The advantages of microwave heating over a traditional oil bath or vapor jacket are speeding up the reaction, milder reaction conditions, higher chemical yields, lower energy consumption and a change in chemical selectivity as compared to conventional heating[8].

The following examples illustrate the application of microwave technology in organic synthesis.

The Heck reaction, coupling aryl halides with a vinyl group, is promoted by microwaves and in situ prepared Pd(II) complex with P(*ortho*-tolyl)$_3$ ligand in ionic liquid 1-butyl-3-methylimidazolium hexafluorophosphate (Bmim).

Use of ionic liquids in microwave-promoted reactions is particularly preferred because of the fast heating of the medium, 10 ℃/s, and very small enhancement of pressure, a problem often encountered in microwave reactors. The reaction outlined in the Scheme 10-10 is very fast, for X = I completed in 5 min and for X = Br in 20 min, and for less effective phosphine ligands it needs up to 45 min. Besides, it was possible to recycle the catalytic system ionic liquid-($PdCl_2$/bmim) five times, maintaining a high reaction rate and yield.

X=Br, I

$PdCl_2/P(o\text{-tolyl})_3$, bmim, Et_3N

MW, 180℃ to 220℃, 5min to 45min

yield 40%~95%

bmim

Scheme 10-10 Heck reaction promoted by microwave radiation

High pressure in microreactors is often solved using dry media or media without solvents. This process starts by adsorption of reagents or catalyst on one or more inorganic supports transparent to microwaves, such as silicates, alumosilicates or clays, or acts as a strong absorber of microwaves such as graphite.

This example demonstrates the promoting effect of microwaves on an asymmetric catalytic reaction. Alkylation of dimethyl malonate carbanion is completed by a racemic allylic alcohol derivative in the presence of the complex of the Mo(III) ion with chiral bidentate nitrogen ligand VII. By this method, the key intermediate in the synthesis of the oral HIV inhibitor tipranavir was prepared in 95% yield and 94% e. e. (Scheme 10-11). Note that the racemic substrate is completely transformed into one enantiomer of the alkylated product revealing catalytic racemization of the non-reactive enantiomer.

$[Mo(CO)_3(C_7H_8)]$

THF, catalyst VII

MW, 180℃, 20min

VII

Tipranavir

Scheme 10-11 Enantioselective alkylation in microwave reactor

As yet, nearly all important synthetic reactions are performed under microwave conditions on a laboratory scale with the aim to become the method of choice in kilo laboratories of the pharmaceutical industry using commercially available microwave reactors[9].

10.5 离子液体中的合成 Syntheses in Ionic Liquids

Ionic liquids are salts in liquid state with a melting point around or just above room temperature. Usually these are the salts between organic cations and complex inorganic anions[10]. As a rule, they are colorless, non-volatile liquids of low viscosity and therefore easy to handle. Moreover, ionic liquids demonstrate high solvation of dissolved molecules, are easy to recycle and can be stored over longer periods without decomposition, rendering them attractive solvents for laboratory and industrial application. Ionic liquids are inconvenient for distillation still their recycling can be elegantly completed. It includes separation of solids by filtration, washing of the collected filter cake with toluene and evaporation of toluene from the ionic liquid, which can be reused[11].

The chemical stability, negligibly low volatility and easy recycling have placed ionic liquids at the top of "green chemistry" solvents. Fig. 10-2 presents the structures of the ionic liquids most frequently used as solvents in organic synthetic chemistry.

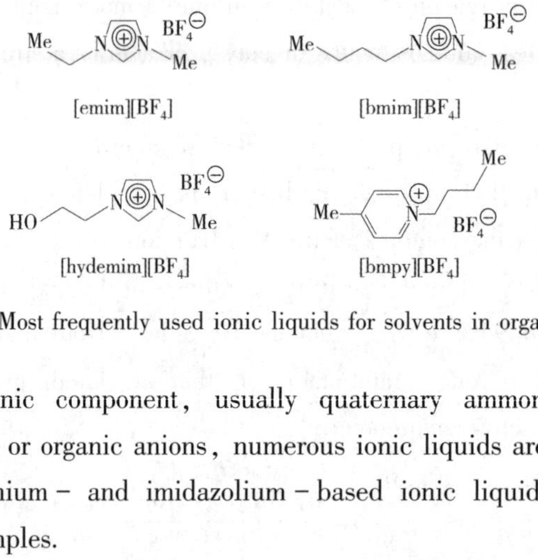

[emim][BF_4] [bmim][BF_4]

[hydemim][BF_4] [bmpy][BF_4]

Fig. 10-2 Most frequently used ionic liquids for solvents in organic synthesis

Combining the cationic component, usually quaternary ammonium cations of nitrogen heterocycles and inorganic or organic anions, numerous ionic liquids are available (Fig. 10-3). The application of pyridinium- and imidazolium-based ionic liquids in organic synthesis is presented in the next examples.

Using pyridinium-based ionic liquid (bmpy) Cl and $AlCl_3$ as the catalyst, an interesting insertion of acetylene into the C—Cl bond of acyl chlorides was achieved (Scheme 10-12).

An alkaline medium is controlled by the formation of [bmpy] $AlCl_4$, which selectively promotes the reaction without the formation of side products. Acetylene is introduced as a continuous gas stream, and products VIII are isolated by distillation in over 90 % yield.

Catalytic asymmetric Michael-type addition of aldehydes to nitroolefins was successfully completed in imidazolium-based ionic liquid XI (Scheme 10-13).

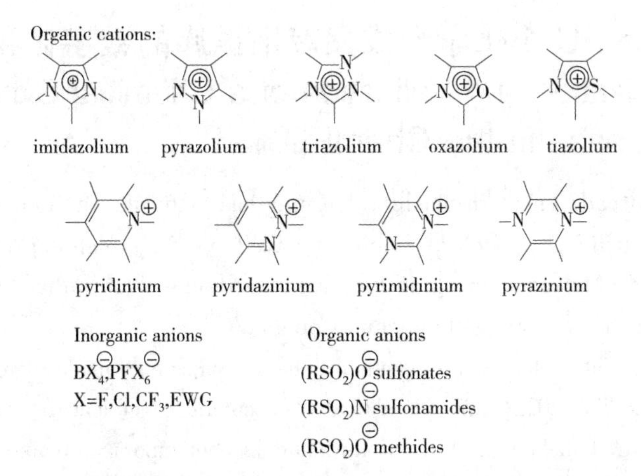

Fig. 10-3 Organic cations and inorganic or organic anions in ionic liquids

R	t/℃
alkyl	60
aryl	0

Scheme 10-12 Preparation of β-chlorovinyl ketones VIII in ionic liquid

Scheme 10-13 Asymmetric Michael addition in ionic liquid

Yields of IX in this reaction vary between 85%~95%, syn/anti ratio 96 : 4, and the e. e. is regularly over 95%. It is peculiar that chiral catalysts act after 12 recyclings without significant loss of enantioselectivity.

In conclusion, a broad application of ionic liquids at the laboratory scale still awaits extension to the industrial scale. Because of their properties described this chapter, ionic liquids are presently being tested as the solvents of choice for many industrial processes compatible with the standards of environmental protection.

10.6 ChemPU 中化学合成文献数据库的数字化和验证 Digitization and Validation of a Chemical Synthesis Literature Database in the ChemPU

During the last decade, modern machine learning has found its way into synthetic chemistry. Some long-standing challenges, such as computer-aided synthesis planning (CASP), have been successfully addressed. Molecular machine learning is a research topic that has great potential to fundamentally change the way synthetic chemists operate[12].

Despite huge potential, automation of synthetic chemistry has only made incremental progress over the past few decades. Till 2022, an automatically executable chemical reaction database of 100 molecules representative of the range of reactions found in contemporary organic synthesis have been found[13]. These reactions include transition *metal*-catalyzed coupling reactions, heterocycle formations, functional group interconversions, and multicomponent reactions. The chemical reaction codes or XDLs (universal Chemical Description Language) for the reactions have been stored in a database for version control, validation, collaboration, and data mining. Of these syntheses, more than 50 entries from the database have been downloaded and robotically run in seven modular chemputers with yields and purities comparable to those achieved by an expert chemist. The automatic purification of a range of compounds using a chromatography module seamlessly coupled to the platform and programmed with the same language are demonstrated (Fig. 10-4).

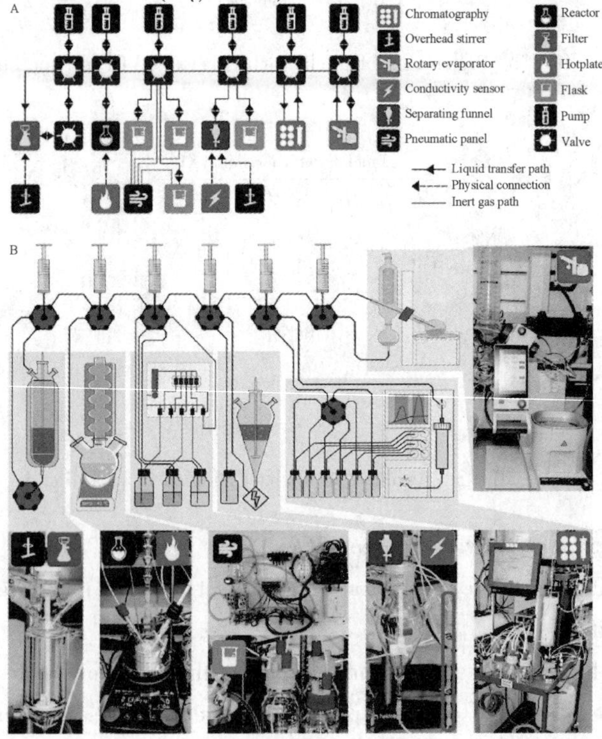

Fig. 10-4 The physical layout of the ChemPU and the available hardware library can be represented as a graph (ChemPU 的物理布局和可用硬件库的示意图形)

Within the context of synthetic design and execution, structural factors that dictate chemical reactivity must be paramount for a total synthesis campaign to be successful. Above all, the consequences of reaction mechanism and thermodynamic parameters determine feasibility in multistep synthesis[14].

10.7 生物分子现代自动化合成 Modern Automated Biomolecule Synthesis

10.7.1 SPPS，多肽固相合成 Solid-phase peptide synthesis

In its modern form, solid-phase peptide synthesis (SPPS) is the most complex and heavily optimized area of synthetic organic chemistry. In practice, SPPS can be used to construct peptides about 50 amino acid residues long. Much larger peptides (that is, proteins) are best prepared using bacterial overexpression. The common term automated peptide synthesis is misleading because it suggests that machines can design and execute peptide synthesis. In fact, in the hands of an amateur, a peptide synthesizer is more likely to deliver a library of compounds than a single product. The diversity of peptide functional groups, relative to that of nucleotides, greatly complicates chemical synthesis. Recall that the ribosome deftly forms peptide amides with perfect chemoselectivity, but chemical peptide synthesis may require the protection of as many as ninedifferent types of functional groups. Protection of the N and C termini prevents polymerization, whereas protection of the side chains prevents branching and other undesired reactions. Successful peptide synthesis requires a deep understanding of structure, reactivity, and mechanism so as to anticipate problems and design conditions accordingly. Protecting groups for the N or C termini must be labile enough to remove without affecting the side-chain protecting groups. The side-chain protecting groups must be robust enough to withstand round after round of acidic or basic reactions.

In theory, it should be possible to couple each amino acid at the N or C terminus of a suitably protected peptide (Fig. 10-5). Unfortunately, however, the condensation of amines with the C terminus of a peptide is accompanied by a competing cyclization that ultimately ruins the growing peptide because of epimerization. When synthesizing a peptide in the N-to-C direction, the entire peptide is ruined. In contrast, when synthesizing the peptide in the C-to-N direction, only the amino acid reagent is ruined. During each round of coupling, it is better to lose 5% of the plentiful amino acid reagent than to ruin 5% of the peptide chains.

Once a carboxylic acid is activated, nearby carbonyl groups can participate in an unwanted cyclization (indicated with a dashed arrow). Depending on the strategy, the cyclization either ruins the amino acid reagent or ruins the growing peptide chain. Ruining the reagents is better than ruining the entire peptide.

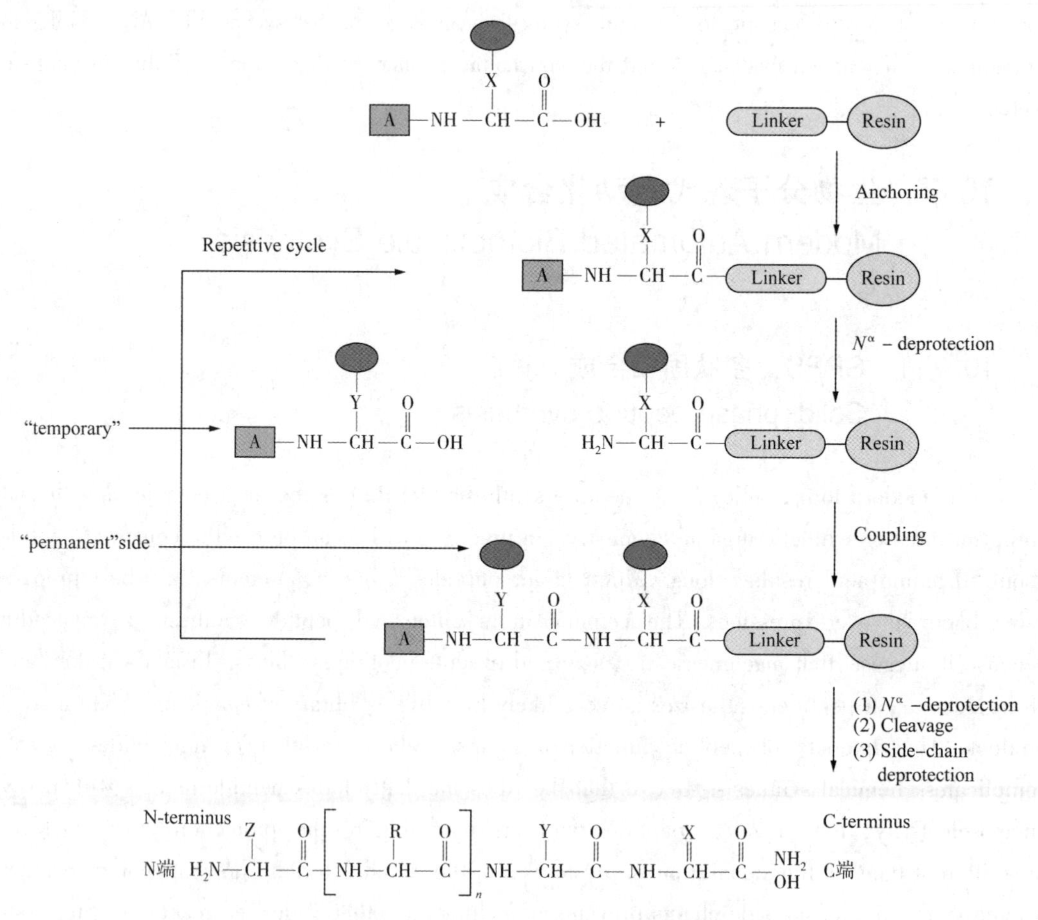

Fig. 10-5 Peptide assembly by Solid-phase peptide synthesis

In modern peptide synthesis, the cyclization of the amino acid reagent is decreased by protecting the $N\alpha$ group as a carbamate. Carbamate protecting groups were found empirically to give lower levels of the cyclization side reaction than amide protecting groups. After amide bond formation, the temporary carbamate protecting group can be removed selectively without disturbing the side chain or the C-terminal protecting groups.

Polystyrene is functionalized with formaldehyde and zinc chloride to give chloromethylated polystyrene, also known as Merrifield resin. The first amino acid is added by an S_N2 reaction, leading to an ester linkage. Thus, the solid support serves as a C-terminal protecting group for the first amino acid. It is generally cheaper to purchase resin with the first amino acid attached than to functionalize it yourself.

The term solid-phase is somewhat misleading in that it implies immobility. Instead of an immobile solid, though, try to think of SPPS beads as liquid spheres. Once resin beads swell with solvent, the peptides and polystyrene chains are highly mobile, allowing reagents to diffuse in and out. Because of this mobility, there is no significant barrier to interactions between peptides bound to a solid-phase resin. Thus, peptides can aggregate during peptide synthesis, which reduces the efficiency of the amino acid coupling reaction.

Peptides are cleaved from Merrifield and PAM resin under strongly acidic conditions. The best conditions for S_N2 cleavage involve anhydrous HF, condensed from the gas phase with special apparatus that is unavailable in most laboratories. HF dissolves both glass and human bones. The weak acidity of HF allows it to penetrate deeply into tissues, making it an extremely dangerous chemical. Trifluoromethanesulfonic acid (TFMSA) is sometimes used as an alternative: CF_3SO_3H/PhSMe/1,2-ethanedithiol (EDT) (2: 2: 1), 25 ℃, 1 hour. The choice of cleavage and deprotection conditions depends greatly on the constitution of the peptide. Typical conditions always employ nucleophilic scavengers to react with benzyl and *t*-butyl cations. The low-high conditions ((i) CF_3SO_3H/TFA/Me_2S/m-cresol (1: 5: 3: 1), 0℃, 3 hours; (ii) CF_3SO_3H/TFA/ PhSMe/EDT (1: 10: 3: 1), 25 ℃, 30 min) prevent alkylation of tyrosine and acylation by glutamine side chains.

A carboxamide group is often desired at the C terminus rather than a carboxylate group. In fact, cleavage and side-chain deprotection usually occur in the same step. Benzhydrylamine resins such as MBHA (methyl benzhydrylamine resin) were developed to allow facile S_N1 cleavage of carbon-nitrogen bonds, providing a carboxamide at the C terminus (Fig. 10-6). A second aryl group provides sufficient activation for efficient ionization using anhydrous HF or TFMSA.

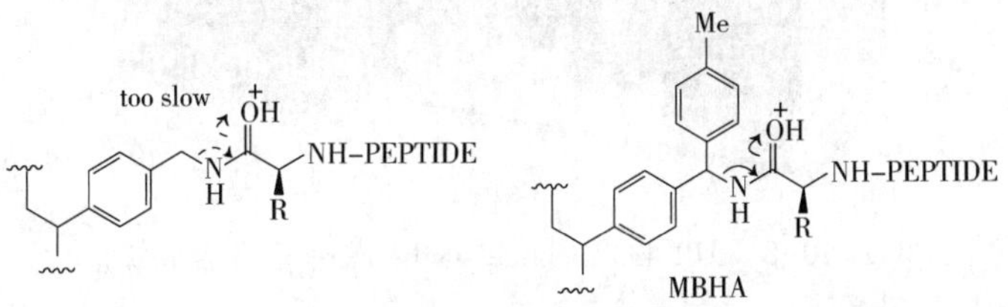

Fig. 10-6 Making C-terminal amides.

When SPPS is performed with base-labile, Fmoc-protected amino acids, peptides go through round after round of coupling and base deprotection; the growing peptide chain does not see acid at any point during peptide elongation. As a result, highly acid-labile linkers can be used that allow the final deprotection and cleavage to be performed under much milder conditions than peptides constructed from Nα-Boc-protected amino acids. Two types of resin that are commonly used for the synthesis of peptides during Fmoc chemistry are chlorotrityl resin and Rink amide resin (Fig. 10-7). Chlorotrityl resin generates a stable triphenylmethyl cation. The chlorine atom subtly attenuates the rate of S_N1 ionization. To facilitate the cleavage of a peptide with a carboxamide terminus, a Rink amide linker is used. The Rink linker generates a highly stabilizedcarbocation, making up for the slow S_N1 ionization of carbon-nitrogen bonds.

Chlorotrityl resin and Rink amide resin are designed to generate highly stable carbocations, resulting in rapid S_N1 cleavage.

Modern automated peptide synthesizer is shown in Fig. 10-8.

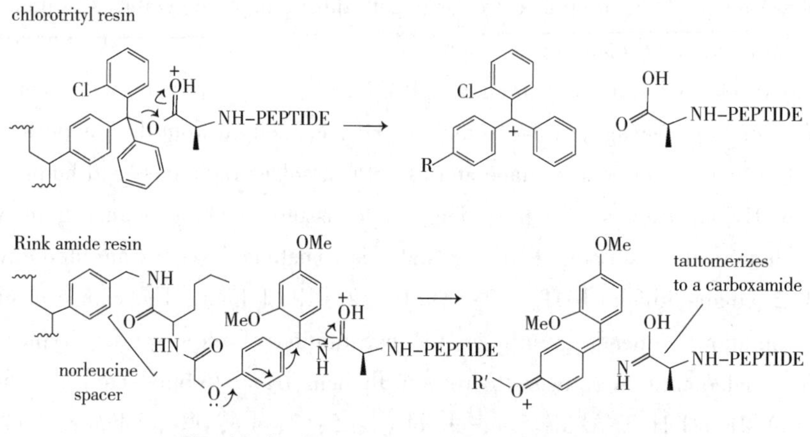

Fig. 10-7 Common resins for Fmoc synthesis.

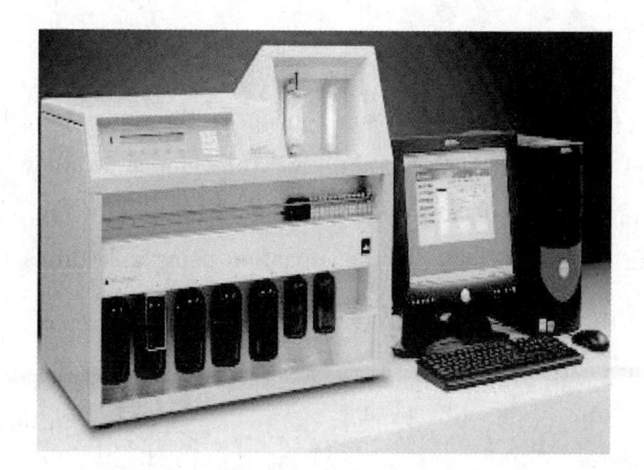

Fig. 10-8 ABI 433Apeptide synthesizer（多肽合成仪）

For SPPS, two major protecting groups for the $N\alpha$ -amino function have been established: Boc (*tert*-butyloxycarbonyl) and Fmoc (Fig. 10-9 and Fig. 10-10). The initial method applied by Merrifield was based on the use of the Boc group as temporary protecting group for the amino function and Bn (benzyl) or related protecting groups for the side chains of trifunctional amino acids. Usually, Boc can be removed by treatment with TFA (trifluoroacetic acid), whereas Bn deprotection requires strong acids such as HF. Hence, this Boc/Bn protecting-group strategy is based on graded acid lability of permanent (also including the linkage to the solid support) and transient protecting groups. Whilst the Boc group has been used exclusively during the first years of SPPS, the introduction of the Fmoc-group opened the path for a novel, more variable synthesis concept. Here, the Fmoc-group, which can be removed by basic conditions, serves as temporary N^{α}-protecting group. Side-chain protecting groups as t-Bu and the linkage of the peptide to the resin are unstable towards TFA-treatment. Nowadays, both protecting group strategies are used for the synthesis of peptides and both methods can be applied for automated synthesis.

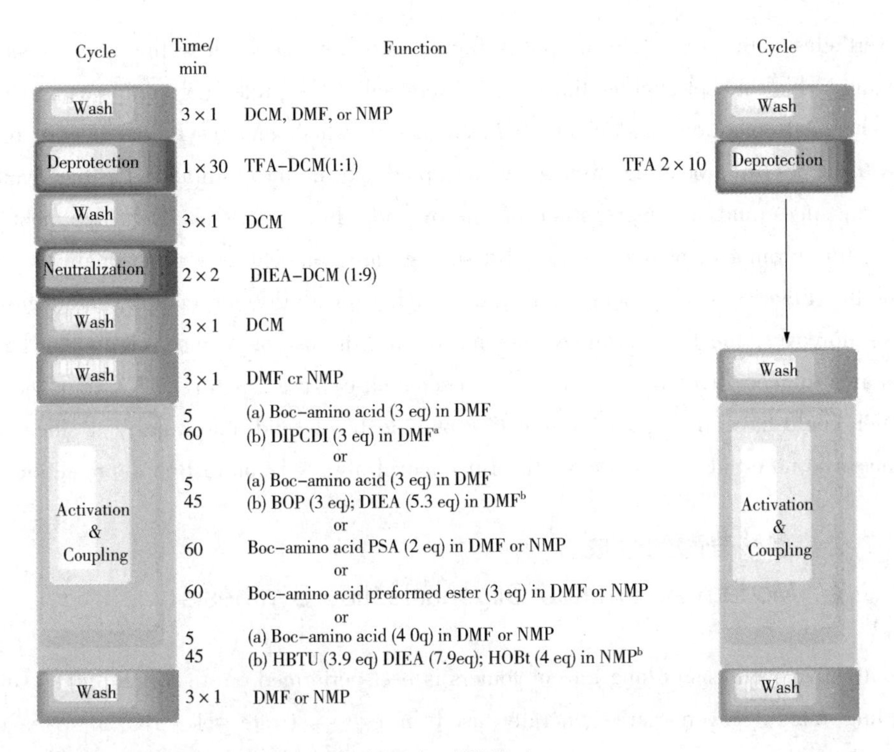

Fig. 10-9 Boc/Bn protecting-group strategy

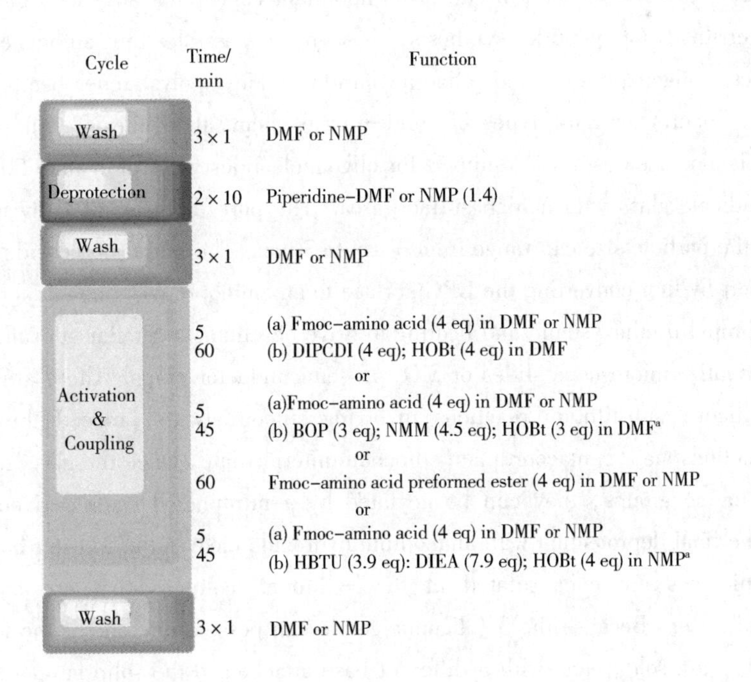

Fig. 10-10 Fmoc/*t*-Bu orthogonal protecting-group strategy

Nevertheless, the *Fmoc/t*-Bu protecting-group approach offers the great advantage of orthogonality. This concept enables the selective removal of the protecting groups using completely different chemical conditions and cleavage mechanisms, which ensures milder overall reactions. Although the Boc/Bn protecting strategy is accepted to be more suitable for the synthesis of difficult sequences and an aggregation of the peptide by repetitious TFA treatment can be prevented, the advantages of the *Fmoc/t*-Bu strategy are notable. The orthogonality is the main benefit of the *Fmoc*-based concept allowing a higher flexibility for complex strategies during synthesis. Moreover, the Fmoc strategy does not require the use of special vessels that have to be stable towards the corrosive and toxic HF and in some cases, the repetitive TFA acidolysis for Boc deprotection could have an impact on sensitive peptide bonds and acid catalyzed side reactions. And, since it is no orthogonal strategy, the Bn removal always leads to Boc deprotection.

10.7.2 寡核苷酸的合成 Modern Automated Oligonucleotide Synthesis

The iterative synthesis of long bio-oligomers is best performed on a solid support. To achieve fast coupling rates, reagents are generally used in excess (rate = k [ROH] [reagent]). Unfortunately, a large excess of chemical reagents is difficult to remove from the desired product. By attaching the substrate to a solid support, the by-products and reagents can be rinsed away from the product, just like washing a car. This important conceptual advance, initially pioneered by Robert Merrifield for peptide synthesis, has made possible the automated synthesis of oligonucleotides, oligopeptides, and oligosaccharides. Tiny polystyrene beads are the most common solid supports for most types of solid-phase chemical synthesis, but controlled-pore glass (CPG) is also used as a solid support for oligonucleotides with fewer than 50 bases. CPG is a porous borosilicate glass with a high surface area. The pore diameter of CPG is usually 500-1000 Å, and the particle size can range from 3 μm to 3 mm. The initial nucleoside is connected to the CPG support by first converting the CPG surface to an aminopropylsilane surface by condensing with 3-aminopropyl triethoxysilane and heating to drive off ethanol. Similar siloxane chemistry can be used to derivatize microscope slides or SiO_2 on semiconductor chips. There are no known S_N2 reactions at silicon; substitution reactions involving silicon always proceed through addition-elimination reactions via a pentacoordinate siliconate intermediate. Once the glass surface has been carpeted with amino groups, they can be acylated by a nitrophenyl ester derived from the first nucleoside. The final deprotection with ammonium hydroxide cleaves the 3′ ester bond. The CPG-nucleoside conjugates are encapsulated in flow-through columns that are connected to an automated synthesizer. Because the 3′ oligonucleotide can possess any one of the four bases, four different columns are sold, each with a different base attached to the solid support.

Modern automated DNA synthesis involves repetitive four-step cycles. The custom-synthesized oligonucleotides are made on automated DNA synthesizers (Fig. 10-11). Each cycle of automated DNA synthesis involves four steps, repeated over and over (Fig. 10-12): 5′ deprotection, phosphoramidite coupling, capping of unreacted 5′-hydroxyl groups, and oxidation of phosphites to phosphates.

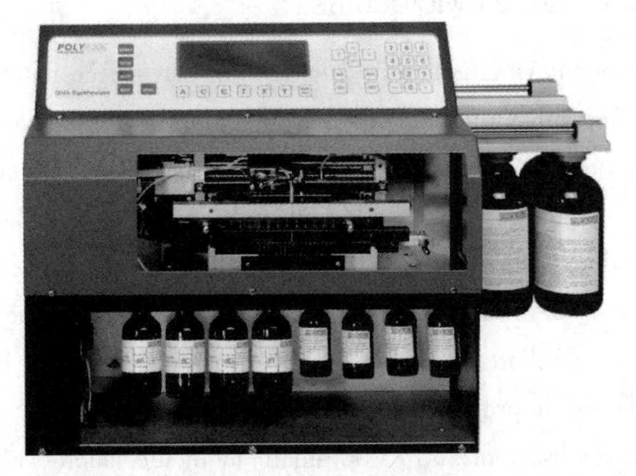

Fig. 10-11 Applied Biosystems DNA Synthesizer(DNA 合成仪)

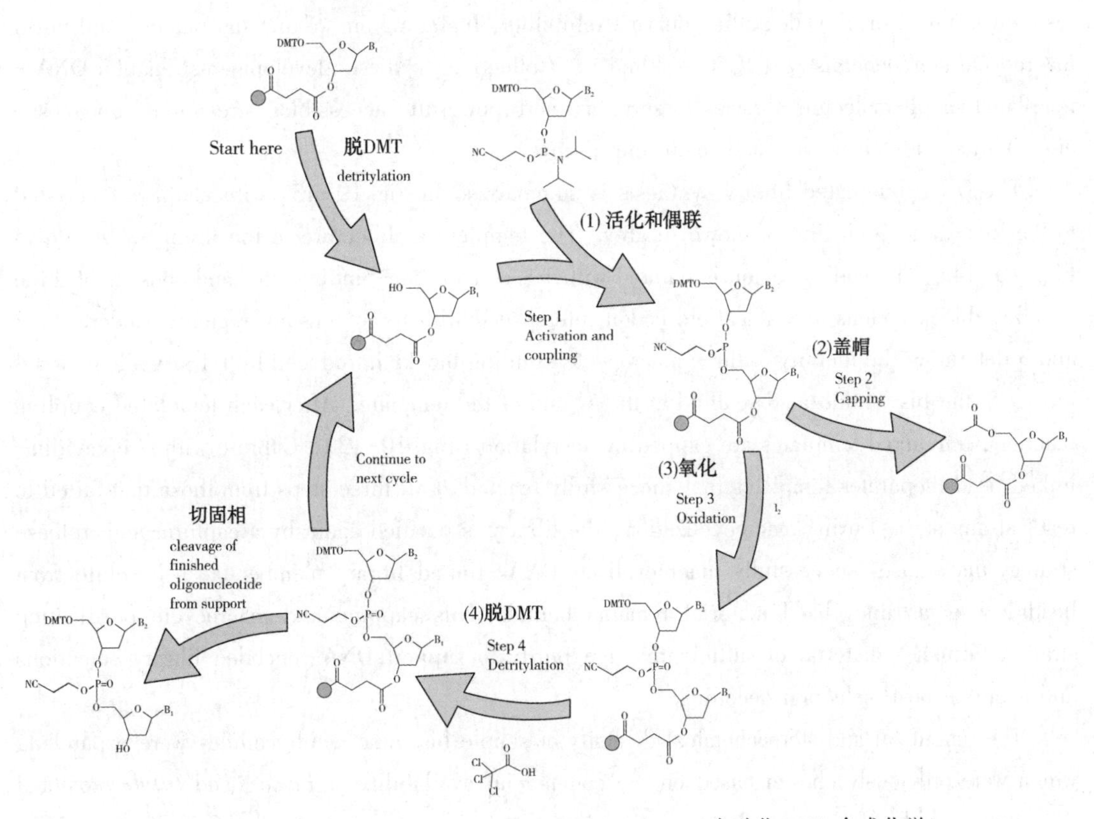

Fig. 10-12 The chemistry of automated DNA synthesis (自动化 DNA 合成化学)

10.7.3 DNA 模板化大环分子库用于发现生物活性小分子 DNA-Templated Macrocycle Libraries for The Discovery of Bioactive Small Molecules

DNA-encoded libraries have emerged as a widely used resource for the discovery of bioactive small molecules, and offer substantial advantages compared with conventional small-molecule libraries. Multiple fundamental aspects of DNA-encoded and DNA-templated library synthesis methodology, including computational identification and experimental validation of a 20×20×20×80 set of orthogonal codons, chemical and computational tools for enhancing the structural diversity and drug-likeness of library members have been developed and streamlined. This is a highly efficient polymerase-mediated template library assembly strategy, and library isolation and purification methods. These improved methods have been integrated to produce a second-generation DNA-templated library of 256, 000 small-molecule macrocycles with improved drug-like physical properties. In vitro selection of this library for insulin-degrading enzyme affinity resulted in novel insulin-degrading enzyme inhibitors, including one of unusual potency and novel macrocycle stereochemistry (IC50 = 40nM). Collectively, these developments enable DNA-templated small-molecule libraries to serve as more powerful, accessible, streamlined and cost-effective tools for bioactive small-molecule discovery.

The DNA-templated library synthesis is summarized in Fig. 10-13, with changes compared to the first-generation library shown in grey. The template architecture of the library is shown in Fig. 10-14a. The coding region is flanked with 10-mer primer-binding sites and consists of three building-block codons and a scaffold codon interspaced with three constant regions. Codons 1, 2 and 3 determine the identity of three macrocycle building blocks introduced by DTS, while codon 4 identifies the bis-amino acid scaffold at the 5′ end of the template. After each templated coupling reaction, unreacted templates are capped by acetylation (Fig. 10-13). Capture with streptavidin-linked beads separates templates that successfully reacted at all three steps from those that failed to react at any step. During macrocyclization, the library is purified again by a capture-and-release strategy that causes successfully macrocyclized DNA-linked library members to self-elute from beads, whereas uncyclized material remains bound. This capping and macrocycle purification strategy furnishes material of sufficiently high purity to support DNA-encoded library selections and accurate post-selection decoding.

The functional and stereochemical diversity of simple bisamino-acid scaffolds were expanded, which were previously chosen based on the commercial availability of *Fmoc*- and *trityl*-protected derivatives suitable for on-bead DNA conjugation. A 1min exposure to 50% trifluoroacetic acid in dichloromethane deprotects Boc-functionalized scaffolds and yields sufficient amounts of intact DNA for subsequent DNA-templated library synthesis steps (Fig. 10-14a). These additions expanded the set of scaffolds from 8 used in the original library to 32 (Fig. 10-14b), and also substantially increased the structural diversity of the resulting library. Building blocks for the second-generation

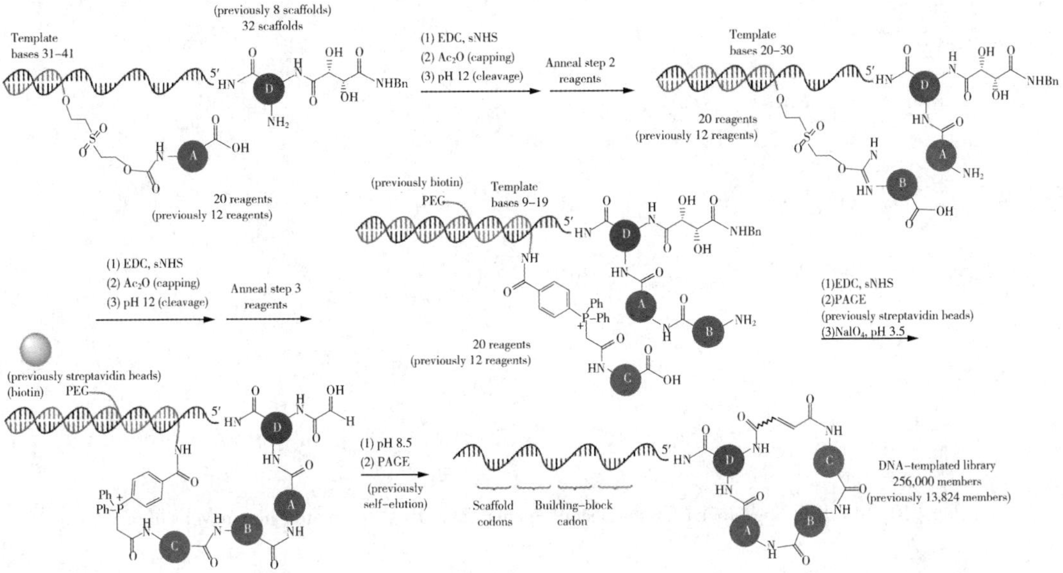

Fig. 10-13 DNA-templated macrocycle library synthesis scheme（DNA 模板大环库的合成方案）

library such that the resulting macrocycles are consistent with Kihlberg rules increased the likelihood of compatibility with cell-based assays and to facilitate subsequent hit-to-lead optimization. A method to calculate the influence of any building-block candidate on the predicted Kihlberg conformity of the resulting library using widely available chemistry software（ChemBioOffice, from CambridgeSoft）were developed. The set of building blocks were iteratively optimized to comply with Kihlberg' s guidelines through minimization of the number of highly polar functional groups and hydrogen bond donors, as well as liberal use of *N*-alkylated amino acids（Fig. 10-14c）.

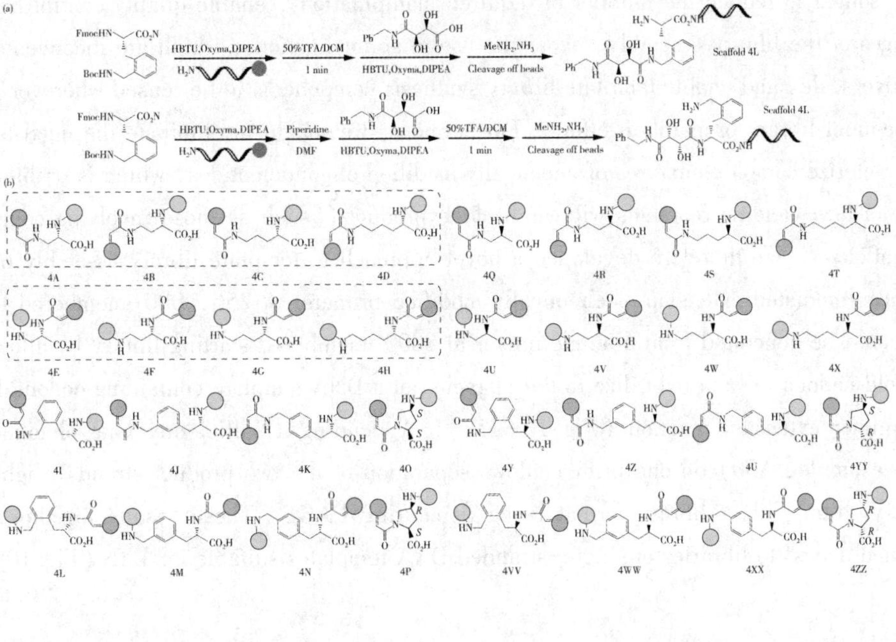

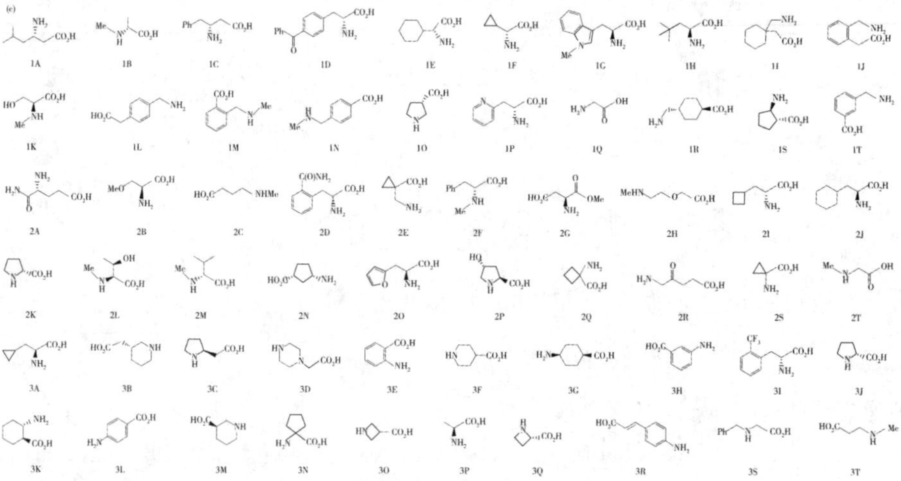

Fig. 10-14 Building blocks for the second-generation DNA-templated macrocycle library

A novel approach to template library assembly based on polymerase-mediated extension of chemically modified primers have been developed. The previously established strategy of assembling the library of DNA templates used split - pool oligonucleotide synthesis of phosphorylated 3′ fragments, followed by enzymatic splint - assisted ligation with chemically modified 5′ fragments. Applying the same approach to the preparation of a 256, 000-membered library would require many more oligonucleotide syntheses and split-pool events; for example, 1, 280 versus 192 oligonucleotide syntheses alone would be required for the preparation of the 3′ fragment (Fig. 10-15a). Splitting the template into three parts rather than two (Fig. 10-15b) could mitigate the problem; however, a more convenient template library assembly to popularize the application of DNA-templated libraries were provided.

We sought to reduce the number of required manipulations, enable quality control before the final stages of the library assembly, avoid the use of splint ligations, which are inconvenient on a preparative scale, and enable template library synthesis components to be reused wherever possible for subsequent library preparation efforts. Furthermore, we sought to eliminate the need to isolate and characterize complex mixtures of chemically modified oligonucleotides, which is problematic in the case of low-yielding reactions with multiple byproducts (such as those involving some of the novel scaffolds). We therefore developed a novel approach to template library assembly based on polymerase-mediated extension of chemically modified primers. A 256, 000-membered template library could be generated from a single universal 8000-membered starting library by allowing the 32 scaffold codons to each hybridize to the I4 region of a DNA template containing codons 1,2 and 3 in a primer extension reaction (Fig. 10-15c). Appending a sufficiently long oligonucleotide tail (for example, A30) on one primer allows separation of the two product strands (light strand and heavy strand) in a library format using denaturing PAGE. These results together provide streamlined access to libraries of single-stranded DNA templates suitable for DTS (Fig. 10-15c).

(a) Synthetic routes enabling incorporation of new scaffold structures into DNA templates, exemplified with scaffolds 4I and 4L.

(b) Scaffolds validated and used in the second-generation library of macrocycles.

(c) Iteratively selected building blocks maximizing overlap of the library with Kihlberg's parameter space for orally bioavailable molecules.

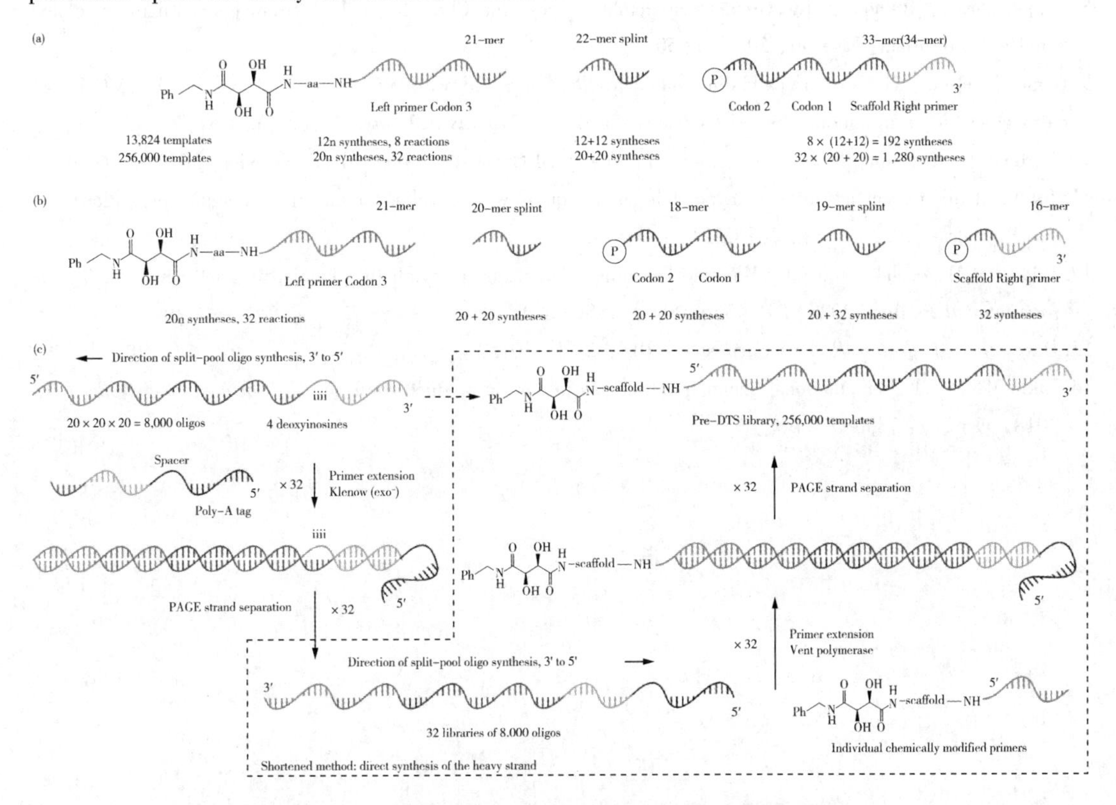

Fig. 10-15 Approaches to the assembly of DNA template libraries

(a) Assembly of the first-generation library of DNA templates.

(b) Modified version of splint ligation assembly for the second-generation DTS library.

(c) Template library assembly strategy via preparative enzymatic primer extensions.

参考文献

References

1. Wang L, et, al. *J. Chinese Journal of Catalysis*. 2022, 43 (7): 1598-1617.

2. Nunes P S G, et, al. Recent advances in catalytic enantioselective multicomponent reactions. *J. Organic & Biomolecular Chemistry*. 2020, 18 (39): 7751-7773.

3. Cargill J F, et, al. New methods in combinatorial chemistry — robotics and parallel synthesis. *J. Current Opinion in Chemical Biology*. 1997, 1 (1): 67-71.

4. Bunin B A, Dener J M, et, al. Annual Reports in Medicinal Chemistry. New York: Academic Press, 1999. 267-286.

5. Ying P, et, al. Liquid - Assisted Grinding Mechanochemistry in the Synthesis of Pharmaceuticals *J. Advanced*

Synthesis and Catalysis. 2021, 363 (5): 1246–1271.

6. Espro C, et, al. Re–thinking organic synthesis: Mechanochemistry as a greener approach. *J. Current Opinion in Green and Sustainable Chemistry*. 2021, 30: 100478.
7. Zhang Y–T, et al. Microwave–assisted synthesis of Zr–based metal–organic polyhedron: Serving as efficient visible–light photocatalyst for Cr(VI) reduction. *J. Inorganica Chimica Acta*. 2022, 543: 121204.
8. Lanjekar K J, Rathod V K. Green Sustainable Process for Chemical and Environmental Engineering and Science. Amsterdam: Elsevier, 2021. 1–50.
9. Demel J, et al. The use of palladium nanoparticles supported with MCM–41 and basic (Al) MCM–41 mesoporous sieves in microwave–assisted Heck reaction. *J. Catalysis Today*. 2008, 132 (1): 63–67.
10. Berthod A, Armstrong D W. Ionic Liquids in Analytical Chemistry. Amsterdam: Elsevier, 2022. 369–394.
11. Jin Y, et al. Recent advances in hypergolic ionic liquids with broad potential for propellant applications. *J. FirePhysChem*. 2022, 2 (3): 236–252.
12. Pflüger P M, et, al. Molecular Machine Learning: The Future of Synthetic Chemistry? *J. Angewandte Chemie International Edition*. 2020, 59 (43): 18860–18865.
13. Davey S G. *J. Nature Reviews Chemistry*. 2018, 2 (1): 0152.
14. Elkin M, et, al. Computational chemistry strategies in natural product synthesis. *J. Chemical Society Reviews*. 2018, 47 (21): 7830–7844.